# Insect Pheromones and their Use in Pest Management

# Insect Pheromones and their Use in Pest Management

P.E. Howse
*School of Biological Sciences, University of Southampton*

I.D.R. Stevens
*Department of Chemistry, University of Southampton*

O. T. Jones
*Chairman, AgriSense-BCS Ltd, Pontypridd*

**CHAPMAN & HALL**
London · Weinheim · New York · Tokyo · Melbourne · Madras

Published by Chapman & Hall, 2–6 Boundary Row, London SE1 8HN, UK

---

Chapman & Hall, 2–6 Boundary Row, London SE1 8HN, UK

Chapman & Hall GmbH, Pappelallee 3, 69469 Weinheim, Germany

Chapman & Hall USA, 115 Fifth Avenue, New York, NY 10003, USA

Chapman & Hall Japan, ITP-Japan, Kyowa Building, 3F, 2-2-1 Hirakawacho, Chiyoda-ku, Tokyo 102, Japan

Chapman & Hall Australia, 102 Dodds Street, South Melbourne, Victoria 3205, Australia

Chapman & Hall India, R. Seshadri, 32 Second Main Road, CIT East, Madras 600 035, India

---

First edition 1998



Typeset in 10/12pt Palatino by Acorn Bookwork, Salisbury, Wiltshire

Printed in Great Britain by T.J. International Ltd, Padstow, Cornwall

ISBN 0 412 80470 0 (Hb) 0 412 44410 0 (Pb)



A Catalogue record for this book is available from the British Library

Library of Congress Catalog Card Number: 97-69425

♾ Printed on acid-free text paper, manufactured in accordance with ANSI/NISO Z39.48-1992 (Permanence of Paper).

† *In Memoriam* I.D.R. Stevens

Ian Stevens, our greatly valued colleague and friend, died while this book was going to press. We dedicate this work to his memory.

*P.E.H., O.T.J.*

# Contents

# Preface

There is now a considerable literature on chemical ecology, which had its beginnings in the study of insect pheromones. This beginning was possible only by combining the disciplines and techniques of biology and chemistry. For a biologist, it is difficult to understand the time-frames of analytical and synthetic chemistry. A compound may take days to characterize and be available in minutes from a bottle on the shelf, or it may take years to characterize and synthesize. Chemists have a similar frustration: after an intense programme of work, the insect in question may not emerge for many months.

The rewards of integrated interdisciplinary study are, however, considerable, because they allow us to understand many facets of insect behaviour and consequently to control that behaviour for our own ends. In this book, we have set out to explain the results of research from chemical and biological perspectives, and see how the knowledge gained has led to novel techniques that can be used in insect pest management and insect control. An important part of understanding insect chemical ecology involves the understanding not only of new concepts but of the vocabularies used by scientists specializing in different fields. It will be clear that the three sections of this book have been written by three different people: an insect behaviourist, an organic chemist and a biologist in industry. The three of us have worked together in the past on various research projects and have attempted here to focus on issues in our own fields that we feel are key to researchers involved in what is now loosely called insect chemical ecology.

*Part One*

# Pheromones and Behaviour

*P.E. Howse*

# 1

# Insect semiochemicals and communication

> The Aurelians take this Moth by Sembling; their manner is, to go out with a live Hen in a Box which is covered down with Gauze or Crape; when they are come to the Appointed place ... they set the Box upon the Ground, and stand ready with their Nets; the Cocks will quickly come and attempt to get at the hen. I have known great Numbers taken in one Hour's time; and it may be depended on, that, if one goes with a Hen, in almost any Place of the Country, they will not fail of Success; not only the Eggers and Vapourers, but any Moth may be taken by Sembling ...
>
> Moses Harris, *The Aurelian* 1766.

In his instructions to insect collectors, Moses Harris did not attempt to explain the senses that male oak eggar moths (*Lasiocampa quercus*) used to find the females. Over a century later the French naturalist, J.-H. Fabre, one of the fathers of insect ethology, carried out simple experiments with the same species of moth and with the great peacock moth, *Saturnia pyri* (Fabre, 1913). One night in May, a female peacock moth emerged from a cocoon in his study. In the eight days that followed, Fabre caught over 150 male peacocks that were attracted into his house, most of which, he concluded, must have come from at least several kilometres away. With both the oak eggar and the great peacock, Fabre observed that moths were attracted to the cages after the virgin females had been removed, and even to a bunch of oak leaves on which they had rested. This led him to the inevitable conclusion that the females emitted a subtle odour which became absorbed by various substrates and which the males detected with their feathery antennae. He made a trap from a narrow-necked bottle baited with a piece of flannel on which a female had been resting. Males entered and could not escape: 'I had devised a trap by means of which I could exterminate the tribe,' he wrote. It was not until later this century, after the gas–liquid chromatograph had been invented, that it was possible to detect and identify the minute traces of volatile chemicals that serve for communication between the sexes of such moths, and to develop traps to aid in control of such insects.

It is perhaps not surprising that insects that are active in the dark, for example, night-flying moths and social insects living in enclosed nests, make great use of odour signals. As it became clear that externally secreted chemicals were involved not only in attraction but in the control of insect development, Karlson and Lüscher (1959), who had been engaged in research on the chemical control of caste development in termites, proposed the term 'pheromone' (a word with Greek roots meaning 'carrier of excitation') to describe a chemical that an animal secretes or excretes that 'releases a specific reaction, for example, a definite behaviour or developmental process' in a member of the same species. Wilson and Bossert (1963) later divided pheromones into **releasers**, which induce an immediate behavioural change, and **primers**, which initiate changes in development, such as sexual maturation, and so do not result in immediate behavioural changes, but predispose to them (note here that a releaser pheromone is not necessarily the same as a 'releaser' in the sense used by ethologists such as Lorenz and Tinbergen to describe a trigger of instinctive behaviour).

The concept of a pheromone is advantageous in one sense, in providing a label that draws attention to a means of communication not previously suspected in most animals, but disadvantageous in another sense, in that it makes it necessary to create other categories of behaviourally active chemicals by exclusion. Chemicals involved in communication are now known as **semiochemicals** (Fig. 1.1). These are further subdivided into pheromones and **allelochemicals** (Whittaker and Feeny, 1971), the latter acting between different species and consisting of two main types: **allomones** and **kairomones** (Brown *et al.*, 1970). Allomones are those chemical signals that give advantage to the emitter (e.g. defensive secretions) and kairomones are those that give an advantage to the receiver (e.g. secretions that can be detected by a parasite or predator). Having created these categories, some consider it necessary to create more and to subdivide existing ones (for a discussion, see Dicke and Sabelis, 1992). **Synomones** are chemical signals that give advantages to both sender and receiver, as in the case of floral scents which indicate a nectar source to insects and ensure pollination of the flowers producing them. **Antimones** are those in which neither benefit. We will keep to the more restricted terminology in order to avoid making too many assumptions about the relative benefits of a given signal. If we accept, as argued below, that a chemical can function as a kairomone in one context and an allomone

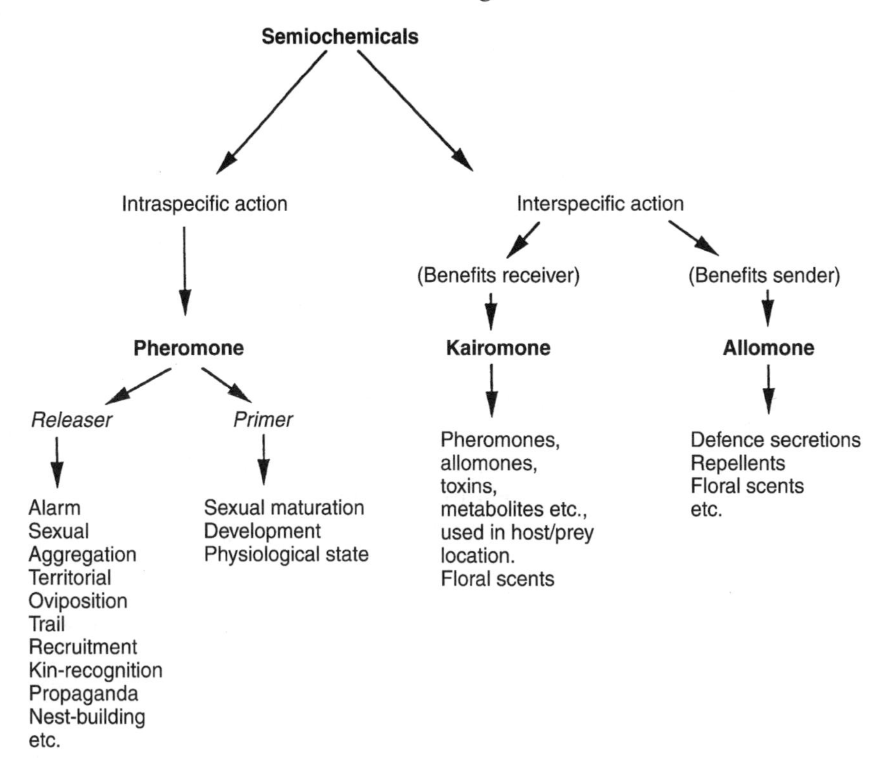


**Fig. 1.1** Categories of semiochemicals, with examples.

in another, then there is no need for a third term to cover both contingencies.

The inconvenience of the classification of semiochemicals becomes clear when we examine, for example, the secretions produced by worker ants when they encounter a threatening situation. They can be called **alarm pheromones** because they generally give rise to a rapid increase in the activity and aggregation of the nest-mates at the site. Some ant species produce formic acid, which is one compound that can have this effect, but it also has a powerful toxic action, especially against other invertebrates. Thus it is either a pheromone or an allomone, according to the context. Indeed it can also function as a kairomone, as it does in the interaction between the carnivorous Australian bulldog ant, *Myrmecia gulosa*, and its ant prey, *Camponotus* spp (Haskins *et al.*, 1973), where the bulldog ants are attracted towards the formic acid that the *Camponotus* release in self-defence. In the New World, blind snakes of the genus *Leptotyphlops* prey upon army ants such as *Neivamyrmex*, which they find by responding to the odour trails that the ants use to mark their trails. These chemicals are therefore either pheromones of kairomones, according to the context in which they act (Bradshaw and Howse, 1984). The 'alarm pheromone' of the ants also serves to repel other ant species, but can act as an attractant to the snakes. It is thus correctly termed either an allomone, a kairomone or a pheromone, according to the circumstances. There are many similar examples, but it is worth noting again that the semantic straitjackets begin to loosen if we speak of allomonal action or kairomonal effects, for example, having first defined the context.

Further difficulties arise from the temptation to categorize pheromones as, for example, 'alarm', 'sexual', 'aggregation', etc., according to the presumed function. Such tags usually involve presumptions that have not been tested experimentally. Furthermore, they are obscurantist: a sexual pheromone may control a diversity of behaviour patterns such as alerting, upwind flight, landing, orientation up an odour gradient and copulation. Similarly, the alarm pheromone of a social insect may induce searching, approach, recruitment, alerting, attack or repellence, depending on the concentration and other conditions.

A number of different behaviour patterns may be released by the same secretion and one reason for this is that pheromones are generally multicomponent. Early workers in the field tended to assume that pheromones could be compared directly with hormones, but acting externally, and terms like 'ectohormones' or 'exohormones' were in prior use, but it is now very clear that this was too simple an analogy. A secretion may be multifunctional as a consequence of being multicomponent. One of the great fascinations of the chemical ecology lies in finding explanations in functional terms for the multicomponent nature of semiochemicals.

It is very tempting to assume that a chemical identified from an animal is, *ipso facto*, a pheromone, particularly if it elicits clearly defined behaviour. This, however, should be proved rigorously. For example, there are many compounds that will elicit forms of alarm in ants, but we cannot assume that each one is used by the insects for this purpose– they may just be responding to it as a foreign odour. When a chemical not found in an animal has a pheromone-like action, it is often referred to as a **parapheromone**. A good example is the synthetic compound trimedlure, which is the best-known attractant for the male Mediterranean fruit fly, *Ceratitis capitata*. This was found by routine screening of compounds and is a chlorine-containing compound not known in nature.

## 1.1 CHEMICAL SIGNALLING

Chemical communication, as we have already noted, is more efficient in certain circumstances than in others. To understand why, it is necessary

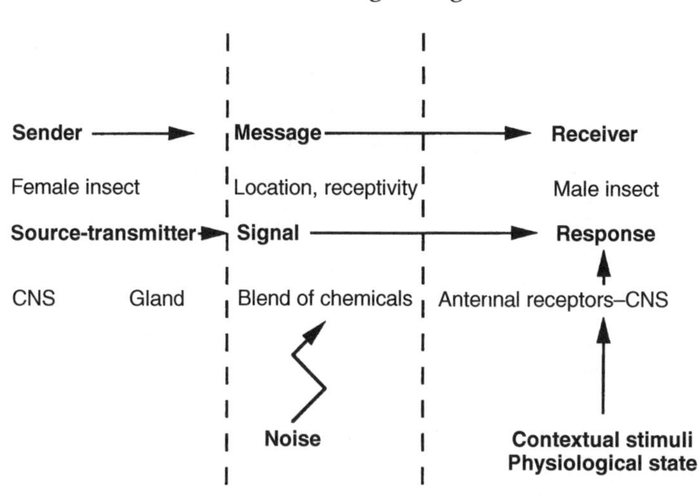


**Fig. 1.2** Elements of a chemical communication system, incorporating the example of sexual communication in a typical moth. (Based on Shannon and Weaver, 1949.)

to look at the components of a communication system (Fig. 1.2). As a result of various internal and external stimuli acting on an animal, a message is transmitted as a signal in coded form through a certain medium to a receiver. It is there registered by sense organs and the code interpreted according to the context, providing a particular 'meaning'. In animals we can only make inferences about the nature of the meaning by observing changes in behaviour of the receiving individual in different circumstances. Looking at the signal alone does not allow us to make any sensible deductions about what the animal will do.

It is believed that, during the course of evolution, many animal signals have become very distinctive to avoid ambiguity, and they tend to occupy an energy band of a certain width which does not overlap too much with that used by other animals. From this arises the concept of **channels of communication**, which can be thought of as broadly analogous to radio or television channels, where signals are transmitted in a narrow frequency band each band carrying, in general, different kinds of information.

Chemical signals, as other types, have certain advantages. They can be used in the dark. They can travel around obstacles without being reflected, so that the emitter can remain hidden from view. Relatively involatile chemicals can be used to mark food resources, routes, territories and home ranges over the long term. Volatile chemicals can travel long distances in the wind (or soluble chemicals can travel in water currents) distributing a signal with no expenditure of energy on the part of the emitter apart from that involved in biosynthesis. Specificity can be built into the signal by adding one or more other chemical compounds, in which case unique blends may result, characterized not only by the compounds composing them but also by the relatively fixed ratios and concentration of ranges in which they occur. A very narrow channel of communication can then be achieved if the sense organs of the receiver are selectively sensitive ('tuned') to the parameters of the signal.

One disadvantage of a chemical signal is that it cannot easily be arrested with the rapidity of, for example, a visual or acoustic signal, and therefore the signal cannot be changed very easily. Concealment from a predator or parasite that hunts by smell is thus very difficult and this no doubt provides pressure for the evolution of narrow communication channels. Neither can a chemical signal be easily modulated in amplitude or changed qualitatively. Hence there is a greater tendency for the same message to have different meanings according to context. In general, chemical messages can be varied in the short term only by the use of different glandular sources, as seen *par excellence* in social insects.

### 1.1.1 Noise

There are a number of advantages in a narrow channel of chemical communication. First, communication is more private; second, energy can be conserved – for example, fewer molecules need to be synthesized; third, selectivity of sensory receptors can be combined with high sensitivity, and only trace amounts of chemical need be released by the sender.

Persistence or repetition of a signal are aspects of **redundancy**. A message is said to be redundant if the receiver can interpret it correctly without it being complete. The advantage of redundancy is that it helps to overcome the problem of **noise** in the communication channel. In acoustic communication, noise can be literally that, but in communication theory it is any phenomenon that tends to compete with or obscure the signal. In chemical communication, 'noise' may be other chemicals that the receiver can detect, or signals in other modalities – visual or acoustic, say – which divert the receiver's attention. Redundancy may be considered to form part of the message, allowing the recipient to 'guess' the details that may be lost in the noise. This certainly happens in human communication – for example, in noisy short-wave radio transmissions where the sender may have to repeat a message a number of times, or increase the content (e.g. 'cat' might be repeated as 'Charlie Alpha Tango'). Chemical signals are generally redundant in that they are emitted continuously over long time periods (and sometimes even in repetitive pulses), and in being multicomponent blends that may require different groups of sensory receptors for their detection.

## 1.2 CHEMICAL COMMUNICATION IN SILK MOTHS

Let us examine the relatively simple case of communication in the courtship of silk moths (Saturniidae), which have proved to be excellent experimental animals because of their large size and the extraordinary sensitivity of the males to the female sex pheromone. It was Butenandt *et al.* (1959) who identified the first sexual pheromone of an insect. The insect was the common silk moth, *Bombyx mori,* in which, as in the majority of nocturnal Lepidoptera, the virgin female produces a volatile pheromone from eversible sac-like glands at the tip of her abdomen. The active component, (*E*)10–(Z)12–hexadecadien–1–ol, christened 'bombykol', was identified from the extract of abdominal tips of 250 000 female moths. This compound is liberated into the wind and is detected by specialized sense organs on the antennae of the male which are extremely sensitive to bombykol. By maintaining a heading into the airstream, the male is then able to find its way to the female (in fact the male *B. mori,* after centuries of domestication, cannot fly, but by flapping its wings it will propel itself across a smooth table-top in a pheromone-laden air stream). The antennal receptors are trichoid sensilla (Fig. 1.3) each innervated (in *B. mori*) by two sensory nerve cells, one of which is selectively responsive to bombykol and the other to the minor component of the pheromone (the corresponding aldehyde) known as 'bombykol' (Kaissling *et al.,* 1978). The antennal flagellum, with its numerous side-branches bristling with trichoid sensilla (Fig. 1.3), is a highly efficient filter, trapping about 30% of the odour molecules in an air stream passing over it (Kaissling, 1971), and radioactive tracer experiments have shown that at least 80% of the molecules are absorbed on the sensilla (Steinbrecht and Kasang, 1972).

In this example, the message is about receptivity for mating. It is transmitted from the abdominal glands as a simple coded signal. The receivers are the antennae of the male and the 'meaning' of the signal to the male moth is demonstrated by its arousal and upwind movement. It is important to realize that the channel of communication – in this case chemical – is always of a defined width, and here is very narrow. That is to say, the message can be carried by only one or very few chemical compounds to which the receiver is selectively sensitive. Another way of expressing this is to say that the receiver is closely tuned to the emitted signal, in the same way that acoustic signals of animals are

a)

b)

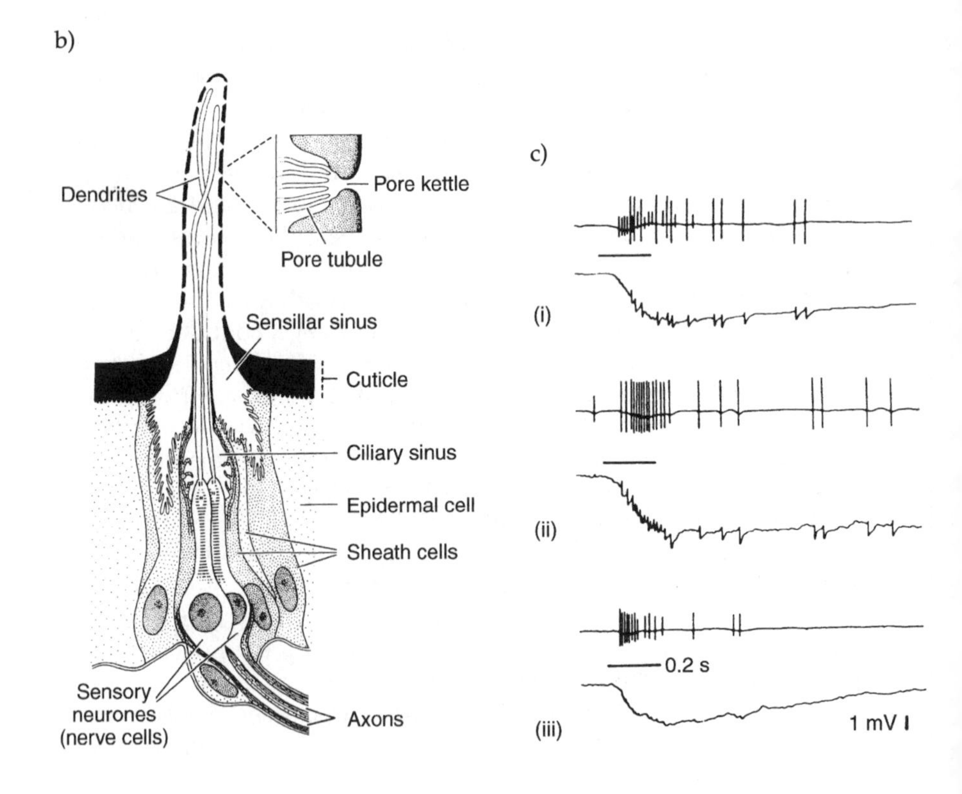
Dendrites
Pore kettle
Pore tubule
Sensillar sinus
Cuticle
Ciliary sinus
Epidermal cell
Sheath cells
Sensory neurones (nerve cells)
Axons
c)
(i)
(ii)
(iii)
0.2 s
1 mV

often confined to a narrow band of frequencies to which the ears of the receiver are selectively sensitive. This tuning may be achieved by a process known as 'sensory filtering', in which the sense organs have a window to only certain frequencies of emitted energy, or to certain chemicals, or to particular wavelengths of light or visual patterns. Clearly, however, the channel of communication is as much a function of the emitter as it is of the receiver. Channels of communication in humans are generally very wide – so wide that it is often difficult for us to accept that the limits are there. This is a conceit that has delayed recognition of the multifarious chemical communication systems in animals, which mostly involve chemicals to which the human nose is insensitive. Because our sense organs assimilate so much information from many different sources, we are not easily deceived. In contrast, it is very easy to deceive insects with simple signals because the insect central nervous system has relatively little capacity to evaluate signals and the context in which they occur. This is a factor of great importance in the control of insect pests by manipulation of their behaviour.

A narrow communication channel has the advantage that it can be largely exclusive or private to the insects concerned. Receptors involved in the detection of pheromones are usually specialized in that they have very low and more or less similar thresholds to one or a series of related molecules. Such receptors have been called 'odour specialists' (Schneider *et al.*, 1964) and can be contrasted with 'odour generalists', which respond to a wide variety of odours, though differing among themselves in their response spectra and thresholds. The ant *Lasius fuliginosus* has antennal receptors that can be divided into 10 groups according to their range of sensitivity (Dumpert, 1970). One cell type can be stimulated only by *n*-undecane, a substance that releases strong alarm behaviour. The burying beetles, *Necrophilus* and *Thanatophilus* spp., have sensilla that are highly sensitive to the odours of carrion (Boeckh, 1962). The effect of specialization is to filter out information that is not of great importance to the survival of an insect. In other words, the signal to noise ratio can be increased by eliminating most of

---

**Fig. 1.3** (a) Antennae of a male saturniid moth. (b) Section through a typical olfactory sensillum from an insect antenna, showing dendritic processes from the sensory cell running into the hair shaft, where they run close to the pores and pore tubules through which odour molecules enter (from Gullan and Cranston, 1994). (c) Action potentials (upper traces) and receptor potentials (lower traces) recorded from an olfactory hair of a male *Antheraea polyphemus* in response to a puff of air carrying the odour of (i) an excised female pheromone gland, (ii) a sample of the synthetic sex pheromone component (*E,Z*)-6,11 hexadecadienyl acetate, and (iii) (*E,Z*)-6,11 hexadecadienal (from Boeckh, 1984, after Kaissling, 1974).

the noise, which also means that relatively high sensitivity can be achieved. In summary, the greater the sensory filtering, the fewer are the demands made on the integrative activity of the nervous system.

Narrowing the communication channel can be achieved, for example, by sensory filtering or by increasing the complexity of the signal. The simplest means of increasing the complexity, analogous to having a key with two teeth that engage a lock instead of one, is to have a bicomponent pheromone. Further specificity can then be encoded in the form of differing ratios of the two components or, to continue the analogy, by having teeth of different length. The design of the molecule is often slightly different in closely related species. Many species of the lepidopteran families Tortricidae, Noctuidae and Pyralidae have female sex pheromones that are straight-chain hydrocarbons, with a functional group at the end of the molecule (Bjostad *et al.*, 1987). Species differences are achieved by changing the chain length (between 10 and 18 carbon atoms), the functional group, the number of double bonds, and their position in the *Z* or *E* configuration (as geometrical isomers).

The high sensitivity of male silk moths to female pheromones is achieved by a combination of design features (Fig. 1.4) all related to the narrow communication channel that exists in silk moths.

The most obvious adaptation lies in the size and geometry of the antennae, which make them highly efficient molecular filters. Commonly (and especially so in many nocturnal Lepidoptera), the male antenna has more and longer side branches than the female antenna, which gives it a large profile. In the male *Antheraea polyphemus*, for example, this covers an area of $85\,mm^2$, while it is only $18\,mm^2$ in the female (Kaissling, 1972). The corresponding figures for the silk moth, *Bombyx mori*, are $6\,mm^2$ in the male and $5.5\,mm^2$ in the female. However, the antenna of the male *B. mori* bears about 17 000 basiconic and trichoid sensilla – about three times as many as in the female. Kaissling has calculated that the antenna of this moth adsorbs 27% of all molecules of bombykol in an air stream with a cross-sectional area of $6\,mm^2$. In this, it is $10^3$ times more efficient at capturing molecules of queen substance than the honey bee drone antenna, and $10^4$ times more efficient at molecule-trapping than a *Drosophila* antenna.

The second design feature is the presence of large numbers of pheromone-specific sensilla on the antennae of the males (Fig. 1.4). Such sensilla are absent from the antenna of female *A. polyphemus* and female *B. mori*. Male *A. polyphemus* have 55 000 sensilla responsive to each of the two isomers of the sexual pheromone, and only about 20 000 'generalists', responsive to a variety of plant volatiles. Incorporated in its two antennae, the insect has a total of nearly a quarter of a million sensory neurones that respond to the female pheromone. This should not, however, be taken as typical, because very few insects have been examined in

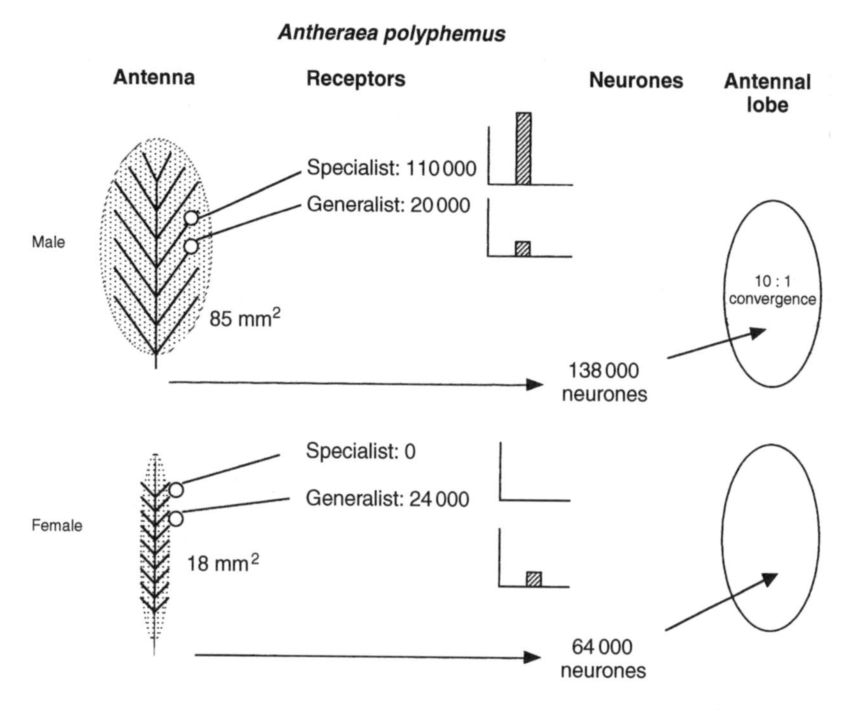


**Fig. 1.4** Chemosensory equipment of *Antheraea polyphemus* male and female moths compared (data from Meng *et al*, 1989, and Kaissling, 1972). The male antennae cover a larger surface area and have a very high density of specialist receptors tuned to the pheromone components of the female. The female has only generalist receptors which facilitate recognition of foodplant volatiles, etc. The 10:1 convergence of receptor axons on to neurones in the antennal glomeruli (right of figure) permits the summation of sub-threshold stimuli from only a few receptors.

such detail. In the moth *Choristoneura fumiferana*, there are fewer than 1000 pheromone specialist cells per antenna (Albert *et al.*, 1974).

Thirdly, the overall sensitivity is affected by the relative sensitivity of the receptor cells and the release rates of the pheromone by the female. The pheromone of *A. polyphemus* differs slightly from that of the related species, *A. pernyi*. Both contain (*E*,*Z*)-6,11-hexadecadienyl acetate ($AC_1$) and the corresponding aldehyde, (*E*,*Z*)-6,11-hexadecadienal (AL), but these are released in different ratios. In *A. polyphemus* the $AC_1$ : AL release rate (which is not necessarily the same as the ratio of compounds present in the intact gland) is estimated to be 90 : 10 (Meng *et al.*, 1989). In *A. pernyi*, a second aldehyde has been identified as (*E*,*Z*)-4,9-tetradecadienyl acetate ($AC_2$), (Kochansky *et al.*, 1975) and the release rate ratio of $AC_1$ : AL : $AC_2$ is estimated by the same authors at

20:100:40. The receptor cells of *A. polyphemus* are more sensitive to $AC_1$ than they are to $AC_2$, and conversely for *A. pernyi* (Meng *et al.*, 1989). These features narrow the communication channel and contribute towards species specificity of response. A mixture of $AC_1$ and AL attracts *A. polyphemus* but does not attract *A. pernyi* in the field, because of the lack of $AC_2$ (Kochansky *et al.*, 1975).

A fourth design feature is the convergence of chemoreceptor neurones on to interneurones in the antennal glomeruli of the brain, which in *Antheraea* is believed to occur at a ratio of 100:1 (Boeckh, 1984). This convergence is thought to explain the lower thresholds of the glomerular interneurones. These respond when the antenna is stimulated by a source of $10^{-6}$ to $10^{-7}$ μg, while individual receptor neurones respond at the higher concentration of $10^{-4}$ to $10^{-5}$ μg at source. This is not because the thresholds of individual receptors are higher, but because the chances of all of, say, 100 receptors being hit by enough molecules to produce a nerve impulse during any one time period diminishes as the concentration falls. For an analogy, imagine rain falling on an array of 100 small pots. If the rain is very intense there will be more water in each pot every time we check. If the rain slows down and almost stops, we might detect another drop of water in only one or two pots during a short interval of time: by checking them all we have increased the sensitivity of our rainfall detection method.

The high efficiency and selectivity of the communication system between male and female silk moths is testimony to a very narrow communication channel. It was surprising, therefore, when Priesner (1968) published an extremely detailed study of 104 species of saturniid moths (about 10% of all the known species) and showed that there was considerable overlap in the responsiveness of male antennal receptors to the sex pheromones of females of related species. He used the technique of electroantennography, or EAG (Chapter 4) in which the responsiveness is measured by the change in potential between the base and tip of the antenna that occurs when sensilla are stimulated along its length. In this way, he measured the responsiveness of males to extirpated female sex pheromone glands in 1900 species combinations.

It proved possible to divide the 104 species into 19 'reaction groups', within each of which there was a complete overlap of EAG responsiveness. Species within each reaction group were all closely related; they were in the same genus or in closely related genera. For example, four species of the Citheroniinae formed one group and three species of *Nudaurelia* formed another, and within each group full EAG responses were elicited in each male antenna from the scent of each female (Fig. 1.5). Species isolation in these silk moths studied by Priesner appears to depend only to a limited extent on pheromone specificity, and a variety of factors which will be considered later (including spatial and temporal

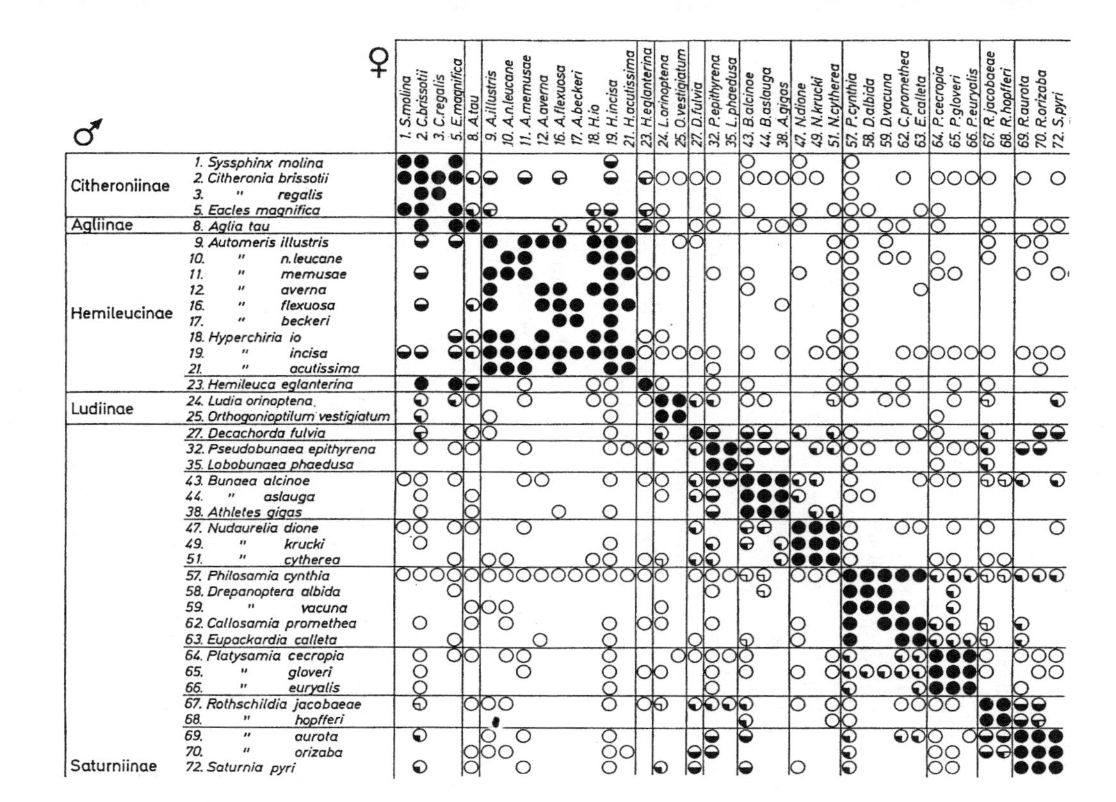


**Fig. 1.5** Reaction spectra of male antennae of various species of saturniid moth. Male antennae were exposed to excised female pheromone glands in a series of species combinations, and the summed electrical response of the antenna (EAG) was recorded. The degree of filling of the circle represents the relative magnitude of the response. (From Priesner, 1968).

separation of the adult moths) may complete the barriers to cross-mating. However, it has become clear that events at the sensory level may also affect behavioural responsiveness. It is important to remember that EAG is not always a good predictor of behavioural responsiveness, for reasons outlined in Chapter 3: for example, some chemicals may stimulate neurones that have inhibitory or synergistic effects within the central nervous system, and this cannot be deciphered from an EAG trace.

## 1.3 CODING BY OLFACTORY RECEPTORS

Let us now examine the role of chemoreceptor mechanisms in the communication process. Molecules that strike an antennal sensillum enter a pore system (Fig. 1.3) that has tubules leading to the dendritic terminal (Steinbrecht and Müller, 1971) where a transduction process occurs leading to a depolarization of the dendrite. Kaissling and Priesner (1970) found that one molecule of bombykol could generate one nerve impulse (action potential) in the appropriate receptor of the male *Bombyx mori* within about half a second. A significant number of moths would then show a behavioural response (wing vibration) when about 1% of all the bombykol-sensitive sensilla were stimulated.

It is believed that odour molecules bind to receptor sites on the membrane of the dendrite. The resulting depolarization (receptor potential) may eventually give rise to nerve impulses (Fig. 1.3c) which are transmitted along the dendrite directly to the antennal glomerulus of the brain. Events at the receptor membrane thus determine the coding of the nerve impulses, and integration in the antennal lobe results in the appropriate change in behaviour. Odour molecules are assumed to fit on to special receptor sites on the membrane like a chocolate sweet fitting into its plastic mould in a box. Essential to the receptor process is a mechanism for freeing these receptor sites as quickly as possible after the change in membrane permeability has been achieved, otherwise the antennal receptors would become blocked and insensitive.

The pore tubules, into which odour molecules enter, appear to be blind-ending, and Vogt and Riddiford (1986) have proposed that the molecules diffuse to the end of the tubule and then pass into the fluid surrounding the dendrite. The fluid has a high concentration of molecules of binding protein, but as the binding is relatively weak, it is suggested that the pheromone molecules migrate to reach the receptor sites across the dynamic bridge that they provide, in much the same way that molecules move through the liquid phase in column chromatography. Once there, the molecules are deactivated by antennal pheromone esterase so that the sites are rapidly freed, but the exact mechanism of this deactivation is not fully understood.

Part of the code of information passed to the antennal lobe is in the form of 'labelled lines'. That is to say, certain dendrites are dedicated to pheromone reception or even reception of particular components of a pheromone. Analysis of a blend of components can then be done by monitoring the traffic in those dendrites, in order to detect cross-fibre patterning. Visser and De Jong (1988) found that patterns of activity were generated in interneurones of the antennal lobe of the Colorado beetle (*Leptinotarsa decemlineata*) in responses to increases in the concentration of leaf odour components (corresponding to the activity of generalist receptors), while other interneurones changed their activity if there was a change in the ratios of the components tested (reflecting the activity of specialist receptors). In this insect, channels of information about odour quantity and odour quality (blends) thus remain distinct within the central nervous system.

Particular regions of the antennal lobe may be devoted to pheromone communication restricting the channel width further. In general, olfactory sensilla sensitive to pheromones go to one part of the antennal lobe and sensilla sensitive to food-related odours go to another part. The male turnip moth, *Agrotis segetum*, has receptors for four components of the female pheromone, and the neurones project to different parts of the antennal lobe (Hansson *et al.*, 1992) Axons from receptor neurones for (Z)-5-decenyl acetate and (Z)-decenol go to the A and B glomeruli, those for (Z)-7-dodecenyl acetate go to the B and C glomeruli, and those for (Z)-9-tetradecenyl acetate go to the B and D glomeruli. The labelled lines thus remain separate to a degree even within the central nervous system. As we have seen (Fig. 1.4), there is a high degree of convergence of neurones on to antennal lobe interneurones in insects such as *Antheraea*. In the American cockroach, *Periplaneta americana*, the convergence ratio is about 4000 : 1 (Boeckh, 1984). There are 180 000 olfactory sensilla on the antenna of the male which connect in the antennal lobe to about 250 interneurones. Of the total, it is estimated that 36 000 receptor cells respond to each of the two components of the female pheromone (periplanone), but these finally converge on to only about 20 interneurones in the macroglomerulus. This is a very narrow channel, from which noise is virtually eliminated, and the central neurones are able to respond to the firing of only a few receptor neurones.

Although pathways from the receptor to the brain appear to remain separate from each other, in some Lepidoptera there is evidence of integration at the receptor level and this may be one mechanism that is involved in blend analysis. The major components of the sex pheromone of the noctuid moth *Spodoptera litura* are (Z)-9,(*E*)-11-tetradecadienyl acetate and its (*E*)-12 positional isomer. Recording from single olfactory cells, Aihara and Shibuya (1977) found that the first compound gave rise to nerve impulses at low frequency but the second had no detectable effect. Applied together in the 9:1 ratio, in which they are produced by the female, nerve impulses were produced in relatively high frequency, suggesting synergy at the level of the receptor cells. Ratios of 5: or 1:9 were less effective than the first compound alone. A similar modulation of receptor responsiveness occurs in the cabbage looper moth, *Trichoplusia ni*, in which the two main components of the female pheromone are (Z)-7-dodecenyl acetate and (Z)-7-dodecenol. O'Connell (1984) investigated the effect of a minor component, dodecyl acetate, on the responsiveness of sensory neurones that were selectively sensitive to one or other of the first two compounds. Although the male antenna appears to have no specialized receptors for dodecyl acetate, when 10% was added to the other two compounds in binary mixtures there was an increased level of response, but the response to ternary mixtures was actually lower than that to the compounds presented singly (Fig. 1.6). Events prior to the generation of action potentials must therefore affect the responsiveness to these mixtures.

The macroglomerular neurones connect with the calyces of the mushroom bodies in the brain (Fig. 1.7) which appear to be concerned with olfactory orientation processes, and with the lateral lobe of the

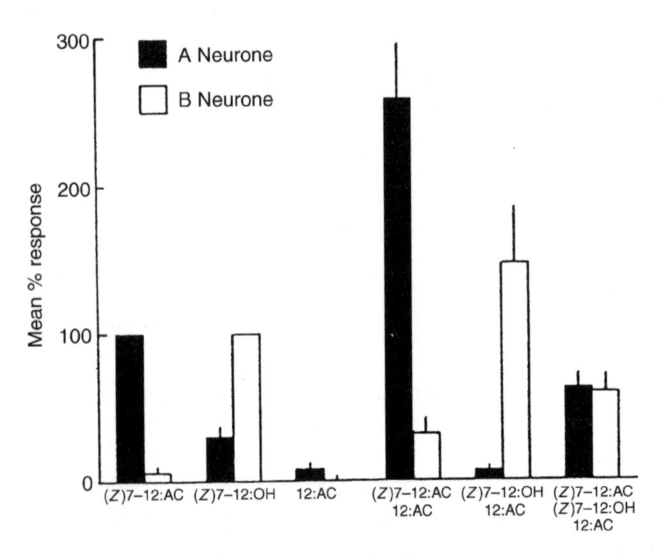


**Fig. 1.6** Electrophysiological responses of olfactory receptor neurones in the moth *Trichoplusia ni* to three components of the female sexual pheromone: (Z)-7-dodecenyl acetate, (Z)-7-dodecenol and dodecyl acetate. Responsiveness is measured relative to two standard amounts of the first two compounds, set at 100%. One of the two receptor neurones in each sensillum is specialized for Z7-12A and the other for Z-12OH. (From O'Connell, 1984.)

protocerebrum, which contains many interneurones that are multimodal (i.e. they integrate signals of various modalities: visual, olfactory, mechanoreceptive, etc.). From the central region of the brain, descending motorneurones go to the thoracic ganglia, where their

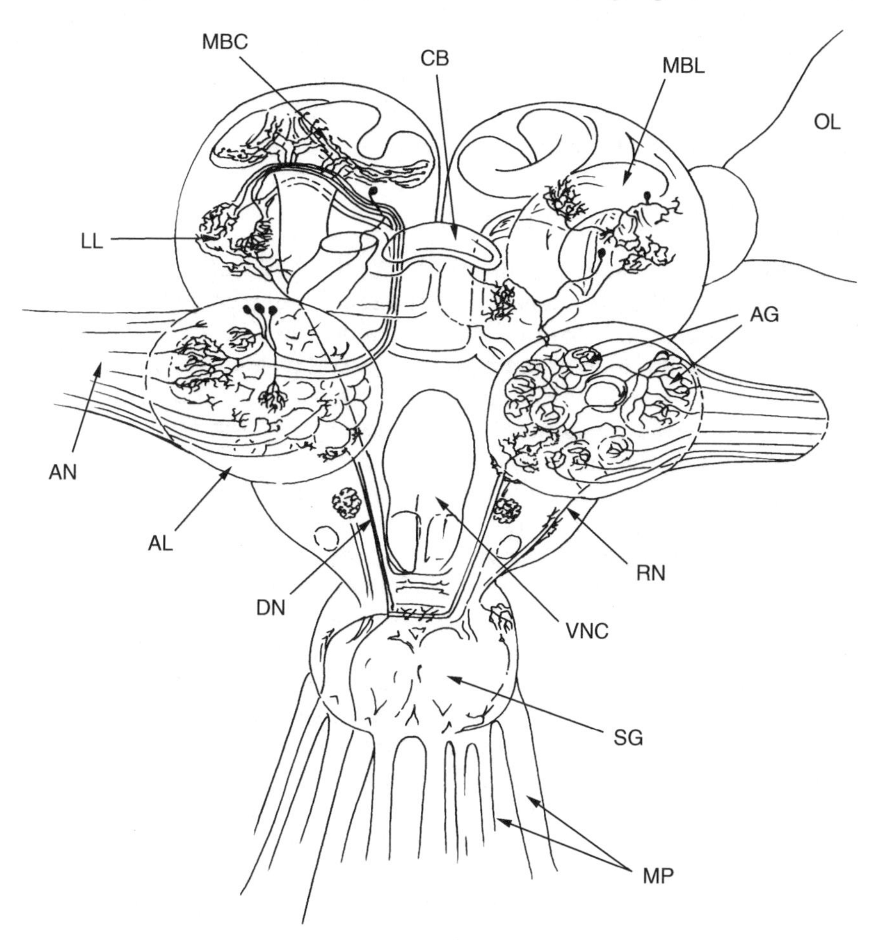


**Fig. 1.7** Frontal view of the brain of the cockroach *Periplaneta americana*, with nerves to mouthparts (MP) in the lower part of the diagram, showing main pathways and connections of olfactory neurones. (Redrawn after Boeckh, 1984, and Boeckh and Ernst, 1987.) Olfactory receptor axons from the antennal nerve (AN) enter the antennal lobe (AL) and terminate in branches in antennal glomeruli (AG). A few (RN) continue to the suboesophageal ganglion (SG). Neurones with cell bodies in the antennal lobe synapse with incoming receptor axons and project to lateral lobes (LL) of protocerebrum, passing lobes of the mushroom body (MBL) and central body (CB), and sending branches to the mushroom body calyces (MBC). Descending neurones (DN) with cell bodies in protocerebrum lead from the lateral lobes to motor centres in the ventral nerve cord (VNC).

activity controls turning movements. Descending interneurones in the ventral nerve cord of *Bombyx mori* have been found (Olberg, 1983) which probably originate in the lateral protocerebral lobe. Their discharge rate switches between high and low stable states, for which reason they are known as 'flip-flopping' interneurones. These interneurones control antennal position and turning movements. They are triggered by air currents containing bombykol and may be part of the mechanism that controls the zigzagging upwind flight movements towards the source of female pheromone.

## 1.4 LEARNING ABILITY OF INSECTS

The theory of instinct was built around the issue of communication. Lorenz and Tinbergen believed that animal signals that they called 'releasers' or 'sign stimuli' acted upon an innate neurological mechanism in the animal brain to elicit a certain pattern of behaviour. They were seen as behavioural keys that fitted a precise neurophysiological lock. This is simply a metaphor for a communication in a narrow channel, but the implication that such behaviour is 'wired in', instinctive and inflexible at all stages of development, is one that has been amply discredited as far as vertebrates are concerned. In insects, the assumption is equally dangerous. It is often overlooked that there is a wealth of information on learning abilities of insects, which can affect their sensitivity to stimuli of all kinds. The ability of honey bees and bumble bees to exploit nectar sources from flowers of certain colours, shapes, scents or opening times is evidence of rapidly acting classical conditioning processes. Erber (1975) and Menzel *et al.* (1974) have shown that a honey bee can be conditioned to associate food with colour stimuli after feeding once. Feeding for 1–1.5 minutes is sufficient for an associated colour signal to be remembered for several days. Floral scents are learnt with even greater facility than colours (Koltermann, 1974). Furthermore, the memory of a scent may depend on time of day (Koltermann, 1971): bees that were allowed to feed at a certain time of day discriminated between the training scent and others only at that time of day, or very close to it. This kind of conditioning process is by no means unique among insects, and is not limited to social insects.

The technique of using models to assess the responsiveness of insects in the field is exemplified by the work of Tinbergen *et al.* (1942) on the grayling butterfly, *Hipparchia (= Eumenis) semele*. The insects showed a preference for colours around the centre of the visual spectrum (yellow, blue) when feeding, but courting males showed different visual 'preferences'. When model butterflies were flown in the air on the end of a fishing line, sexually motivated males made the greatest number of approach flights to models that resembled the female only in being of similar size, black and moving through the air with a jerky motion. This work ignored the possibility that previous experience may have shaped any of the responses but, unlikely as it may seem in certain instances, this cannot be ignored. Swihart and Swihart (1970) showed that the apparent preferences of the passion vine butterfly, *Heliconius charitonius*, could be changed if butterflies were given sugar water in association with certain colours. Pioneering experiments were carried out by Dudai *et al.* (1976) on olfactory conditioning in *Drosophila melanogaster*, prompted by an interest in behavioural mutants. Flies were conditioned by placing them in vertical tubes and allowing them to fly upwards towards light. In the presence of a chosen volatile, the flies received a mild electrical shock on landing on a grid at the top of the tube. Flies that had been shocked then tended to avoid entering tubes containing the test volatile but entered others which had a different odour.

A similar ability to associate certain odours with prey or with stimuli associated with the food or prey after a period of conditioning has been demonstrated in some parasitoids. Thorpe and Jones (1937) first showed that the ichneumon wasp *Venturia (Nemeritis) canescens*, which parasitizes the flour moth (*Ephestia kuehniella*), will orientate towards the odour of peppermint if this is added to the diet of the wasp as larva. Associative learning has been found to be involved in the location of the larvae of the cotton earworm, *Heliothis zea*, by the braconid parasitoid *Microplitis croceipes* (Lewis and Tumlinson, 1988), The parasitoid was strongly attracted to the odour of cowpea plants if it was reared on larvae that had fed on cowpea (Herard *et al.*, 1988). Females reared from hosts fed on artificial diet were not attracted. A volatile component from *H. zea* caterpillar faeces attracted female wasps that had already been exposed to faeces, and a non-volatile component detected by antennal contact induced the wasps to fly to host-related odours. Even when an alien odour such as vanilla was presented along with this non-volatile component, the insects would subsequently respond to vanilla. There are apparently possibilities here for improving biological control by pre-training of parasitoids. There is evidence that chemicals from the plant are retained on the cocoon of the parasitoid and that contact with these during eclosion affects the responsiveness of the adult insect to odours of the host plant. The eucloid hymenopteran *Leptopilina heterotoma* is a parasitoid of *Drosophila*. Papaj and Vet (1990) investigated the effects of two hours' exposure of the female parasitoids to *D. melanogaster* and *D. stimulans* larvae in artificial diets of apple-yeast or decaying mushroom. Experienced females were subsequently more likely than naive females to find the microhabitat they had experienced, and they took less time to find it.

Workers of the ant *Formica polyctena* become imprinted to the odour of their nest-mates as a result of experience immediately after emergence from the pupa (Jaisson, 1975). In the ant *Cataglyphis cursor* there is another learning phase in which the larvae imprint on cues from the adult workers (Isingrini *et al.*, 1985). Learning shown by the establishment of preferences can extend to other odour stimuli: young *Camponotus vagus* and *Formica polyctena* workers were exposed to the odour of thyme, to which they are normally aversive, near a water reservoir. Following this, the workers demonstrated a preference for thyme (Jaisson, 1980).

It is difficult to generalize about the role of pheromones in insects, because this depends much on their way of life, relationship with host plants, level of sociality, etc. However, certain orders of insects appear to share common features. Nocturnal Lepidoptera mostly produce $C_{10}$ to $C_{21}$ unsaturated straight-chain aliphatic alcohols, ketones, acetates and aldehydes. These are usually liberated by the female and serve to attract males upwind. Species differ in terms of the blend of compounds they produce, and the compounds differ in terms of the number and position of double bonds, the chirality and the functional group. Pheromones of this type have been studied intensively because of their potential use in new methods of pest control, but many other kinds of pheromone are found in Lepidoptera. Males of both butterflies and moths possess scent glands that produce relatively involatile compounds acting at short range and acting as key elements in the courtship sequences that are a prelude to mating. For this reason, they are sometimes referred to as aphrodisiac pheromones. Oviposition pheromones have been identified from various Lepidoptera – for example, from the large cabbage white butterfly, *Pieris brassicae*, where they deter other females from laying in the vicinity (Schoonhoven, 1990). The gregarious larvae of lackey moths or 'tent caterpillars' (*Malacasoma* spp.) lay a pheromone trail from a sternal secretory gland when searching for new feeding sites on the host plant (Fitzgerald, 1976). This is usually placed on silk threads that the exploring caterpillars deposit between their silken nest and the leaves they are feeding on.

Because of their economic importance as herbivores or wood-borers, Coleoptera have also been intensively studied. Their sexual pheromones are very diverse in chemical structure, including aliphatic and aromatic compounds, some of which are terpenes and many of which show optical isomerism. Aggregation pheromones occur widely in weevils and serve to attract members of the same sex, in general, to host plants.

Hemiptera are more conservative in their pheromones. Many scale insects produce mono- and sesquiterpenes as sexual attractants. Others, such as the green stink bug, *Nezara viridula*, produce aggregation and alarm pheromones in addition to sexual attractants, and aphids use farnesenes as alarm pheromones which stimulate rapid dispersion.

The pheromones of Diptera are little understood. The early stages of sexual attraction may depend on visual pursuit, as they do, for example, in house flies and tsetse flies in which, at very close range, aliphatic monoenes act as arrestants and aphrodisiacs. The pheromones of tephritid fruit flies are extraordinarily complex: over 50 compounds (including terpenes, nitrogenous compounds and aliphatics) have been isolated from the male Mediterranean fruit fly, *Ceratitis capitata* (Jang *et al.*, 1989). These are implicated in the attraction of virgin females to males and the formation of male aggregations or leks prior to courtship, but their precise role and interrelationship with host plant volatiles are still shrouded in mystery.

## 1.5 SOCIAL INSECTS: THE HONEY BEE

Communication by pheromones finds its apotheosis in eusocial insects (bees, wasps, ants and termites). In such insects many categories of pheromone have been described, amongst which are alarm, recruitment, sexual, aphrodisiac, aggregation, brood, queen, trail, leaf-marking (leaf cutting ants), territorial, lekking (bumble bees), nest-building (termites) and thermoregulatory pheromones (wasps, bumble bees). Such a list has limited utility. It draws attention to the many kinds of chemical communication that may exist, but behaviour often does not fit neatly into preconceived categories. As a corollary, such labels tend to appear as explanations, like many terms used in ethology, and can inhibit empirical investigation. Perhaps the most tendentious term is 'alarm'. To us, this word signifies some impending hazard. Insects, like humans, may show many different types of behaviour to cope with alarm, which may include summoning help and deploying a number of defensive strategies. We have to take great care, therefore, to describe insect behaviour precisely; for each insect species we should ask ourselves what we mean by terms such as 'alarm', 'sexual' and so forth, when describing pheromonal communication, and, if necessary, redefine the term for the species and context. Although there are many types of pheromone in the honey bee, *Apis mellifera*, we see a principle at work which Blum (e.g. 1977) has termed 'pheromone parsimony': the use of a single chemical compound for carrying different messages or different parts of messages.

### 1.5.1 Queen-produced pheromones

The queen honey bee is the focus of activity in the hive and is constantly surrounded by workers. When the queen of a colony dies, the workers construct queen cells from existing worker cells containing female larvae or eggs, and at the same time the ovaries of the adult workers begin to enlarge. These are responses to the lack of a multicomponent pheromone commonly referred to as 'queen substance', which is produced by the queen's mandibular glands, coats her body and is transferred to workers by trophallaxis (the mutual licking, feeding and grooming that occurs continuously among members of the colony). The main components of the queen substance (Fig. 1.8) are 9-oxo-(*E*)-2-decenoic acid (9-ODA) (Callow and Johnstone, 1960) and 9-hydroxy-(*E*)-2-decenoic acid (9-HDA) (Butler and Fairey, 1964). It was found only

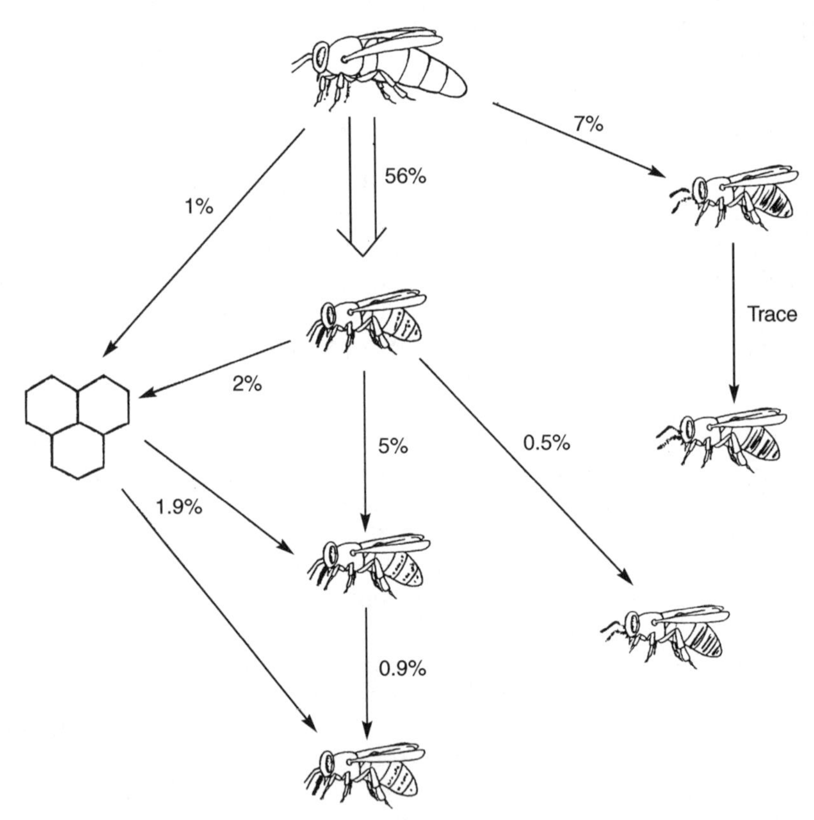


**Fig. 1.8** Diagram of circulation of queen pheromone within a honey bee colony, construed by following fate of radiolabelled 9ODA. (From Winston and Slessor, 1992.) Messenger bees gather pheromone from queen using their antennae or mouthparts. Most pheromone (56%) is taken up by licker bees: antennating bees take up 7%. Figures give percentages of pheromone produced by the queen and transmitted in direction of arrows. A large proportion is taken into bodies of bees. Wax of the comb acts as a minor store from which the pheromone is redistributed.

recently (Kaminski *et al.*, 1990; Slessor *et al.*, 1988) that both enantiomers, the (+) and (−) forms, of 9-HDA are present, along with methyl *p*-hydroxybenzoate and 4-hydroxy-3-methoxyphenylethanol. A mixture of these five compounds elicits **retinue behaviour**, in which workers cluster around the queen, making frequent contact with her. The individual compounds by themselves have no such effect.

The movement of queen pheromone around the colony has been studied by Naumann *et al.* (1991). The queen produces about 500 μg of pheromone per day. She swallows about 36% of this, but the majority (48.5%) is picked up by the 10% of workers that spend much of their time licking the queen. A licker bee plays a key role in the distribution of pheromone (Fig. 1.8). About 40% of the pheromone it receives appears to be swallowed and is removed from circulation, while the rest finds its way within a few minutes to the surface of the cuticle by grooming and passive transport.

Overcrowding of the hive, which occurs when a colony builds up in numbers during the summer, leads to the construction of new queen cells, apparently because queen substance is less efficiently distributed in larger colonies. The existing queen may then leave with an attendant swarm. Winston *et al.* (1990) supplemented the titre of queen substance by adding synthetic equivalents of the total mixture to hives. After removal of the queen from the hive, the equivalent of the pheromone of one queen per day delayed rearing of new queens.

Treatment of queen-right colonies (i.e. with queens) with 10 female equivalents delayed swarming for an average 25 days (Winston *et al.*, 1991). Swarming bees that have lost their queen are attracted upwind by the odour of the 9-ODA, but do not form a stable cluster unless the 9-HDA is also present (Butler and Simpson, 1967). As queen substance is not attractive within the hive and, outside, rarely attracts workers from queen-right colonies, it appears that environmental and physiological factors associated with swarming operate a motivational switch for the workers, making them sensitive to the pheromone. The 9-ODA also serves as a sexual attractant for drones, but again is effective only in certain conditions outside the hive: queens must be flying at between 4 m and 25 m above the ground (depending upon windspeed) before drones respond to them in appreciable numbers. The drones then approach the odour source by flying upwind (Butler and Fairey, 1964) and will approach a source of the 9-ODA flown from a balloon. The amazing persistence of this pheromone component was shown in an unplanned experiment: Butler accidentally contaminated the mast of his yacht with the acid, and found that the mast still attracted drones seven years later.

A drone is attracted to a virgin queen by visual stimuli when he is close to her, but he will mount her after making contact with her in flight only if she carries the odour of the 9-ODA: at short range this appears to act as an aphrodisiac that is essential to the courtship sequence. At short range, also, secretion from tergal glands attracts drones and stimulates copulatory activity (Renner and Vierling, 1977).

A further effect that appears to be intermediate between that of the releaser and a primer is the ability of supplemental pheromone to stimulate foraging and brood-rearing. Higo *et al.* (1992) found that pheromone application increased the harvesting of pollen by 80% and the rearing of brood by 18%.

Queen substance thus has a number of separable functions according to the context or circumstances in which it is used, acting as a primer pheromone in suppressing ovarial development in workers and as a releaser pheromone in most other respects. It can be produced in small quantities by workers, particularly in egg-laying false queens that develop in the absence of a queen.

### 1.5.2 Worker-produced pheromones

Workers do not produce queen pheromone; their main mandibular gland component is (*E*)-10-decenoic acid, which is used in royal jelly (Crewe and Velthuis, 1980). Worker bees produce pheromones from

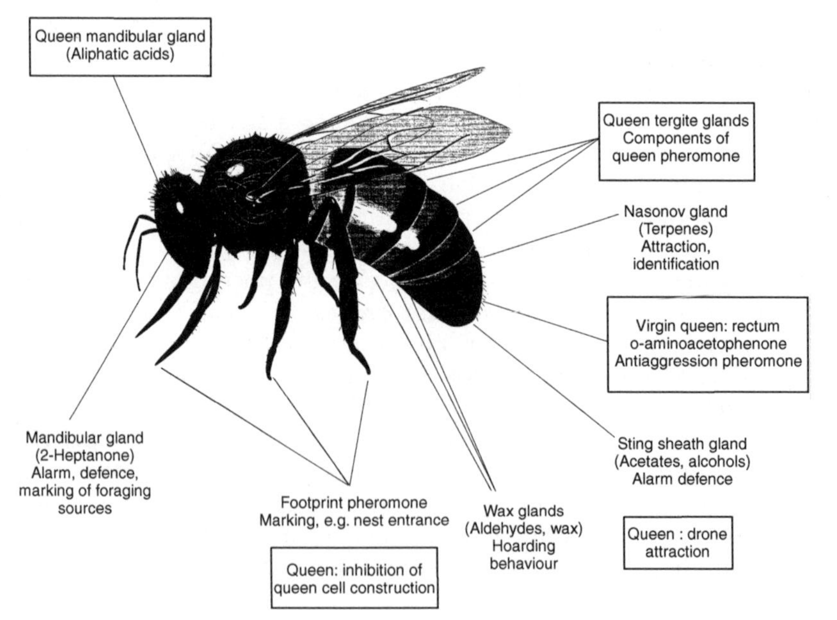


**Fig. 1.9** Honey bee worker showing some main sources of pheromones. Boxed information refers to queen pheromones only. (After various authors: see text.)

several different glands, including the Nasonov gland (Fig. 1.9) between the abdominal tergites. The major component of the gland is geraniol, but this is converted enzymatically in the gland to (*E*)-citral and geranic acid (Pickett *et al.*, 1980), which are very attractive to other workers. The Nasonov gland scent is released when foragers have found water, and sometimes when they have found food (Free and Williams, 1972). It is also released at the hive entrance and serves to guide returning workers: this behaviour can be stimulated by odours from the comb and by 9-HDA (Free, 1987). In addition, the Nasonov pheromone is involved – with queen pheromone – in swarming, and in marking the entrance to a new nest site (Free *et al.*, 1983).

In an established hive or nesting site, the entrance is marked with a 'footprint' pheromone, the glandular origin of which is uncertain (Butler *et al.*, 1969). This may also be used, along with the Nasonov pheromone, to mark foraging sites. In addition, Ferguson and Free (1979) have found evidence for a forage-marking pheromone produced from glands on the dorsal surface of the abdomen.

A variety of pheromones controls alarm and defensive behaviour. When a worker stings, the barbed sting usually remains anchored in position and the poison gland and associated tissues remain with it and are torn from the abdomen. Alarm pheromone is then released from the sting sheath gland. This spreads alarm and incites other workers to sting in the same area. The major component is isopentyl acetate (Boch *et al.*, 1962), but more than 30 other components have been identified from the sting and associated glands, many of which also elicit strong alarm (reviewed by Blum and Fales, 1988), including (Z)-11-eicosen-1-ol (Pickett *et al.*, 1982), which also attracts foraging workers.

The mandibular gland of worker honey bees produces another alarm pheromone consisting mainly of 2-heptanone (Boch and Shearer, 1966) which is used mainly by guard bees at the nest entrance, possibly to repel robber bees (Simpson, 1966). It also appears to be part of a repellent scent mark that forager bees leave on depleted flowers (Vallet *et al.*, 1991).

New pheromones continue to be discovered in honey bees, as in other social insects, including a pheromone emanating from the brood that stimulates foraging and affects ovary development, and pheromones that are involved in recognition of nest-mates (reviewed by Winston, 1987). Capping of worker brood cells is stimulated by the presence of fatty acid esters (methyl palmitate, oleate, linoleate and linolenate) on the surface of the worker and the drone larvae shortly before pupation (Le Conte *et al.*, 1990). The brood pheromone glyceryl-1,2-dioleate-3-palmitate stimulates the nurse bees to incubate the pupae, producing heat by muscular contraction (Koeniger and Vieth, 1983). Virgin queens, which risk being killed within the hive, have been found to produce a pheromone in the rectum that repels workers and reduces their tendency to attack (Post *et al.*, 1987). The main component is *o*-aminoacetophenonone (Page *et al.*, 1988).

In many aspects of honey bee reproductive behaviour, then, the channel of communication for a rich panoply of behaviour patterns is a very narrow one, in which two decenoic acids play a major role. Blum *et al.* (1971) synthesized 19 alkenoic acids closely related to 9-ODA and tested them by flying samples from a balloon in areas where drones were present. None of the synthesized compounds attracted drones: any structural modification of the molecule means loss of the message. The message is therefore very stereotyped and the same for many types of behaviour, but it is clear that the meaning changes in a radical way according to the context. We also learn that pheromonal mixtures from different glandular sources can act in concert to control social behaviour.

## 1.6 EXPLOITING PHEROMONES

Insect control by pheromones alone has limitations, but pheromones can be used in integrated control along with many other methods, as will be explained in detail later. Insecticide use generally precludes the concurrent use of other methods, including most kinds of biological control. In contrast, pheromones can be used in such a way that natural enemies thrive.

One of the first attempts at controlling insects by pheromones was made against the gypsy moth, *Porthetria dispar*. This insect was introduced into Maryland in the United States in the late nineteenth century, with the hope of hybridizing it with the common silkworm to produce an oak-feeding silk-producer. Not surprisingly, this failed, but the insect thrived on native oaks and spread very rapidly throughout the western United States. Attempts were made in 1896 to control it using traps baited with virgin females (Forbush and Fernald, 1896). Later, in the 1940s, crude extracts of female abdominal tips were used in monitoring traps (Holbrook *et al.*, 1960). A research programme was then started to identify the active principle. The first candidate was an alcohol named gyptol, which was identified using a bioassay that measured arousal of males in the laboratory. Although this was active in concentrations approaching 1 pg (Jacobson *et al.*, 1960), it was ineffective in the field. Subsequently (Z)-7,8-epoxyoctadiene (disparlure) was isolated and synthesized (Bierl *et al.*, 1970) and later successfully tested in the field, the (+) enantiomer eventually proving 10 times more effective (Carde *et al.*, 1977).

Earlier in the century, pheromone identification was hampered by the fact that only minute amounts of chemicals are present in a single insect (usually a few nanograms or picograms), and analytical methods of sufficient sensitivity did not exist. In view of this, and of the complexity and diversity of chemical communication systems, it is not surprising that initial attempts to develop insect control methods based on the use of pheromones met with little success. The reasons, more precisely identified, are as follows.

- Mistakes were made in identification.
- Inadequate behavioural bioassays were employed that validated the wrong compounds.
- Synthetic pheromones were not sufficiently pure. (Even minor impurities in the form of the wrong stereo- or optical isomers can greatly reduce male responsiveness.)
- The multicomponent nature of pheromones was overlooked, and hence the possibility that different components may control separate phases of attraction and that they must be present in the correct ratios.
- The orientation of free-flying insects towards an odour source is more complex than originally realized, and so the design and positioning of traps can greatly affect the efficiency of a chemical lure.
- Physiological factors, including sexual maturity, temperature, humidity, previous exposure to pheromone ('habituation') and circadian rhythms of threshold change, were found to affect trap catches.

### 1.6.1 Monitoring

The use of sexual pheromones as lures in monitoring traps is now widespread. Although they usually trap only males, they are particularly useful where large areas have to be surveyed, such as forests and orchards. Monitoring serves four functions: detection of outbreaks, establishment of emergence times of adult insects, distribution mapping, and assessment of changes in abundance. Pheromone monitoring systems can thus provide vital intelligence for the timing of insecticidal control measures (Chapter 9). However, there are no known cases of control by trapping using female sexual pheromones alone. Pheromone monitoring traps have great advantages over light traps: they are cheap to produce and transport, do not require a power source and, above all, are highly selective; no sorting or identification of the catch is needed.

### 1.6.2 Mass trapping

Population reduction by mass trapping becomes feasible with attractants that can outcompete sexual attractants, or when females can be captured. This has rarely been possible (Chapter 10) and perhaps the most successful examples are those of the bark beetle (Scolytidae) trapping programmes carried out in Scandinavia and North America. In these insects, communication involves both aggregation and sexual pheromones (Chapter 2), and so both males and females can be lured into traps.

### 1.6.3 Mating disruption

This technique depends on blocking the communication channel between male and female insects by flooding the medium with sexual pheromone. This effectively increases the noise level so that signals from individual insects are swamped. The exact way in which mating disruption is achieved is not well understood; various phenomena may be involved and it is quite possible that mechanisms differ among different insect species. It may include masking of aerial trails by the persistent cloud of odour, the formation of false trails from dispensers which act as female (or male) mimics, greatly outnumbering the calling insects. It may occur by raising the threshold of chemoreceptors or by habituation of responsiveness of the central nervous system as a result of prolonged exposure to the pheromone. Mating disruption can also be achieved by blocking of antennal receptors through the use of modified pheromones: fluorinated compounds are particularly effective.

### 1.6.4 Inhibition of oviposition

Anti-oviposition pheromones are known to occur in various Lepidoptera and Diptera. They are also known as **epideitic** pheromones, in reference to their effects of reducing intraspecific competition. Female cabbage white butterflies (*Pieris brassicae* and *P. rapae*) add an anti-oviposition pheromone to the eggs during egg-laying. This is produced by glands at the tip of the abdomen (Behan and Schoonhoven, 1978) and inhibits egg-laying by conspecific females. Females of the cherry fruit fly (*Rhagoletis cerasi*) and the apple maggot (*R. pomonella*) mark the fruit around the oviposition puncture by dragging the ovipositor across the surface. A *Rhagoletis pomonella* female drags her ovipositor over the surface of the fruit in which she has just oviposited, leaving a trail which contains a pheromone produced in the midgut and released with other gut contents (Prokopy, 1972). This host-marking pheromone deters other females of the same species from ovipositing on the same fruit. A formulation of oviposition pheromone of cherry fruit fly applied to whole cherry trees has been shown to reduce the number of fruit attacked by this insect by up to 90% (Katsoyannos and Boller, 1980).

### 1.6.5 Alarm pheromones

A component of the alarm pheromone of a number of aphid species is (*E*)-β-farnesene (Bowers *et al.*, 1972), which causes groups of aphids to stop feeding and disperse rapidly. Gibson and Pickett (1983) found that glandular hairs on leaves of the wild potato, *Solanum berthaultii*, released this compound, which repels the aphid *Myzus persicae*. This pheromone has been used to mobilize feeding aphids so that they are more easily targeted by insecticide sprays and fungal pathogens (Griffiths and Pickett, 1980).

### 1.6.6 Use of toxic baits

There has been little use of pheromones in control by baiting, though food attractants and parapheromones have been used extensively. Methyl eugenol, which is a component of many flower fragrances, is strongly attractive to males of certain tephritid Diptera, including the oriental fruit fly, *Bactrocera (Dacus) dorsalis*. It has been successfully used in eradication campaigns in various Japanese islands (Koyama *et al.*, 1984), added to small fibre-board bait blocks impregnated with insecticide. No naturally occurring compound of similar attractiveness has been found for the more economically important Mediterranean fruit fly, *Ceratitis capitata*. 'Trimedlure', a male attractant for this species that is in wide use, is a synthetic parapheromone containing chlorine atoms: the reasons for its properties are not understood. This has been used in bait sprays, but protein-based mixtures such as hydrolysed yeast are more commonly used because they attract insects of both sexes.

The trail pheromones of ants and termites have been explored as potential attractants for use in bait particles that can be harvested and spread throughout the colony, but their use has proved to be uneconomical. In leaf-cutting ants, trail pheromone has been shown to increase the radius of attraction of baits by a small amount (Robinson *et al.*, 1982) but not to increase harvesting.

### 1.6.7 Stimulo-deterrent methods

As we understand more about the action of different kinds of pheromone within the natural environment, possibilities open up of moving populations of insects out of crops into places where they do less harm, or can be destroyed. This may involve deterring insects from colonizing certain plants and, at the same time, attracting them to other areas; hence the term 'stimulo-deterrent'. This has antecedents in the use of trap crops, which have been used for control of the boll weevil (*Anthonomus grandis*) in the United States (Lloyd, 1986). Strips of cotton are planted early in the season and baited with grandlure, the synthetic pheromone of the boll weevil. Weevils are then killed with insecticide before they can spread to the commercial crop, which fruits later. The potential for the use of stimuli-deterrent strategies is increasing as our knowledge of the role of kairomones and allomones increases. For example, kairomones can be used to guide parasitoids and predators to their hosts or to areas of crops containing the hosts. Allomones can be used as antifeedants; components or extracts of the neem tree, *Azadirachta indica*, have been used in this way against various insect pests (reviewed by Jermy, 1990). The greater use of semiochemicals in insect control is compatible with biological control and it is to be hoped that it will herald a new era in which the use of chemical pesticides will decline.

### 1.6.8 Bioelectrostatic methods

Potential now exists for the use of electrostatically charged dry powders as carriers for insecticides and biologically active chemicals (Howse, unpublished). Powders can be designed which adhere to the insect cuticle by electrostatic forces and can be used as slow-release substrates for semiochemicals, microbial insecticides, etc. In the future, pest control may be achieved by attracting insects to bait stations with sex pheromones, where they pick up an inoculum of slow-acting pesticides or pathogen, which they then pass on to mates during the mating process.

## REFERENCES

Aihara, Y. and Shibuya, T. (1977) Response of single olfactory receptor cells to sex pheromones in the tobacco budworm moth, *Spodoptera litura. J. Insect Physiol.*, **22**, 779–784.

Albert, P.T., Seabrook, W.D and Paim, U. (1974) Isolation of a sex pheromone receptor in males of the eastern spruce budworm, *Choristoneura fumiferana* (Clem.) (Lepidoptera: Tortricidae). J. *Comp. Physiol.*, **91**, 79–89.

Behan, M. and Schoonhoven, L.M. (1978) Chemoreception of an oviposition deterrent associated with eggs in *Pieris brassicae. Ent. Exp. Appl.*, **24**, 163–179.

Bierl, B.A., Beroza, M. and Collier, C.W. (1970) Potent sex attractant of the gypsy moth: its isolation, identification and synthesis. *Science, NY*, **170**, 87–89.

Bjostad, L.B., Wolf, W.A. and Roelofs, W.L. (1987) Pheromone biosynthesis in lepidopterans: desaturation and chain shortening. In *Pheromone Biochemistry* (eds G.D. Prestwich and G.J. Blomquist), Academic Press, New York, pp. 77–120.

Blum, M.S. (1977) Behavioural responses of Hymenoptera to pheromones, allomones and kairomones. In *Chemical Control of Insect Behaviour, Theory and Applications*, (eds H.H. Shorey and T.J. McKelvey), Wiley, New York and London, pp. 149–168.

Blum, M.S. and Fales, H.M.(1988) Eclectic chemisociality of the honeybee: a wealth of behaviours, pheromones, and exocrine glands. *J. Chem. Ecol.*, **14**, 2099–2107.

Blum, M.S., Boch, R., Doolittle, R.E., *et al.* (1971) Honey bee sex attractant: conformational analysis, structural specificity, and lack of masking activity of congeners. *J. Insect. Physiol.*, **17**, 349–364.

Boch, R. and Shearer, D.A. (1966) 2-Heptanone and 10-hydroxy-trans-dec-2-enoic acid in the mandibular glands of honey bees of different ages. *Z. vergl. Physiol.*, **54**, 1–11.

Boch, R., Shearer, D.A. and Stone, B.C. (1962) Identification of iso-amyl acetate as an active compound in the sting pheromone of the honey bee. *Nature*, **195**, 1018–1020.

Boeckh, J. (1962) Elektrophysiologische Untersuchungen an einzelnen Geruchsrezeptoren auf den Antennen des Totengrabers (*Necrophorus*, Coleoptera). *Z. vergl. Physiol.* **46**, 212–248.

Boeckh, J. (1984) Neurophysiological aspects of insect olfaction. In *Insect Communication. 12th Symposium, Royal Entomological Society of London* (ed. T. Lewis), Academic Press, London,

Boeckh, J. and Ernst, K.-D. (1987) Neurophysiology of insect olfaction. *J. Comp. Physiol. A.*, **161**, 550.

Bowers, W.S., Nault, L.R., Webb, R.E. and Dutky, S.R. (1972) Apid alarm pheromone: isolation, identification, synthesis. *Science*, **117**, 1121–1122.

Bradshaw, J.W.S. and Howse, P.E. (1984) Sociochemicals of ants. In *Chemical Ecology of Insects* (ed. W.J. Bell and R.T. Carde), Chapman & Hall, London and New York, pp. 429–474.

Brown, W.L., Eisner, T. and Whittaker, R.H. (1970) Allomones and kairomones: transpecific chemical messengers. *BioScience*, **20**, 21–22.

Butenandt, A., Beckmann, R., Stamm, D. and Hecker E. (1959) Uber den Sexuallockstoff des Seidenspinners *Bombyx mori*. Reindarstelling und Konstitution. *Z. Naturforsch.*, **14b**, 283–284.

Butler, C.D. and Fairey, E.M. (1964) Pheromones of the honeybee: biological studies of the mandibular gland secretion of the queen. *J. Apicult. Res.*, **3**, 65–76.

Butler, C.D. and Simpson, J. (1967) Pheromones of the queen honeybee (*Apis mellifera* L.) which enable workers to follow her when swarming. *Proc. R. Ent. Soc. Lond. (A)*, **142**, 149–154.

Butler, C.D., Fletcher, D.J.C. and Walter, D. (1969) Nest entrance marking with pheromones by the honey bee *Apis mellifera L.* and by a wasp *Vespa vulgaris L. Anim. Behav.*, **17**, 142–147.

Callow, R.K. and Johnstone, N.C. (1960) The chemical constitution and synthesis of queen substances of honeybees (*Apis mellifera L.*). *Bee World*, **41**, 152–153.

Carde, R.T., Doane, C.C., Balzer, T.C. *et al.* (1977) Attractancy of optically active pheromone for male gypsy moths. *Environ. Ent.*, **6**, 768–772.

Crewe, R.M. and Veltuis, H.H.W. (1980) False queens: a consequence of mandibular gland signals in worker honey bees. *Naturwiss.*, **67**, 467–469.

Dicke, L. and Sabelis, M.W. (1992) Costs and benefits of chemical information conveyance: proximate and ultimate factors. In *Insect Chemical Ecology* (eds B.D. Roitberg and M.S. Isman), Chapman & Hall, New York and London, pp. 122–155.

Dudai, Y. (1976) Properties of learning and memory in *Drosophila melanogaster. J. Comp. Physiol.*, **114**, 69–89.

Dudai, Y., Jan. J.-N., Byers, D. *et al.* (1976) *Dunce*, a mutant of *Drosophila* deficient in learning. *Proc. Natl Acad, Sci., USA*, **73**, 1984–1688.

Dumpert, K. (1970) Alarmstoffrezeptoren auf der Antenne von *Lasius fuliginosus* (Latr.) (Hymenoptera, Formicidae). *Z. vergl. Physiol.*, **76**, 403–425.

Erber, J. (1975) The dynamics of learning in the honey bee (*Apis mellifica carnica*). The time dependence of the choice reaction. *J. Comp. Physiol.*, **99**, 231–242.

Fabre, J-H. (1913) *The Life of the Caterpillar*. Translated from *Souvenirs Entomologiques* by A.T. de Mattos. Hodder & Stoughton, London and New York, 382 pp.

Ferguson, A.W. and Free, J.B (1979) Production of a forage marking pheromone by the honeybee. *J. Apic. Res.* **18**, 128–135.

Fitzgerald, T.D. (1976) Trail marking by larvae of the eastern tent caterpillar. *Science, NY*, **194** 961–963.

Forbush, E.H. and Fernald, C.H. (1896) *The Gypsy Moth*, Wright & Potter, Boston.

Free, J.B. (1987) *Pheromones of Social Bees*, Comstock Pub., Ithaca, NY.

Free, J.B. and Williams, I. (1972) The role of the Nasonof gland pheromone in crop communication by honeybees. *Behaviour*, **41**, 314–318.

Free, J.B., Ferguson, A.W. and Al-Sa'ad, B.N. (1983) Effect of the honeybee Nasanov and alarm pheromone components on behaviour at the nest entrance. *J. Apic. Res.*, **22**, 214–223.

Gibson, R.W. and Pickett, J.A. (1983) Wild potato repels aphids by release of aphid alarm pheromone. *Nature, Lond.*, **302**, 608–609.

Griffiths, D.C. and Pickett, J.A. (1980) A potential application of aphid alarm pheromones. *Ent. Exp. Applicata*, **27**, 199–201.

Gullan, P.J. and Cranston, P.S. (1994) *The Insects: an Outline of Entomology*, Chapman & Hall, London and New York.

Hansson, B.S., Ljungberg, H., Hallberg, E. and Lofstedt, C. (1992) Functional specialization of olfactory glomeruli in a moth. *Science, Wash.*, **256**, 1313–1315.

Haskins, C.P., Hewitt, R.E. and Haskins, F. (1973) Release of aggressive and capture behaviour in the ant *Myrmecia gulosa* F. by exocrine products of the ant *Camponotus*. *J. Ent. (A)*, **47**, 125–139.

Herard, F., Keller, M.A., Lewis, W.J. and Tumlinson, J.H. (1988) Beneficial arthropod behaviour mediated by airborne semiochemicals. IV. Influence of host diet on host-oriented flight chamber responses of *Microplitis demolitor* Wilkinson. *J. Chem. Ecol.* **14**, 1597–1606.

Higo, H.A., Colley, S.J., Winston, M.L. and Slessor, K.N. (1992) Effects of honey bee queen mandibular gland pheromone on foraging and brood rearing. *Can. Entomologist*, **124**, 409–418.

Holbrook, R.F., Beroza, M. and Burgess, E.D. (1960) Gypsy moth (*Porthetria dispar*) detection with the natural female sex lure. *J. Econ. Ent.*, **53**, 751.

Isingrini, M., Lenoir, A. and Jaisson, P. (1985) Preimaginal learning as a basis of colony-brood recognition in the ant *Cataglyphis cursor*. *Proc. Natl. Acad. Sci. USA*, **82**, 8545–8547.

Jacobson, M., Beroza, M. and Jones, W.A. (1960) Isolation, identification, and synthesis of the sex attractant of the gypsy moth. *Science*, **132**, 1011–1012.

Jaisson, P. (1975) L'impregnation dans l'ontogenese du comportement desoins aux cocons chez la jeune formi rousse (*Formica ployctena* Forst.). *Behaviour*, **52**, 1–37.

Jaisson, P. (1980) Environmental preference induced experimentally in ants (Hymenoptera: Formicidae). *Nature, Lond.*, **286**, 388–389.

Jang, E.B., Light, D.M., Flath, R.A., *et al.* (1989) Electrantennogram responses of the Mediterranean fruit fly, *Ceratitis capitata*, to identified volatile constituents from calling males. *Ent. Expl. et Appl.*, **50**, 7–9.

Jermy, T. (1990) Prospects of antifeedant approach to pest control – a critical review. *J. Chem. Ecol.*, **16**, 3151–3166.
Kaissling, K.-E. (1971) Insect olfaction. In *Handbook of Sensory Physiology* (ed. L.M. Beidler Vol.4, Springer, Berlin, and NewYork, pp. 351–431.
Kaissling, K.-E. (1972) Kinetic studies of transduction in olfactory receptors of *Bombyx mori.* In *Olfaction and Taste* (ed. D. Schneider), Wiss, Stuttgart, pp. 207–213.
Kaissling, K.-E. (1974) Sensory transduction in insect olfactory receptors. *In Biochemistry of Sensory Function* (ed. L. Jainicke), Springer Verlag, Berlin and London, pp. 243–273.
Kaissling, K.-E. and Priesner, E. (1970) Die Riechswelle des Seidenspinners. *Naturwiss.*, **57**, 23–28.
Kaissling, K.-E., Kasang, G., Bestmann, H.J. *et al.* (1978) A new pheromone of the silkworm moth *Bombyx mori*, sensory pathway and behavioural effect. *Naturwiss.*, **65**, 382–384.
Kaminski, L.-A., Slessor, K.N., Winston, M.L. *et al.* (1990) Honey bee response to queen mandibular pheromone in a laboratory bioassay. *J. Chem. Ecol.*, **16**, 63–64.
Karlson, P. and Lüscher, M (1959) Pheromone, ein Nomenklatur-vorschlag fur ein Wirkstoffklasse. *Naturwiss.*, **46**, 63–64.
Katsoyannos, B.E. and Boller, E.F. (1980) Second field application of oviposition deterring pheromone marking pheromone of the European cherry fruit fly, *Rhagoletis cerasi* L. (Diptera: Tephritidae). *Z. angew. Ent.*, **89**, 278–281.
Kochansky, J., Taschenberg, E.F., Carde, R.T. *et al.* (1975) Sex pheromone of the moth *Antheraea polyphemus. J. Insect Physiol.*, **21**, 1977–1983.
Koeniger, N.I. and Vieth, M.J. (1983), Glyceryl-1,2-dioleate-3-palmitate, a brood pheromone of the honeybee (*Apis mellifera* L.). *Experientia*, **39**, 3137–3150.
Koltermann, R. (1971) Rassen-bzw. artspezifische Duftbewertung bei der Honigbiene und oekologische Adaptation. *J. Comp. Physiol.*, **85**, 327–360.
Koltermann, R. (1974) Periodicity in the activity and learning of the honeybee. In *Experimental Analysis of Insect Behaviour*, (ed. L. Barton-Browne), Springer, Berlin and New York, pp. 228–236.
Koyama, J., Teruya, T. and Tanaka, K. (1984) Eradication of the oriental fruit fly (Diptera: Tephritidae) from the Okinawa islands by a male annihilation method. *J. Econ. Ent.*, **77**, 468–472.
Le Conte, Y., Arnold, G., Trouiller,T. and Masson, C. (1990) Identification of a brood pheromone in honeybees. *Naturwiss.*, **77**, 334–336.
Lewis, W.T. and Tumlinson, J.H. (1988) Host detection by chemically mediated associative learning in a parasitic wasp. *Nature, Lond.*, **331**, 257–259.
Lloyd, E.P. (1986) The boll weevil: research developments and progress in the USA. *Agricultural Zoology Review*, **I**, 109–135.
Meng, L.Z., Wu, C.H., Wicklein, M. *et al.* (1989) Number and sensitivity of three types of pheromone receptor cells in *Antheraea pernyi* and *A. polyphemus. J. Comp. Physiol.* A, **165**, 139–146.
Menzel, R., Erber, J. and Mashur, T. (1974) Learning and memory in the honeybee. In *Experimental Analysis of Insect Behaviour*, (ed L. Barton-Browne), Springer, Berlin and New York, pp. 195–217.
Naumann, K., Winston, M.L., Slessor, K.N. *et al*, (1991) The production and transmission of honey bee queen (*Apis mellifera* L.) mandibular gland pheromone. *Behav. Ecol. and Sociobiol.*, **29**, 321–332.
O'Connell, R.J. (1984) Electrophysiological responses to pheromone blends in

single olfactory receptor neurones. In *Insect Communication. 12th Symposium, Royal Entomological Society of London* (ed. T. Lewis), Academic Press, London.

Olberg, R.M. (1983) Pheromone-triggered flip-flopping interneurons in the ventral nerve cord of the silkworm moth, *Bombyx mori. J. Comp. Physiol.*, **152**, 297–307.

Page, R.E., Blum, M.S. and Fales, H.M. (1988) *o*-Aminoacetophenone, a pheromone that repels honeybees (*Apis mellifera* L.). *Experientia*, **44**, 270–271.

Papaj, D.R. and Vet, L.E.M. (1990) Odor learning and foraging success in the parasitoid, *Leptopilina heteroma. J. Chem Ecol.*, **16**, 3137–3150.

Pickett, J.A, Williams, I.H., Martin, A.P. and Smith, M.C. (1980) The Nasonov pheromone of the honeybee, *Apis mellifera* L. (Hymenoptera, Apidae). Part 1. Chemical characterization. *J. Chem. Ecol.*, **6**, 425–434.

Pickett, J.A., Williams, I.H. and Martin, A.P. (1982 (Z)-11-Eicosen-1-ol, an important new pheromonal component of the sting of the honeybee, *Apis mellifera* L. (Hymenoptera, Apidae). *J. Chem. Ecol.*, **8**, 163–175.

Post, D.C., Page, R.E and Erickson, E.H. (1987) Honeybee (*Apis mellifera* L.) queen feces: source of a pheromone that repels worker bees. *J. Chem. Ecol.*, **13**, 583–591.

Priesner, E. (1968) Die interspezifischen Wirkungen der Sexuallockstoffe der Saturniidae (Lepidptera). *Z. Vergl. Physiol.*, **61**, 263–297.

Prokopy, R.J. (1972) Evidence for a marking pheromone deterring repeated oviposition in apple maggot flies. *Environ. Ent.*, **1**, 326–332.

Robinson, S.W., Jutsum, A.R., Cherrett, J.M. and Quinlan, R.J. (1982) Field evaluation of methyl-4-methyl pyrrole-2-carboxylate, an ant trail pheromone, as a component of baits for leaf cutting ant (Hymenoptera: Formicidae) control. *Bull. Ent. Res.*, **72**, 345–356.

Schneider, D., Lacher, V. and Kaissling, K.-E. (1964) Die Reaktionsweise und das Reaktionsspektrum von Riechzellen bei *Antheraea pernyi* (Lepidoptera, Saturniidae). *Z. vergl. Physiol.*, **48**, 632–662.

Schoonhoven, L.M. (1990) Host-marking pheromones in Lepidoptera, with special reference to two *Pieris* spp. *J. Chem. Ecol.*, **16**, 3034–3052.

Shannon, C. and Weaver, W. (1949) *The Mathematical Theory of Communication*, Illinois University Press.

Simpson, J. (1966) Repellency of the mandibular gland scent worker honey bees. *Nature, Lond.*, **209**, 531–532.

Slessor, K.N., Kaminsky, L.-A., King, G.G.S. *et al.* (1988) Semiochemical basis of the retinue response to queen honey bees. *Nature, Lond.*, **332**, 354–356.

Steinbrecht, R.A. and Kasang, G. (1972). Capture and conveyance of odour molecules in an insect olfactory receptor. In *Olfaction and Taste*, Vol. IV (ed. D. Schneider), Wiss, Verlagesellschaft, Stuttgart, pp. 193–199.

Steinbrecht, R.A. and Muller, B. (1971) On the stimulus conducting structures in insect olfactory receptors. *Z. Zellforschung*, **117**, 570–575.

Swihart, C.A. and Swihart, S.L. (1970) Colour selection and learned feeding preference in the butterfly, *Heliconius charitonius* Linn. *Anim. Behav.*, **18**, 60–64.

Thorpe, W.H. and Jones, F.G.W. (1937) Olfactory conditioning in a parasitic insect and its relation to the problem of the host selection. *Proc. R. Ent. Soc. B*, **124**, 56–81.

Tinbergen, N., Meeuse, B.J.D., Boerma, L.K. and Varossisan, W.W. (1942). Die Balz des Samtfalters, *Eumenis (= Satyrus) semele* (L.).*Z. Tierpsychol.*, **5**, 182–226.

Vallet, A., Cassier, P. and Lensky, Y. (1991) Ontogeny of the fine structure of the mandibular glands of the honeybee (*Apis mellifera* L.) workers and the pheromonal activity of 2-heptanone. *J. insect Physiol.* **37**, 789–804.

Visser, J.H. and De Jong, R. (1988) Olfactory coding in the perception of semiochemicals. *J. Chem. Ecol.*, **14**, 2019–2028.

Vogt, R.G. and Riddiford, L.M. (1986) Pheromone reception, a kinetic equilibrium. In *Mechanisms in Insect Olfaction*, (eds T.L. Payne, M.C. Birch and C.E.Kennedy), Clarendon Press, Oxford, pp. 157–162.

Whittaker, R.H. and Feeny, P. (1971) Allelochemicals: chemical interactions between species. *Science*, **171**, 757.

Wilson, E.O. and Bossert, W.H. (1963) Chemical communication among animals. *Recent Progress in Hormone Research*, **19**, 673–716.

Winston, M.L. (1987) *The Biology of the Honey Bee*, Harvard Univ. Press. Cambridge, Mass.

Winston, M.L. and Slessor, K.N. (1992) The essence of royalty: honey bee queen pheromone. *American Scient.*, **80**, 374–385.

Winston, M.L., Higo, H.A. and Slessor, K.N. (1990) Effect of various dosages of queen mandibular pheromones on worker attraction to swarm clusters and inhibition of queen rearing in the honey bee (*Apis mellifera* L.). *Insectes Soc.*, **36**, 15–27.

Winston, M.L., Higo, H.A., Colley, S.J. *et al.* (1991) The role of queen mandibular pheromone and colony congestion in honey bee (*Apis mellifera* L.) reproductive swarming. *J. Insect Behav.*, **4**, 649–659.

# 2

# The role of pheromones in insect behaviour and ecology

## 2.1 PRIMER PHEROMONES

Primer pheromones may act upon sexual maturation of adult insects (as in the honey bee, Chapter 1) or on development. In locusts both kinds of effect occur. At low densities the desert locust, *Schistocerca gregaria,* is cryptically coloured, but in high densities the insects undergo a startling colour change and they aggregate, sometimes forming swarms of hundreds of millions of insects. This phase change involves graded changes in colour, morphology, behaviour and physiology. Crowding of hoppers is one factor that leads to gregarization as a result of the experience of increased tactile stimulation (Ellis, 1959). The existence of a gregarization pheromone has been demonstrated in nymphal faeces, changing behaviour and colour in seven days (Gillett, 1983). Exposure to adult faeces, on the other hand, leads to solitary behaviour within four days without affecting colour. There has been controversy over the chemical nature of the gregarization pheromone, but Fuzeau-Braesch *et al.* (1988) found that the air from around gregarious *Locusta migratoria* and *Schistocerca gregaria* was rich in phenol, guaiacol and veratrole. A mixture of these three compounds causes locusts to aggregate and they may be part of the gregarization pheromone, which produces a 'cohesive' effect.

Phase change colour can occur within the life span of an individual or can progress across generations through a process of epigenetic maternal inheritance (Islam *et al.,* 1994a,b). By using behavioural changes as an index of phase change, Islam *et al.* discovered a maternally transmitted agent (semiochemical) which acts on developing embryos and influences their behaviour on hatching. It appears that the female locust is using her own experience of crowding, right up to the time of egg-laying, as an indication that her offspring will emerge into a population of high density. She then triggers changes in their colour and behaviour by adding a primer pheromone to the foam surrounding the egg pod.

Desert locusts that are sexually mature produce a pheromone which accelerates the development of sexual maturity in other males, resulting in them rapidly turning from the typical pink colour to the yellow colour of the mature insects. Ovarian development is accelerated in sexually immature females in the presence of mature males, even if copulation is prevented, again as a result of pheromonal action (Norris, 1968).

## 2.2 COURTSHIP IN LEPIDOPTERA

Day-flying Lepidoptera depend very much upon vision for orientation, either to nectar sources or to mates. Communication by visual signals is rapid over relatively long distances; the visual elements may include colour, ultraviolet reflectance patterns and patterns of flight. Nevertheless, male butterflies will often respond to leaves, flower petals or pieces of paper blowing in the wind. There have been various studies on the nature of such visual stimuli using models., for example, Tinbergen *et al.* (1942) showed that male grayling butterflies (*Hipparchia semele*) responded to objects towed through the air on the end of a fishing rod (Chapter 1), demonstrating that motion was more important than colour eliciting courtship approach flights from females. Magnus (1958) studied the silver-washed fritillary, *Argynnis paphia*, placing models on the end of a carousel apparatus. Again, it was found that the males were attracted more by a flapping model, which presented a flickering image, than by colour or wing pattern. These insects have a basic brown or orange ground pattern to the wing, respectively. Crane (1955) investigated courtship of the passion vine butterfly, *Heliconius erato*, which has blackish wings with a splash of bright red across the forewing. She painted over the red with a variety of colours and found that the red band acted as a sexual attractant, and that colours close to red, such as yellow and orange, produced more responses from males than blue or green.

We cannot assume that flicker is generally more important than any other feature of a visual signal in courtship. Patterns of ultraviolet reflection and absorbance on the wings which govern the responsiveness of the males towards females have been shown to be important as releasers of courtship behaviour and species-isolating mechanisms in some butterflies. In one of the cabbage white butterflies (*Pieris rapae crucivora*) courtship depends on visual stimuli alone (Obara, 1970). Males can discriminate between resting females and other males, and respond equally well to a dead female in a closed glass box. The undersurfaces of the female's hindwing reflect 40% of near UV, compared with only 10% in the male, and a paper model with similar reflectivity to the female wing strongly attracted males, which also attempted copulation with it. Males of the sympatric yellow pierids *Colias eurytheme* and *C. philodoce* have different UV reflectance patterns in their wings, the former reflecting and the latter absorbing (Silberglied and Taylor, 1978). Female *C. eurytheme* respond to the UV reflectance on male wings, but not to the reflectance in visual spectrum, while female *C. philodoce* respond neither to 'visual' nor UV reflectance patterns. Male pheromones differ between the two species (see below) and it appears that the females depend more upon these in discriminating between males.

However, it is generally true that pheromones produced by males interact with, and supplement, visual stimuli at close range. In the accounts of butterfly collectors over the past century there are many records of distinctive odours of male butterflies, but research in this field has been limited to very few species. It appears that male-produced pheromones are used at close range in courtship as gate-keepers to mating, or 'aphrodisiacs'. In the grayling butterfly, Tinbergen *et al.* (1942) described a courtship ritual (Fig. 2.1) which begins with the female placing her antennal clubs against scent scales on the forewing of the male. There is analogous behaviour in the silver-washed fritillary, in which the male has distinctive black forewing bands packed with scent scales, also known as *androconia*. *Colias philodice* females select males on the basis of pheromones produced by androconial scales on the wing base, which spread across the wing surface (Grula *et al.*, 1980;

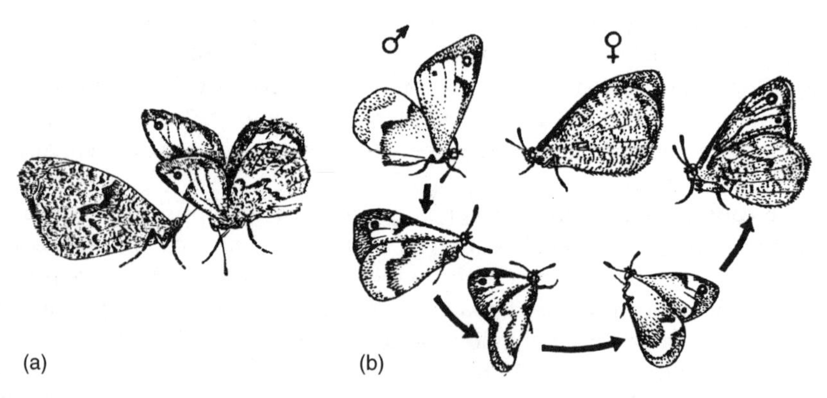


**Fig. 2.1** Courtship ritual of grayling butterfly, *Hipparchia semele*: (a) bowing (male on right); (b) half-circle performed by male after bowing. (After Tinbergen *et al.*, 1942.)

Sappington and Taylor, 1990). The main components of the pheromone are *n*-hexyl palmitate, myristate and stearate, while *C. eurytheme* males produce 13-methylheptacosane, *n*-heptacosane and *n*-nonocosane. It is suggested that such high molecular weight compounds that are relatively involatile are suited to use at very close range during courtship and small amounts equip a male with a lifelong pheromone supply. This supply builds up to a steady level 12 hours after eclosion from the pupa, and thereafter very little is lost from the wings at normal field temperatures. Courtship involves the female making antennal contact with the wings of the male, and the male beats his wings against a perched female during courtship. UV is reflected strongly from the dorsal surface of the wings of the male of the small sulphur butterfly, *Eurema lisa*, but not from the wings of the female. This lack of reflectance is one factor that initiates courtship (Rutowski, 1978), which involves a rapid sequence of events. The wing pheromones of the male elicit abdominal extension of the female, but they are not absolutely essential to the courtship sequence, as some females will extend their abdomens in response to the visual stimulus of an approaching male.

The courtship of pierid butterflies is a brief affair: in *Eurema lisa* it is over in three to four seconds (Rutowski, 1978). It appears that the insects are vulnerable when their attention is diverted by the opposite sex. Distasteful butterflies, on the other hand, can afford a more leisurely approach. Brower *et al.* (1965) described the courtship behaviour of the queen butterfly *Danaus gilippus berenice*, analysing it in terms of a stimulus–response sequence. The courtship is eye-catching and takes an average of 32 seconds. Again, the first approach is visual. The male overtakes the female in the air and extrudes from the tip of its abdomen a pair of brushes, or 'hair pencils', which it manoeuvres above the antennae of the female (Fig. 2.2). This has an arrestant effect on a receptive female, who lands, at which juncture the male continues hair pencilling while the female is perched (ground hair pencilling). The male then alights to the side of the female and copulation begins.

The hair pencils (Fig. 2.3) are androconial organs which dispense pheromone on particles of cuticular dust that adhere to the antennae of the female. These pheromone transfer particles are extremely fine (around 1–5 microns in diameter), and their forms differ according to species. They detach from the hairs and fall under gravity, adhering to the antennae of the female by virtue of an oily hydrocarbon component. A pheromone, given the trivial name danaidone, stimulates chemoreceptors on the antennae of the female (Meinwald *et al.*, 1969; Myers and Brower, 1969).

Amateur entomologists have been perennially frustrated by the apparent inability to breed danaid butterflies in cages in captivity;

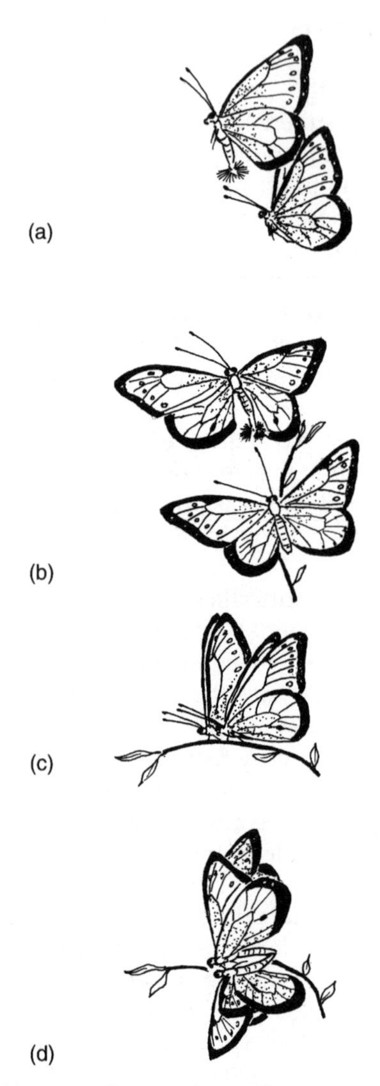


**Fig. 2.2** Courtship behaviour of queen butterfly, *Danaus gilippus berenice.* (After Brower *et al.*, 1965). Courtship begins with aerial pursuit, followed by hair pencilling, in which male stays just above female, extrudes his hair pencils and dusts her antennae with pheromone. A responsive female slows her flight and alights on vegetation while male flies in a narrow arc around her. If she flutters her wings, he withdraws his hair pencils and strikes down on her with his body. She closes her wings; he then alights alongside and attempts copulation.

courtship is very rarely successful. The reason for this came to light as a result of observations carried out by Edgar *et al.* (1973) on Australasian Danaidae. They saw males in the field sucking juices from decaying heliotrope plants, and showing sexual behaviour towards other males in the vicinity. It was found that the males extract poisonous pyrrolizidine alkaloids (PAs) from the plants and convert them to pyrrolizines, such as danaidone. Recently, it has been found that this extraction of pheromone precursors occurs in a wide range of Lepidoptera, including monarchs (*Danaus plexippus*), arctiid moths such as *Creatonotus*, ctenuchiid moths, and ithomiine butterflies (Boppré, 1990; Schneider *et al.*, 1982; Eisner and Meinwald, 1987).

Three possible types of interaction occur between Lepidoptera and PA-containing plants. The latter include Compositae, such as *Senecio*, Boraginaceae, such as *Heliotropium*, and Leguminosae such as *Crotalaria* (Fig. 2.4). In some Arctiidae, such as *Utetheisa* (Culvenor and Edgar, 1972), the larval food plants contain PAs which are sequestered also in the pupal and adult stages and the sexual pheromone is then biosynthesized from them. Adult male danaid butterflies, such as *Danaus*, *Amauris* and *Euploea* species and various ithomiine butterflies, are attracted to dead and decaying PA-containing plants and extract the PAs in liquid they regurgitate from their probosces (Boppré, 1990).

In the Danaidae, the pyrrolizidines are believed to accumulate in a small gland near the centre of each hindwing of the male, into which the folded hair pencils are inserted at regular intervals. Neotropical Ithomiinae dispense pheromones from hair pencils on the front margin of the hindwings, which are used in courtship during hovering flight. Pursued males cant their hindwings in flight, exposing the hair pencils, and the pheromone repels conspecific males and the males of other species that share the same pheromone components (Pliske, 1975). Many species of ithomiine fly together in Müllerian mimicry rings, and the repellent action of the male pheromone prevents time being wasted in unsuccessful courtship. However, other studies suggest that in certain circumstances the scent organs are used to aggregate males of the same species which form mating leks, and can attract other species forming part of a Müllerian mimicry ring (Haber, 1978).

Species of *Danaus* and *Amauris* extract the PA lycopsamine (Fig. 2.4), which is simply converted into danaidone, while *Euploea* converts it mainly into the corresponding aldehyde. Ithomiines studied by Pliske (1975) use the other part of the lycopsamine molecule, which forms a lactone. A third variant is shown by *Utestheisa*, which synthesizes aldehydes from the lycopsamine that it sequesters as a larva.

PA compounds have been shown to have a variety of possible functions. For example, they act as aphrodisiac pheromones or pheromone precursors. The volatile element, hydroxydanaidal, may be

a)
b)
c)
d)

termed a kairomone because it indicates plant sources from which the adult or the larvae can feed. The butterfly orchid, *Epidendrum paniculatum*, uses PAs as allomones, attracting arctiid and ctenuchiid moths to its flowers (DeVries and Stiles, 1990). In the moth *Creatonotus*, PAs have a primer effect; the males of this insect have enormous coronemata (Fig. 2.3). These are inflatable and extensible structures. Their size is related to the amount of PAs consumed by the larva from its foodplant (Schneider, *et al.* 1982). Adult butterflies sequester PAs and use them as defensive allomones (Brown, 1984): in ithomiines they find their way to the cuticle and the reproductive organs (from where they may be transferred to eggs). Brown showed that the PAs protect the butterflies against the predatory web-spinning spider *Nephila*, which cuts the insects out of its web rather than feeding on them. Other species of Lepidoptera use PAs to defend themselves against vertebrates such as toads and lizards (Boppré, 1990).

### 2.2.1 Aggregations and leks

Harvesting of chemicals for pheromone synthesis occurs also in solitary euglossine bees, which are conspicuous and often brightly coloured elements of the neotropical fauna. The males have organs on their hind tibiae consisting of a brush with a cavity (reminiscent of a dustpan) which they use to collect scents from flowers such as orchids, and from fruit and other sources (Williams and Dobson, 1972; Dodson, 1975; Dressler, 1982). According to Dodson, each species collects a different selection of floral fragrances. Certain bees collect a sufficiently varied cocktail to become 'attractors' for other male bees of the same species, which then form leks that attract females, probably by the visual stimuli from a cloud of brightly coloured males. Whitten *et al.* (1989) found that the males of *Eulema cingulata* secrete the contents of their labial glands, which consist of lipids, on to the surface of flowers. The lipids act as solvents and apparently aid in mopping up the scents, which are then transferred to the tibial organs of the hind legs.

Lek formation has also been described in the saltmarsh caterpillar moth, *Estigmene acrea* (Willis and Birch, 1982). Like some other members of the Arctiidae, the male is able to inflate enormous coronemata, which

---

**Fig. 2.3** Scent organs of Lepidoptera. (a) Hair pencils of a male Ithomia butterfly (from Pliske, 1975). (b) Male *Creatonotus gangis* showing inflated coronemata, used to disperse pheromone (redrawn from Schneider *et al.*, 1982). (c) Hair pencil of a danaid butterfly (photo: T. Eisner). (d) Hair pencil of angle shades moth, *Phlogophora meticulosa* (from Birch, 1970). Male (right) responds in flight to a calling female (left) by everting a pair of abdominal hair pencils and fanning the scent (dotted lines) towards the female with his wings.

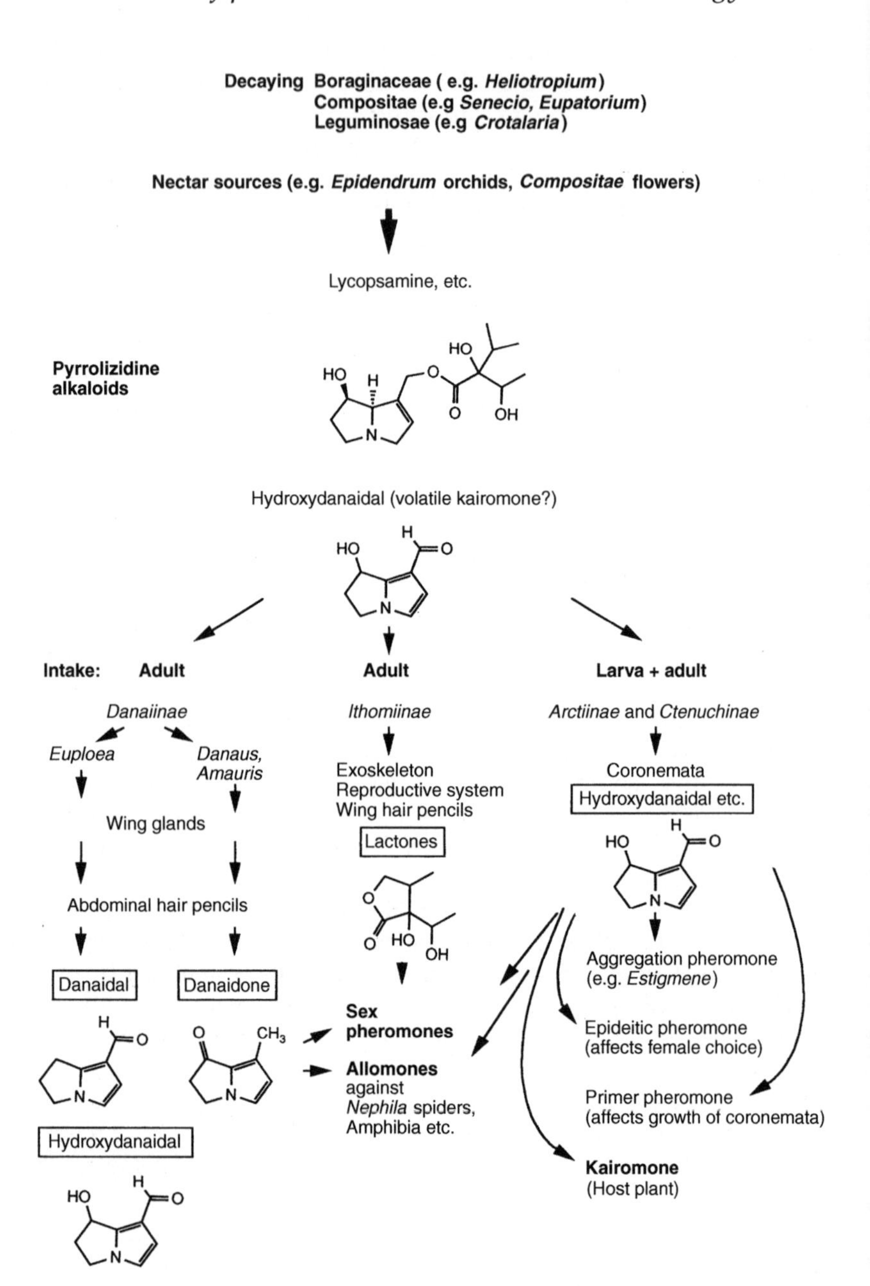


**Fig. 2.4** Interactions between plants containing pyrrolizidine alkaloids (PAs) and Lepidoptera. (Based on various authors: see text.)

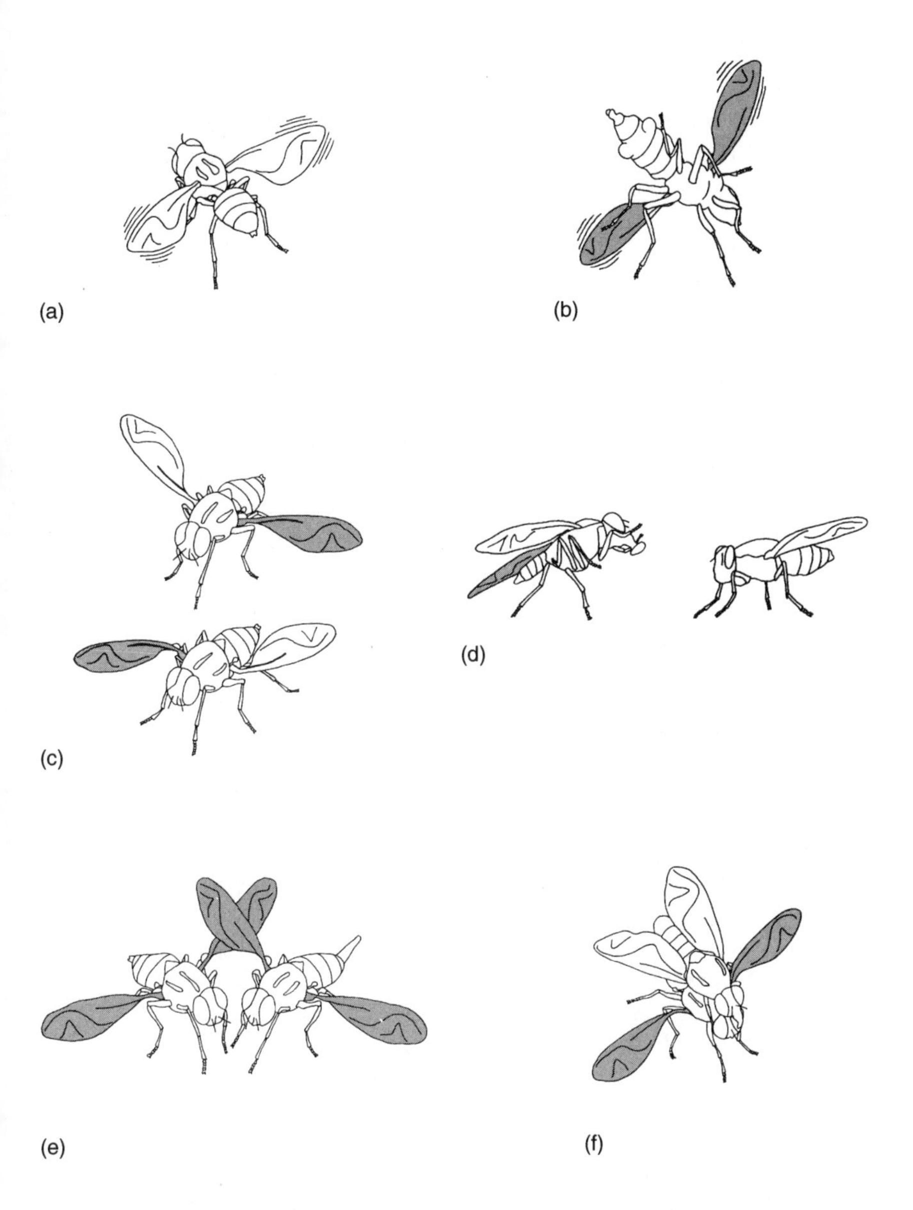


**Fig. 2.5** Courtship behaviour of tephritid fruit fly *Anastrepha fraterculus* (courtesy of I. S. de Lima). (a) Male begins calling by vibrating his wings and then (b) everting lateral abdominal pouches, which contain volatile sex pheromone. Fanning is alternated with (c) wing-waving display, in which first one wing and then the other is moved forwards and rotated on its axis (underside shaded). (d) Defensive posture of male (left) towards another male: proboscis and forelegs extended forwards and abdomen contracted. (e) When a receptive female approaches a male, both rotate their wings; (f) if female contacts male with her proboscis, he rotates on to her back and attempts copulation. An unreceptive female escapes by falling to the ground.

are air-filled tubes covered with hairs (Fig. 2.3). The pheromone that is released attracts both males and females. The males remain close by and also begin to call, forming leks, with an average of nine males in each, while females that arrive select males and mate with them. Curiously, the more usual system of sexual attraction also occurs in this species: females begin calling from fixed positions later in the evening, after lekking has ceased, and the males fly towards them. The value of this dual communication system is obscure.

Lekking is also commonly found in tephritid fruit flies, such as medfly, *Ceratitis capitata*, and the South American fruit fly, *Anastrepha fraterculus*. Males establish leaf territories at certain times of day, and call by inflating pleural glands on the sides of the abdomen and everting the rectal sac (Fig. 2.5). Males and females are attracted, but they defend their leaf territories against other males, which may join the lek, and court females that arrive.

### 2.2.2 Pheromones of nocturnal Lepidoptera

The courtship behaviour of male night-flying moths has been studied more than that of any other group of insects. One reason is that many of the world's major crop pests are moths, but another is that the attraction of the sexes is relatively easy to study because it depends almost exclusively in many cases on the use of pheromones. The relatively long-distance visually guided approach of the male to the female is replaced by upwind orientation up an odour 'plume'. At close range, however, the males dispense pheromone in a manner analogous to that of male butterflies. While males of some species of nocturnal Lepidoptera have eversible coronemata, others have hair pencils or brushes (Fig. 2.3).

A much studied example is that of the oriental fruit moth, *Grapholitha molesta*. The female dispenses a pheromone mixture from a gland at the tip of the abdomen, which contains four main components: the *Z* and *E* isomers of the 8-dodecenyl acetate, (Z)-8-dodecenyl alcohol and dodecyl alcohol. This cocktail of scents is carried in air currents, and sexually competent males orientate upwind, in a manner that will be discussed at greater length in Chapter 3. Initially, it was shown that when the two acetates are present in a 93:7 ratio they induce activation (raising of the antennae and preparation for take-off) followed by upwind flight (Carde *et al.*, 1975). As the male approaches the source, these compounds elicit counter-turning, in which the insect continues to fly upwind but tracks from side to side, producing a typical zigzag flight pattern. The insect may also show casting flight close to the source, in which it makes wide cross-wind excursions in an apparent attempt to relocate the odour plume. When the dodecyl alcohol is also present, it stimulates a high frequency of landing, and after landing the males walk towards the odour source fanning their wings. The dodecyl alcohol also elicits a precopulatory display in which the male everts abdominal scent brushes, fanning the scent towards the female with his wings. The puffs of pheromone-laden air then initiate a final series of interactions in which the female is first attracted to the male and walks towards his hair pencils until she touches them. The contact stimulus induces the male to face away from her and attempt copulation.

This brief description conceals a host of questions about the action of the female pheromone components, which require separate detailed discussion (Chapter 3). They include the following.

- Is each component of the behaviour of males triggered by a particular compound or do they respond to the critical ratios of the blend of compounds?
- How are the ratios measured by the antennal receptors?
- What is the nature of the plume of female pheromone?
- How does the male adjust its orientation when flying into the wind?
- How is counter-turning controlled?

## 2.3 PHEROMONES OF DIPTERA

The courtship of many Diptera is similar to that of butterflies in that the approach from a distance is visually guided, and only at close range do chemical signals intervene. In the majority of cases that have been investigated the pheromones are relatively involatile chemicals of high molecular weight. For example, the sexual pheromone of the common house fly, *Musca domestica*, is a mixture of several components found in the cuticular wax of the female, of which the most important are Z-tricosene and heneicosene (Carlson *et al.*, 1971). The pheromone of the tsetse fly, *Glossina morsitans*, also a constituent of female cuticular wax, consists mainly of the long-chain hydrocarbon 15,19,23-trimethylheptatriacontane (Carlson *et al.*, 1978). These compounds have a mainly arrestant effect, and so they have potential use in control measures in conjunction with contact insecticides. In tsetse flies, the pheromone-receptive sensilla are found on the legs (tibia and tarsus): removal of the antennae has no effect on responses to the pheromone (Langley *et al.*, 1987). For long range attraction, volatile kairomones (constituents of the exhaled breath of cattle) are used for luring tsetse flies to traps and treated surfaces (Vale *et al.*, 1988).

Studies of pheromone communication in *Drosophila* fruit flies have provided us with a remarkably complex picture, summarized in Fig. 2.6, which adds a new layer of understanding to the well-described chain sequence of visual, acoustic and tactile interactions (reviewed in Scott *et al.*, 1988).

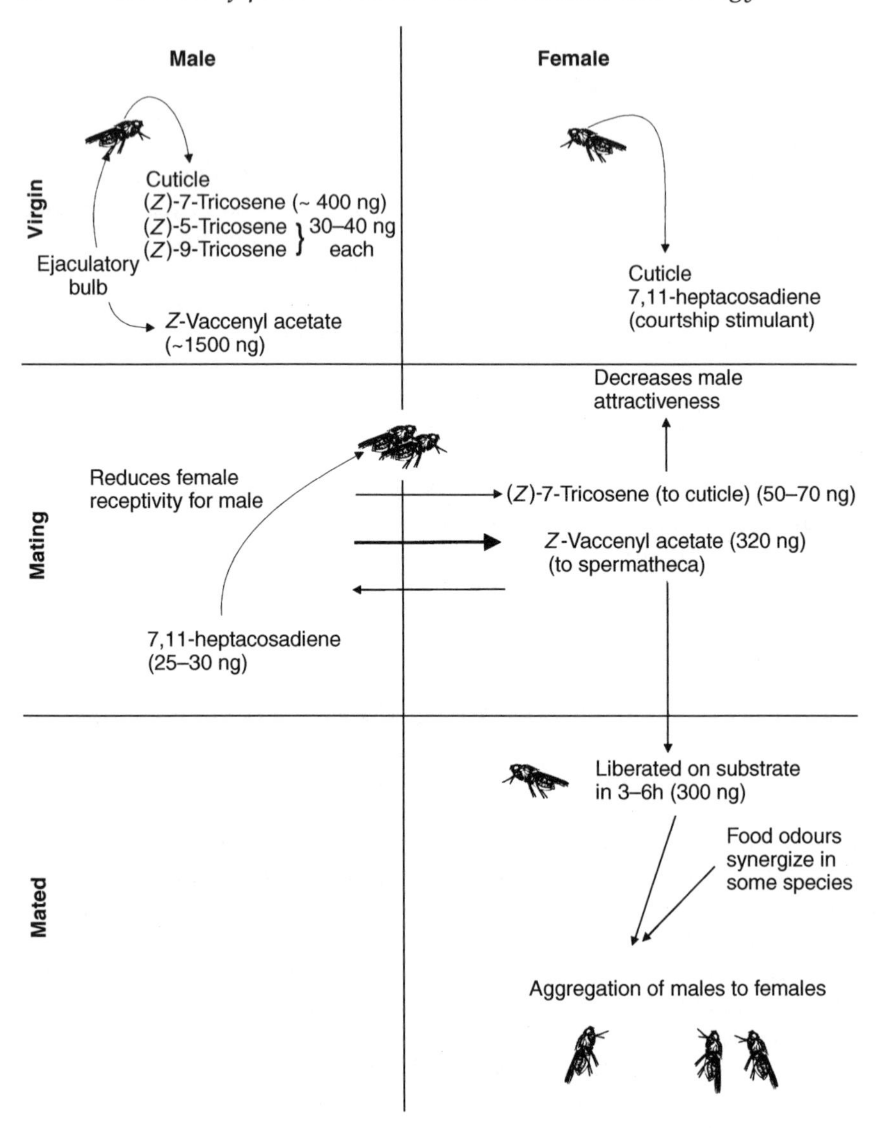


**Fig. 2.6** Semiochemicals in courtship of *Drosophila melanogaster*. (Details from various authors, summarized in Scott *et al.*, 1988, and Schaner *et al.*, 1989.)

Acoustic stimuli resulting from wing vibration of the males prime the insects for courtship activity. *Drosophila melanogaster* has two distinct pheromonal systems: male-produced and female-produced. The male pheromone consists of compounds such as tricosenes which are found on the cuticle and gradually attain a titre of up to 1000 ng during the first five days of adult life (Bartelt *et al.*, 1988). In addition, about 1500 ng of *Z*-18-11 acetate is found in the male ejaculatory bulb. The female produces small amounts of 7,11-heptacosadiene, which acts as a courtship stimulant (Scott and Richmond, 1987).

During mating, there is an exchange of pheromone between the sexes (Scott and Richmond, 1987). About 50–70 ng of the 7-tricosene is transferred to the female, and about 320 ng of *Z*-vaccenyl acetate passes into the spermatheca of the female. At the same time, 25–30 ng of 7,11-heptacosadine is transferred from the female to the male. Almost all (300 ng) of the *Z*-vaccenyl acetate is then released by the female on to the substrate (Scott and Richmond, 1987), where, in conjunction with food odours, it acts as an aggregation pheromone, attracting both males and females. All the acetate disappears from the female in 3–6 hours: the very curious result is that the pheromone which attracts both sexes from a distance is produced by the male and liberated by the female. This means, however, that more insects are attracted to sites where mating is taking place and where there is a food resource. In some species (e.g. *D. borealis*) the pheromone is synergized by volatiles from the host plant, in this case fermenting aspen bark (Bartelt *et al.*, 1988). The 7-tricosene is also used by the mated female. It rapidly disappears from her cuticle, but is secreted again in the presence of a male, when it acts as a courtship inhibitor. The 7,11-heptacosadine transferred to the male also acts as a courtship inhibitor, discouraging both females and other males from making courtship attempts. As males mature, they increase their complement of 7-tricosene and so become less attractive to females, a fate which is accentuated by marking with 7,11-heptacosadiene. This is believed to help the females to choose more vigorous virgin males.

Other kinds of pheromone are known in Diptera. For example, several species of tephritid fruit flies have been shown to produce oviposition inhibitor pheromones (Chapter 1), which the female puts on to fruit after oviposition. An oviposition attractant pheromone is produced by female *Culex quinquefasciatus* mosquitoes, and is released from apical droplets on the eggs (Laurence and Pickett, 1982). This attracts other females of this and related species to the vicinity of the egg raft.

## 2.4 AGGREGATION AND COURTSHIP IN BARK BEETLES

In bark beetles of the genera *Ips* and *Dendroctonus* (Scolytidae), colonization of host trees and courtship are interdependent activities (Byers, 1989). Adult *Dendroctonus brevicomis* females are attracted to their host plant, normally a *Pinus ponderosa* tree, by defensive allomones of the tree, such as the monoterpene, myrcene, released by stress or injury

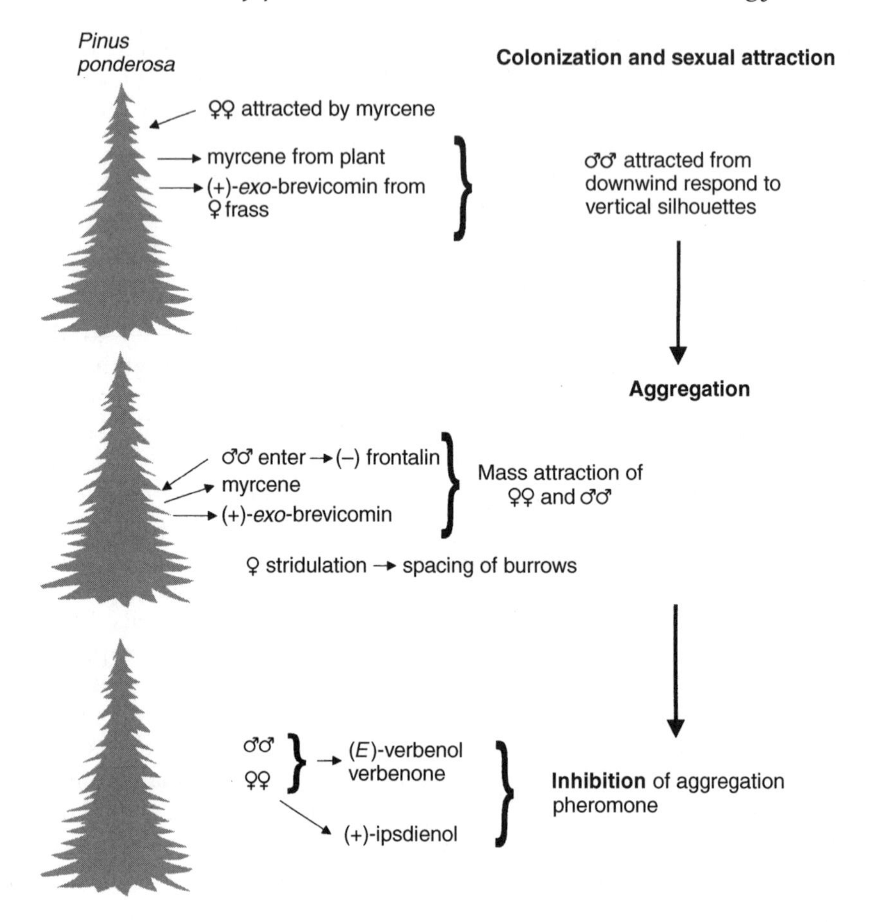


**Fig. 2.7** Semiochemicals in host tree colonization by bark beetle *Dendroctonus brevicomis*. (Details from various authors, summarized in Byers, 1989.)

(Fig. 2.7). Females boring into the tree release a pheromone, (+)-*exo*-brevicomin. This can be found in the frass, and its production is stimulated by myrcene. The two compounds attract males from downwind, which orientate partly visually, steering towards upright silhouettes. The males that enter the burrows start to liberate their own pheromone, (−)-frontalin, which is produced in the hindgut. This synergizes the action of the *exo*-brevicomin and the myrcene, and the mixture attracts large numbers of both sexes, thereby deserving the name of an aggregation pheromone. The females that have attracted males to their burrows stridulate, which repels other insects arriving at the same spot and spaces out the attack.

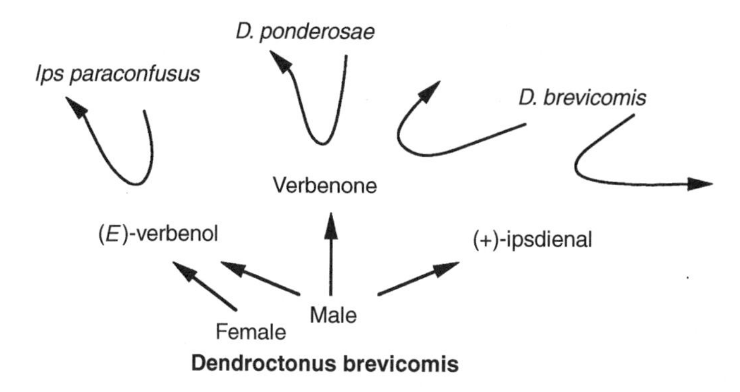


**Fig. 2.8** Specificity of bark beetle pheromones: repellent action of *Dendroctonus* pheromone components. (Details from various authors, summarized in Byers, 1989.)

Once established as pairs in their burrows, the males and females both begin to produce verbenone and (*E*)-verbenol, and the males also liberate (+)-ipsdienol. These three components inhibit the action of the aggregation pheromone and so prevent overcolonization of the tree. After several days, the amounts of brevicomin and frontalin diminish below the threshold for long-range upwind attraction, and colonization ceases.

The pheromones of *D. brevicomis* are also used by other species of bark beetle: for example, frontalin takes its name from *D. frontalis*, and ipsdienol from the other common genus, *Ips*. Specificity of response is achieved by the use of different mixtures of the compounds (Fig. 2.8).

## 2.5 PHEROMONES OF SOCIAL INSECTS

Pheromones of social insects form the body of an intricate communication network within a colony. Figures 1.9 and 2.9 show the sources of some of the main semiochemicals produced by honeybees, and by leaf-cutting ants respectively. These can be grouped into chemicals concerned in the communication of food finding and exploitation, alarm and defence, and colony and individual identification.

### 2.5.1 Alarm and defence

The so-called alarm pheromones of social insects rarely have a single action upon the species that produce them. The behaviour they elicit may include withdrawal, approach or disorientation, and a variety of responses concerned with defence and aggression (agonistic behaviour). A worker wood ant (*Formica rufa*) that is suddenly disturbed opens its

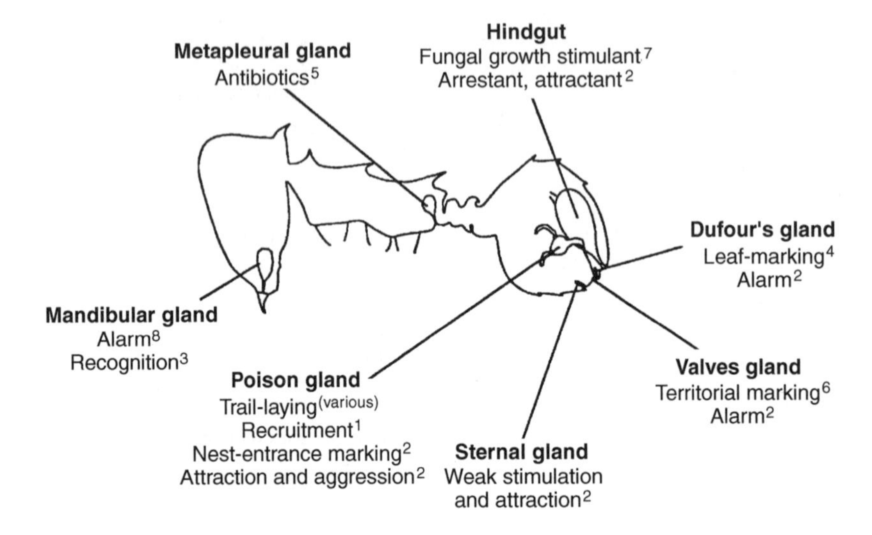


**Fig. 2.9** Sources of semiochemicals in leaf-cutting ants (Attini). Details from various authors, summarized in Howse (1992): (1) Jaffé and Howse, 1979; (2) Hölldobler and Wilson, 1986; (3) Jaffé, 1983; (4) Bradshaw *et al.*, 1979; (5) Schildknecht and Koob, 1970; (6) Jaffé *et al.*, 1979; (7) Weber, 1966; (8) various.

mandibles wide, swings its gaster forwards beneath the thorax, and sprays a secretion containing formic acid at any suitable target that presents itself. This behaviour may be described variously as aggression, defence, threat or alarm, but such terms are subjective and imply that the motivation of the insect is understood. Alarm pheromones are now generally recognized to be multicomponent mixtures, and a new approach to understanding their action and significance involves determining the role of each of the main components.

Alarm pheromones are ubiquitous among social insects and in many instances they are spread by diffusion rather than by air currents. This must generally be the case in ants in which pheromone is released into the boundary layer close to the surface. The depth of the boundary layer will vary according to the air velocity above the surface, but within the layer the air will be effectively stationary and pheromone released from a point source will diffuse outwards at a fixed rate depending on its molecular weight and diffusion coefficient. Bossert and Wilson (1963) gave a method of calculating diffusion rates of volatile chemicals, and showed that a given chemical can be pictured as diffusing out from a point source in a hemispherical cloud, progressively enlarging in diameter. At a given time, the concentration of pheromone at the boundaries of this cloud will be just sufficient to overcome the sensory threshold of the ants for the substance concerned.

There is thus a hemispherical active space, within which ants can detect the pheromone. Where the pheromone is released in an instantaneous 'puff', the radius of the active space will extend with time and then contract as diffusion continues and the concentration of odour molecules falls.

The mechanism of communication of alarm can be understood with reference to the African weaver ant, *Oecophylla longinoda*. Some aspects of alarm/defence behaviour are controlled by the mandibular gland secretion, but in dire emergencies abdominal secretions are used. In major workers, the mandibular gland secretion has over 30 components (Bradshaw *et al.*, 1979a). Of these, hexanal has an alerting function, the less volatile 1-hexanol is an attractant, and 2-butyl-3-octenal and 3-

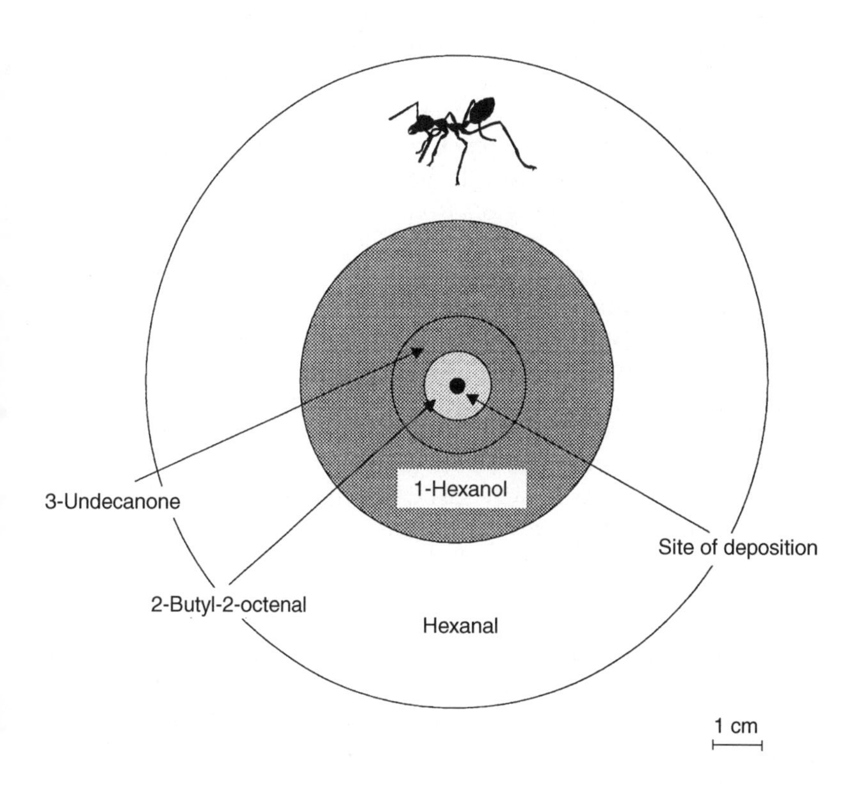


**Fig. 2.10** Active spaces of main components of mandibular gland secretion of major worker of weaver ant, *Oecophylla longinoda*, 20 seconds after deposition at a central point of a flat surface in still air. Circles show approximate limits of detection of components. Insect is alerted when it detects hexanal and is then able to orientate up the odour gradient provided by other components. Biting responses are elicited by 2-butyl-2-octenal and, to a lesser extent, by 3-undecanone. (After Bradshaw *et al.*, 1975.)

undecanone, both of low volatility, elicit localized biting (Fig. 2.10). Certain other components can act as synergists.

Figure 2.10 shows the active spaces, determined by behavioural methods, for four components of the mandibular gland secretion of *O. longinoda* major workers, approximately 20 seconds after release from a central point. The pheromone mixture supplies a sequential message: ants coming within range of the most volatile component (hexanal) are first alerted and rush around rapidly with their mandibles opened. If they enter the zone of 1-hexanol they orientate up the odour gradient towards the source. They then encounter the less volatile 3-undecanone, which in addition to supplying directional information also lowers the threshold for biting. The least volatile of the four compounds is 2-butyl-2-octenal, which has the greatest effect on biting thresholds. The overall effect is that ants orientate towards a source of pheromone and tend to bite at any moving object contaminated with pheromone.

Weaver ants overcome their opponents by pinning them down: they surround other insects, grab their legs and antennae and pull on them. In these circumstances the pheromone communication system is adequate for defensive purposes. If, however, they cannot contain their attackers, they resort to spraying from the tip of the gaster a poison gland secretion containing formic acid. They direct the spray forwards by rotating the gaster upwards, but this is a high-risk strategy because the formic acid mixture is toxic to their own nest-mates, which can easily become contaminated. The acid alone cannot penetrate the cuticle, but it is produced in admixture with a series of non-polar hydrocarbons, such as undecane and tridecane, which facilitate its passage through the wax layer of the arthropod cuticle. The defensive mixture also has a pheromonal action (Bradshaw *et al.*, 1979b), attracting other ants; indeed there is synergistic activity between undecane and formic acid. An arthropod adversary that is coated with the spray may not be immediately immobilized, but more weaver ant workers will be recruited either up the odour gradient or along small odour trails left by the moving target.

Further complexity of the alarm communication system is revealed by caste differences and by inter-colony variation in the mandibular gland components. The minor workers have components that differ qualitatively and quantitatively from those of the majors; for example, they are rich in nerol, geraniol and primary alcohols such as 1-hexanol. These may be substances that are more suitable for communication within the relatively closed nest atmosphere, and less toxic than those used in quantity by the major workers. Nerol is relatively involatile and repels other minor workers, while attracting majors. 1-Hexanol does the same, but diffuses out more rapidly from the source and is probably mainly used as an alerting pheromone. Between individuals of the same colony there is a consistency in ratios of mandibular gland components, but there can be marked differences between individuals of different colonies such that the major components are not even the same ones. This has the intriguing implication that individuals of each colony interpret the pheromone message in different ways.

The other ant species in which pheromone communication has been studied in detail are leaf-cutting ants (Attini) and fire ants (*Solenopsis* spp.). Both groups are of immense economic importance in the New World; hence there is great interest in ways of controlling their behaviour.

The attines forage for living leaves which they cut and transport back to their subterranean nest, where they use the triturated vegetable material as a substrate for their symbiotic fungus. Exocrine secretions are involved in many aspects of their social life (Fig. 2.9). The first ant trail pheromone to be identified was that of the leaf-cutting ant *Atta texana*. Tumlinson *et al.* (1971) extracted 3.7 kg of ants and found that the main component of fractions that elicited trail-following was methyl 4-methylpyrrole-2-carboxylate (Fig. 2.9). The synthetic compound was found to be active against several other *Atta* species. Its activity is astonishingly high: it was calculated that the threshold of responsiveness is $3.48 \times 10^8$ molecules per cm of trail. Theoretically, therefore, 0.33 mg should be sufficient to draw a detectable trail around the circumference of the Earth. Mass recruitment along a foraging trail can be achieved in *Atta cephalotes* by the amount of pheromone present in the poison gland of one worker (Jaffe and Howse, 1979).

As a worker cuts a leaf disc, it marks the surface with a leaf-marking pheromone from the Dufour's gland. This pheromone potentiates leaf-carrying behaviour and makes it more likely that dropped discs are picked up (Bradshaw *et al.*, 1986). Leaf discs are incorporated into the fungus garden by the smaller workers (minima), which add to it material from the hindgut containing enzymes, arrestants and attractants. Nest hygiene is accomplished physically by removal of pathogens and fungal spores, to which the workers are highly sensitive, and chemically by the use of allomones from the metapleural glands of workers which function as antibiotics and 'herbicides', inhibiting the growth of alien fungi (Schildknecht and Koob, 1970).

The system of alarm communication has many similarities with that of *Oecophylla*. Some attines, including *Acromyrmex octospinosus*, appear to have overcome the problem of rapid loss of the odour gradient that occurs after the release of a quantity of alarm pheromone. In *A. octospinosus* there are two main pheromone components in the mandibular gland: 3-octanone and 3-octanol. Both attract ants up the odour gradient, but the former induces rapid running and alerting of workers, particularly in the peripheral zone (Fig. 2.11). The 3-octanol stimulates

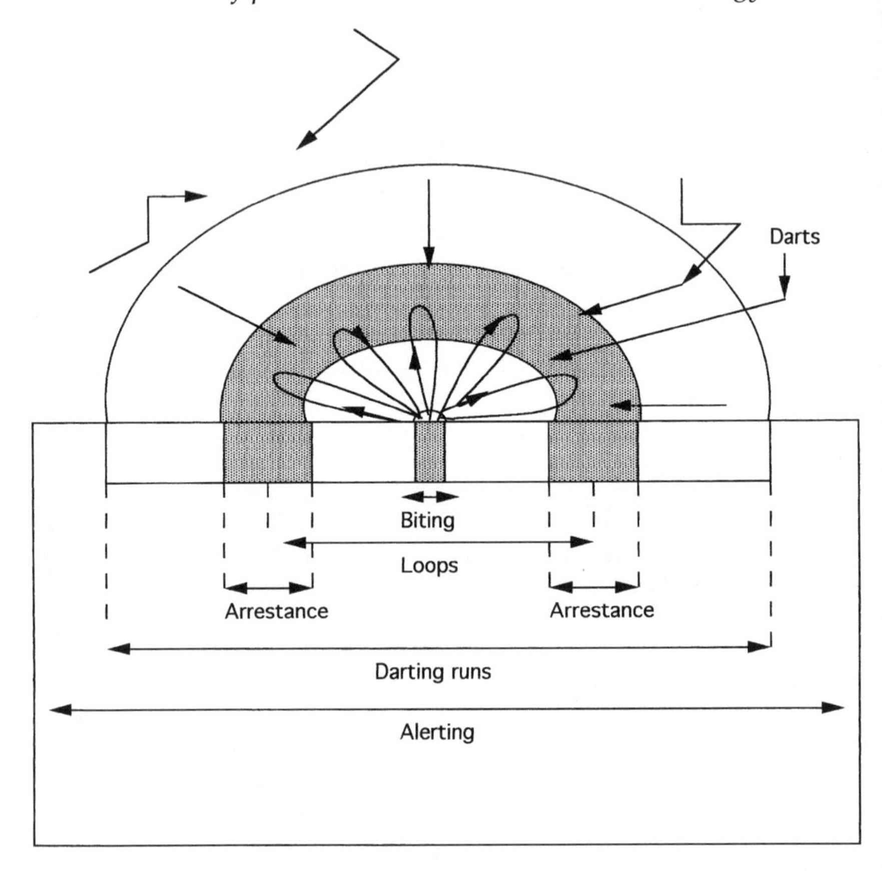


**Fig. 2.11** Scheme of alarm/recruitment system of the leaf-cutting ant *Acromyrmex octospinosus*. Alarm pheromone released at central point diffuses out rapidly in still air. At outer edges of zone, insects are alerted; as they detect higher concentrations they indulge in darting runs in seemingly random directions. Others are able to orientate up the odour gradient and are arrested a few centimetres from source. As concentration falls with time, they tend to bite at the source and make looping runs centred on the release point, which serve to mark position of disturbance with a more persistent trail pheromone. (Howse, unpublished.)

biting at the source of the odour, and induces some of the ants to lay peculiar looping pheromone trails passing through the source. This appears to be a means of 'signposting' a source of danger for nestmates, which are then able to locate the source after the odour gradient has been lost or has become very shallow (Howse, 1997).

### 2.5.2 Marking and identification

Various insects, including cockroaches, crickets and Collembola, are known to mark the substratum with a pheromone that is attractive to other members of the same species and so mediates aggregation. Some species of bees, including carpenter bees and bumble bees, patrol a small territory in which they mark regular settling areas with cephalic gland secretions. These serve to attract females and increase the sexual motivation of females. The secretions are very specific: two distinct forms of *Bombus lucorum* have been detected in Scandinavia, one of which has ethyl decanoate as the principal component of the marking secretion and the other ethyl tetradec-(Z)-9-enoate (Bergstrom *et al.*, 1973).

Major workers of the weaver ant *Oecophylla longinoda*, mark their territories with droplets of rectal sac fluid. They distinguish their own droplets from those of other colonies (Holldobler and Wilson, 1977) to which they show aggressive responses. Territorial marking also occurs in the leaf-cutting ant *Atta sexdens* (Vilela and Howse, 1984): ants on the territory of an alien species of leaf-cutting ant, or on the territory of another colony of the same species, tend to flee, while ants on their own territories are very aggressive towards aliens.

Cuticular hydrocarbons have been implicated in recognition of nest-mates in many social insect species, different species often having differing and characteristic cocktails of cuticular lipids (Blomquist and Dillwith, 1985). Workers of the wasp *Polistes annularis* have as the main components of their cuticular lipids 13,17-dimethylhentriacontane, 3-methylnonacosane and 3-methylheptacosane (Espelie and Hermann, 1990). These same compounds are found on the cuticles of the larvae, on the eggs and on the surface of the nest and nest pedicel. It appears that these compounds aid the wasps in recognising their own nest. Workers of a related species, *Polistes metricus*, selected their own nest in laboratory tests 66% of the time when given a choice between that and two others (Espelie *et al.*, 1990). After the surface hydrocarbons (mainly $C_{27}$ and $C_{29}$ *n*-alkanes and $C_{31}$ and $C_{33}$ methylalkanes) had been extracted, they chose their own nest only 8% of the time, but after the surface hydrocarbons had been restored the choice went up to 47%. The nest pedicel hydrocarbons, however, were very similar in composition in different nests, and there is evidence from other species that they serve as repellents to predatory ants. For example, methyl hexadeconate from the sternal gland of *P. fuscatus* workers, which is smeared on to the nest pedicel, repels ant predators (Henderson and Jeanne, 1989).

Using a conditioning process, Getz *et al.* (1988) showed that honey bee workers could discriminate between cuticular waxes from different adult nest-mates, eggs from the same and different hives, and larvae of similar age from the same hive.

### 2.5.3 Recruitment and orientation

Social insects of many species lay chemical trails which are used in foraging and recruitment to food, and sometimes in transference of the colony, or part of it, to a new nest site. The nature of the trail, and its usage, may depend on motivational factors affecting the whole colony: for example, neotropical army ants (Ecitonini), which are nomadic, form raiding trails which differ according to the reproductive state of the colony (Schneirla, 1971). Colonies of *Eciton hamatum* develop a quite simple raiding trail when they are in the stationary phase.

In his now classical study, Wilson (1962) showed that the foraging behaviour of the fire ant, *Solenopsis saevissima*, is organized by a trail pheromone which is secreted by the Dufour's gland that opens into the sting. This secretion is exuded through the sting and the trail is highly species-specific. A worker searching for food orientates with respect to visual stimuli (Fig. 2.12). When she has found a source of food, she returns to the nest laying a trail of the highly volatile pheromone. At first sight this appears paradoxical, because a trail half a metre long will begin to be lost by volatilization before a trail-laying worker is able to return with recruited nest-mates. However, Wilson showed that the volatility is essential to the simple and elegant method of controlling recruitment.

A returning forager recruits other workers within the nest which follow the trail to the food source. Those that arrive and cannot attain the food because there are already too many ants in the way, or those

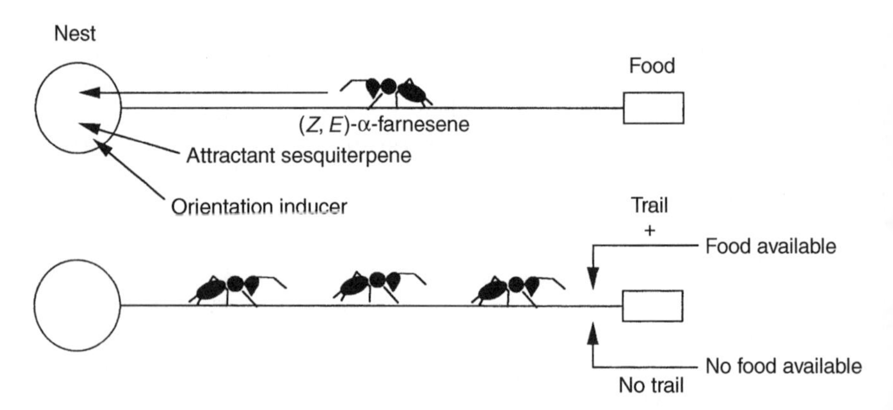


**Fig. 2.12** Use of semiochemicals in trail-following and recruitment in fire ants (*Solenopsis* spp.). Forager finding food lays trail back to nest. Recruitment along trail depends on presence of high concentration of Dufour's gland secretion. (Data from Vander Meer *et al.*, 1988, 1989; Hangartner, 1969; Wilson, 1962.)

that find the food has all been taken, return to the nest without reinforcing the trail. There is thus a simple feedback mechanism in which the quantity of food is measured. This feedback takes several minutes to operate, so that if the trail is suddenly removed during recruitment there is an overshoot of workers arriving for several minutes. At distances greater than 0.5 m this overshoot becomes smaller, as the trail loses its effectiveness by volatilization unless it is heavily populated.

Trail-laying at distances greater than 0.5 m is achieved by a system of multiple trails: ants coming to the end of an existing trail tend to continue onwards. If they find food they then lay a new trail back to the terminus which attracts other workers from the existing trail. The interval between laying of individual trails gets progressively shorter and a new continuous trail comes into being. The maximum length of the trail is dictated by, among other factors, the nature of the terrain and the size of the colony.

*Solenopsis geminata* can vary the amount of pheromone on the trail by changing the pressure of the sting on the substratum (Hangartner, 1969). Although the trail itself appears to act as a releaser for trail-following in *S. saevissima,* meticulous studies on the related species *S. invicta* (Vander Meer *et al.*, 1988, 1989) have shown that a trail formed of Dufour's gland secretion acts as a guide to workers already foraging, but does not function in recruitment. The component responsible for orientation is (*Z,E*)-α-farnesene. There are two other compounds, a sesquiterpene and an orientation inducer pheromone. The former acts as an attractant, but the latter stimulates trail-following. However, for them to exert these linked functions, they need to be present at a concentration over 250 times that on the trail. A forager that, returning to the nest, releases the Dufour's gland secretion in high concentration thereby stimulates some of the reserve force of foragers to follow her return trail, or to join in the traffic along an existing trail. The interest of this mechanism lies in the discovery that, as in communication of alarm among ants, it may also be that different components of the same pheromone control different subcategories of behaviour.

Many species of ant have two kinds of trail: one or more trunk trails leading out from the nest, which are in more or less permanent use, and more ephemeral trails leading off from the trunk trails. A trail made from an extract of the poison gland of the harvester ant *Pogonomyrmex badius* is very attractive to foragers and is followed accurately, but begins to lose its attractiveness after 25 minutes (Holldobler and Wilson, 1970). An extract of the Dufour's gland is effective for far longer. The secretion of this gland is used for setting up long-lasting trails and in homing to the nest, while the poison gland pheromone is used for recruitment to new food sources.

The mechanism of orientation in trail-following has been investigated in *Lasius fuliginosus* by Hangartner (1967). Workers following a straight trail make alternate drumming movements with each antenna and compensate for any slight lateral deviations from the trail. If one antenna is removed, the compensatory movement on one side does not occur until the ant has left the scent field, which suggests that it is orientating tropotactically, according to the balance of stimulation on each antenna.

A primitive stage of trail-following is represented by the phenomenon of 'tandem running', described by Möglich *et al.* (1973) in the ant *Camponotus sericeus*. A returning forager uses a tactile display inside the nest which induces a nest-mate to follow her in tandem. The forager grasps the other ant between her mandibles, shakes her, and then reverses and offers her gaster to the accosted ant. The latter then touches the forager with her antennae and is led along the trail in the tandem so formed. The leader maintains her position as a result of mechanical stimuli resulting from regular drumming by the antennae of the follower on her gaster and legs, and the follower requires a chemical stimulus from a pheromone on the surface of the cuticle of the leader.

Pheromone trails are also used by many termites (Howse, 1984) and neotropical stingless bees (Lindauer and Kerr, 1958). A forager of the bee *Trigona postica* that finds food makes several direct flights between the nest and the food source before alighting at intervals of 2–3 m along her path to deposit droplets from her very large mandibular glands on to vegetation, stones or earth. Inside the nest, she alerts nest-mates with an acoustic signal and flies back in front of the alerted bees to the food. To demonstrate the importance of the trail, a food source was placed on the opposite side of a lake from the nest. Bees that found the food attempted vainly to deposit pheromone on the water surface, but there was no recruitment until the lake was spanned with a rope on which vegetation was suspended.

### 2.5.4 Brood pheromones

Some brood pheromones stimulate heat production by workers, which is achieved with abdominal pumping movements. Such pheromones are termed **thermoregulatory pheromones**. Ishay (1972) made alcoholic extracts of the pupae of various wasps and hornets (Vespinae). When these extracts were put on filter paper strips they elicited abdominal pumping behaviour of workers that came into contact with them. The queen bumble bee *Bombus vosnesenskii* sits astride her brood and can raise her thoracic temperature to 34.5–37.5°C (Heinrich, 1974). She is able to warm the temperature of her brood by 20°C when the external temperature is 5°C. This behaviour is controlled by a pheromone which the queen herself deposits on the brood clump.

## 2.6 PHEROMONES IN PREDATOR–PREY INTERACTIONS

Any communication channel is open to exploitation by predators and parasites, which can either intercept the message and use it for their own ends, or counterfeit a signal that is then used to mislead.

When a pheromone signal is intercepted by a predator, it is used as a kairomone, and there are numerous examples of the use of kairomones by parasitoids (reviewed by Lewis and Martin, 1990). Before the present terminology was adopted, Tinbergen (1932) analysed the stimuli involved in prey capture by the hunting wasp, *Philanthus triangularum*, which catches and stings live honey bees in flight. *Philanthus* approaches bees upwind, but it will not pounce on them and sting them unless they carry 'bee odour'. Dead honey bees from which the odour had been extracted in solvent were not captured when hung on threads in the wind, but pieces of stick contaminated with bee odour were captured and attempts were made to sting them.

Some predators appear to conceal their own 'odour' in order to gain acceptance into social insect colonies. In most known examples this is achieved through the use of cuticular hydrocarbons, which often exist as a species-specific complex, modifiable to a greater or lesser extent through incorporation of elements from the environment. Staphylinid beetles of the genus *Tricopsenius* live by soliciting food from subterranean termite hosts (*Reticulitermes flavipes*). They synthesize a mixture of cuticular hydrocarbons, which matches that of their host, and are therefore accepted as nest-mates (Howard *et al.*, 1980). The workerless cuckoo ant, *Leptothorax kutteri*, is an obligate parasite of *Leptothorax acervorum*, gaining control of the workers and queens. The parasite grooms the host queen at an exceptionally high frequency (Franks *et al.*, 1990) and so acquires the host mixture as a disguise, which means she is tolerated within the nest both by the queen and by the *L. acervorum* workers who have also obtained the pheromone mixture by mutual grooming. *Ectatomma ruidum* is a common neotropical ant in which individual workers gain entry to colonies other than their own and steal the incoming food, which they take back to their own nest. They acquire the odour of the host nest by close contact and use this as camouflage to gain entry (Breed *et al.*, 1992).

Alarm pheromones are used by some insects to exploit others. The slave-making ants *Formica pergandei* and *F. subintegra* have outsize Dufour's glands which contain decyl, dodecyl and tetradecyl acetates (Regnier and Wilson, 1971). These are also components of the alarm pheromones of other species of formicine ant that they enslave, after killing the queen. When the ants enter a host nest, they release large amounts of the pheromone, which causes panic and breakdown of communication among defenders. For this reason, they have been described as **propaganda pheromones** by Regnier and Wilson (1971). Similar behaviour is found in some neotropical stingless bees (Meliponini). *Lestrimelitta limao* gets its name from the odour of lemons it produces, due to citral which is contained in hypertrophied mandibular glands (Blum *et al.*, 1970). Scouts seek out the nests of certain other stingless bees (e.g. *Trigona subterranea*) that use citral as a trail pheromone. They then recruit hundreds of nest-mates that plunder the food stores of the *Trigona* workers, which results in more citral being liberated, producing a 'propaganda' effect: complete disruption of the social organization of the nest in which the field and guard bees abandon the colony completely, leaving the way clear for the robber bees. Not all species of *Trigona* have this panic alarm response to citral: Wittmann *et al.* (1990) have described the remarkable defensive behaviour of the bee that is able to resist the *Lestrimelitta. Trigona angustula* nests are guarded by groups of workers that hover to one side of the nest entrance. These guards detect *L. limao* workers by the visual stimuli they provide in conjunction with components of their mandibular gland pheromone, principally consisting of citral and 6-methyl-5-hepten-2-one. These kairomonal components alone trigger immediate and very rapid attack, but at the same time, the kairomonal stimulus is translated into a pheromonal message by release of benzaldehyde, which is the main component of the species-specific alarm pheromone in the *Trigona*, and hundreds of nest-mates are recruited to defend the entrance.

The use of sexual pheromones as lures is known in bolas spiders of the genus *Mastigophora* (Eberhard, 1977; Stowe, 1988). *M. corniger* produces (Z)-9-tetradecenyl acetate, (Z)-9-tetradecal and (Z)-11-hexadecenal, which are common components of female lepidopteran sex pheromones and attract at least 19 species of male moth. The spiders catch the approaching males by flicking a thread at them with a sticky droplet (bola) on the end. There is evidence that similar deception is employed by orchids (Kullenberg and Bergström, 1975). Bee orchids (*Ophrys* spp.), common in southern Europe, are pollinated by solitary bee and wasp males, which attempt copulation with the flowers and in the process have pollinia attached to them.

## REFERENCES

Bartelt, R.B., Schaner, A.M. and Jackson, L.L. (1988) Aggregation pheromones in *Drosophila borealis* and *Drosophila littoralis*. *J. Chem. Ecol.*, **14**, 1319–1327.

Bergström, G., Kullenberg, B. and Stallberg-Stenhagen, S. (1973) Studies on natural odiferous compounds. VII. Recognition of two forms of *Bombus lucorum* L. (Hymenoptera, Apidae) by analysis of the volatile marking secretion from individual males. *Chemica Scripta*, **2**, 9–11.

Birch, M.C. (1970) Pre-courtship use of abdominal brushes by the nocturnal moth *Phlogophora meticulosa* (L.) (Lepidoptera: Noctuidae). *Anim. Behav.*, **18**, 310–316.

Blomquist, G.J. and Dillwith, J.W. (1985) Cuticular lipids. In *Comprehensive Insect Physiology, Biochemistry and Pharmacology*, (eds. G.A. Kerkut and L.I. Gilbert), Vol. 3, Pergamon, Oxford, pp. 117–154.

Blum, M.S., Crewe, R.M., Kerr, W.E. *et al.*(1970) Citral in stingless bees: isolation and function in trail laying and robbing. *J. Insect. Physiol.*, **16**, 1637–1648.

Boppre, M. (1990) Lepidoptera and pyrrolizidine alkaloids. Exemplification of complexity in chemical ecology. *J. Chem. Ecol.*, **16**, 165–185.

Bossert, W.H. and Wilson, E.O. (1963) The analysis of olfactory communication among animals. *J. Theor. Biol.*, **5**, 443–469.

Bradshaw, J.S.W., Baker, R. and Howse, P.E. (1979a) Multicomponent alarm pheromones in the mandibular gland of the African weaver ant, *Oecophylla longinoda. Physiol. Ent.*, **6**, 395–412.

Bradshaw, J.W.S., Baker, R. and Howse, P.E. (1979b) Chemical composition of the poison apparatus secretions of the African weaver ant, *Oecophylla longinoda*, and their role in behaviour. *Physiol. Ent.*, **4**, 39–46.

Bradshaw, J.W.S., Howse, P.E. and Baker, R. (1986) A novel autostimulatory pheromone regulating transport of leaves in *Atta cephalotes. Anim. Behav.*, **34**, 234–240.

Breed, M.D., Snyder, L.E., Lynn, T.L. and Morhart, J.N. (1992) Acquired chemical camouflage in a tropical ant. *Anim. Behav.*, **44**, 519–523.

Brower, L.P., Brower, J.V.Z. and Cranston, F.P. (1965) Courtship behavior of the queen butterfly, *Danaus gilippus berenice* (Cramer). *Zoologica (N.Y.)*, **50**, 1–39.

Brown, K.S. (1984) Adult-obtained pyrrolizidine alkaloids defend ithomiine butterflies against a spider predator. *Nature*, **309**, 707–709.

Byers, J.A. (1989) Chemical ecology of bark beetles. *Experientia*, **45**, 271–283.

Carde, R.T., Baker, T.C. and Roelofs, W.L. (1975) Ethological function of components of a sex attractant system for oriental fruit moth males, *Grapholitha molesta* (Lepidoptera: Tortricidae). *J. Chem. Ecol.*, **1**, 475–491.

Carlson, D.A., Mayer, M.S., Silhacek, D.L., *et al.* (1971) Sex pheromone attractant of the house fly: isolation, identification and synthesis. *Science*, **174**, 76–78.

Carlson, D.A., Langley, P.A. and Huyton, P.M. (1978) Sex pheromone of the tsetse fly: isolation, identification and synthesis of contact aphrodisiacs. *Science*, **210**, 750–753.

Crane, J. (1955) Imaginal behaviour of a Trinidad butterfly, *Heliconius erato hydrata* Hewitson, with special reference to the social use of colour. *Zoologica*, **40**, 167–196.

Culvenor, C.C.J. and Edgar, J.A. (1972) Dihydropyrrolizine secretions associated with coronemata of *Utetheisa* moths (family Arctiidae). *Experientia*, **28**, 627–628.

DeVries, P.J. and Stiles, F.G. (1990) Attraction of pyrrolizidine alkaloid seeking Lepidoptera to *Epidendrum paniculatum. Biotropica*, **22**, 290–296.

Dodson, C.H. (1975) Coevolution of orchids and bees. In *Coevolution of Animals and Plants*, (eds L.E. Gilbert and R.H. Raven), University of Texas Press, Austin, pp. 91–99.

Dressler, R.L. (1982) Biology of the orchid bees (Euglossini). *A. Rev. Ent.*, **13**, 121–146.

Eberhard, W.G.G. (1977) Aggressive chemical mimicry by a bolas spider. *Science*, **198**, 1173–1175.

Edgar, J.A., Culvenor, C.C.J. and Robinson, G.S. (1973) Hairpencil dihydropyrrolizidines of Danainae from the New Hebrides *J. Aust. Ent. Soc.*, **12**, 144–150.

Eisner, T. and Meinwald, J. (1987) Alkaloid-derived pheromones and sexual selection in Lepidoptera. In *Pheromone Biochemistry (eds G.D. Prestwich and G.D. Blomquist), Academic Press, New York, pp. 215–269.*

Ellis, P.E. (1959) Learning and social aggregation in locust hoppers. *Anim. Behav.*, **7**, 91–106.

Espelie, K.E. and Hermann, H.R. (1990) Surface lipids of the social wasp *Polistes annularis* (L.) and its nest and nest pedicel. *J. Chem. Ecol.*, **16**,1841–1852.

Espelie, K.E., Wenzel, J.W. and Chang, G. (1990) Surface lipids of the social wasp *Polistes metricus* Say and its nest and nest pedicel and their relation to nestmate recognition. *J. Chem. Ecol.*, **16**, 2229–2241.

Franks, N., Blum, M., Smith, R-K, and Allies, A.B. (1990) Behaviour and chemical disguise of cuckoo and *Leptothorax kutteri* in relation to its host *Leptothorax acervorum*. *J. Chem. Ecol.*, **16**, 1431.

Fuzeau-Braesch, S., Genin, E., Jullien, R. *et al.* (1988) Composition and role of volatile substances in atmosphere surrounding two gregarious locusts, *Locusta migratoria* and *Schistocerca gregaria*. *J. Chem. Ecol.*, **14**, 1023–1033.

Getz, W.M., Bruckner, D. and Smith, K.B. (1988) Variability of chemosensory stimuli within honeybee (*Apis mellifera*) colonies: differential conditioning assay for discrimination cues. *J. Chem. Ecol.*, **14**, 253.

Gillett, S. (1983) Primer pheromones and polymorphism in the desert locust. *Anim. Behav.*, **31**, 221–230.

Grula, J.W., McChesney, J.D. and Taylor, O.R. (1980) Aphrodisiac pheromones of the sulfur butterflies *Colias eurytheme* and *C. philodoce* (Lepidoptera, Pieridae). *J. Chem. Ecol.*, **6**, 241–256.

Haber, W.A. (1978) Evolutionary ecology of tropical mimetic butterflies (Lepidoptera: Ithomiinae). PhD dissertation, University of Minnesota, Minneapolis.

Hangartner, W. (1967) Spezifitat und Inaktivierung des Spurpheromons von *Lasius fuliginosus* Latr. und Orientierung der Artbeiterinnen im Dutfield. *Z. vergl. Physiol.*, **57**, 103–136.

Hangartner, W. (1969) Structure and variability in the individual odor trail in *Solenopsis geminata* Fabr. (Hymenoptera, Forrmicidae) *Z. vergl. Physiol.*, **75**, 123–142.

Henderson, G. and Jeanne, R.L (1989) Response of aphid-tending ants to a repellent produced by wasps (Hymenoptera: Formicidae, Vespidae). *Ann. Ent. Soc. Am.*, **82**, 516–519.

Heinrich, B. (1974) Pheromone induced brooding behaviour in *Bombus vosnesenskii* and *B. edwardsii* (Hymenoptera: Bombidae). *J. Kansas Ent. Soc.*, **47**, 396–404.

Holldobler, B. and Wilson, E.O. (1977) The multiple recruitment systems of the African weaver ant *Oecophylla longinoda* (Latrielle) (Hymenoptera: Formicidae), *Behav. Ecol. Sociobiol.*, **3**, 431–433.

Howse, P.E. (1984) Semiochemicals of termites. In *Chemical Ecology of Insects* (eds W.J. Bell and R.T. Carde), Chapman & Hall, London and New York, pp. 475–520.

Howse, P.E. (1992) Pheromonal control of behaviour in leaf-cutting ants. In *Applied Myrmecolocy. A World Perspective* (eds R.K. Vander Meer, K. Jaffe and A. Cedeno), Westview Press, Boulder, CO, pp. 427–437.

Ishay, J. (1972) Thermoregulatory pheromone in wasps. *Experientia*, **28**, 1185–1187.

Islam, M.S., Roessingh, P., Simpson, S.J. and McCaffery, A.R. (1994a) Parental effects on the behaviour and coloration of nymphs of the desert locust, *Schistocerca gregaria*. *J. Insect Physiol.*, **40**, 173–181.

Islam, M.S., Roessingh, P., Simpson, S.J. and McCaffery, A.R.(1994b) Effects of population density experienced by parents during mating and oviposition on the phase of hatchling desert locusts. *Proc. Roy. Soc. Lond.,B*, **257**, 93–98.

Jaffe, K. and Howse, P.E. (1979) The mass recruitment system of the leaf-cutting ant *Atta cephalotes*. *Anim. Behav.*, **27**, 930–939.

Kullenberg, B. and Bergström, G. (1975) Chemical communication between living organisms. *Endeavour*, **34**, 59–66.

Langley, P., Huyton, P.M. and Carlson, D.A. (1987) Sex pheromone production by males of the tsetse fly *Glossina morsitans morsitans*. *Physiol. Ent.*, **42**, 425–433.

Laurence, B.R. and Pickett, J.A. (1982) Erythro-6-acetoxy-5-hexadecanolide the major component of a mosquito oviposition attractant pheromone. *J. Chem. Soc. Chem. Commun.*, **1982**, 59–60.

Lewis, W.J. and Martin, W.R. Jr (1990) Semiochemicals for use with parasitoids: status and future. *J. Chem. Ecol.*, **16**, 3067–3090.

Lindauer, M. and Kerr, W.E. (1958) Die gegenseitige Verstandigung bei den stachellosen Bienen. *Z. vergl. Physiol.*, **44**, 405–434.

Magnus, D. (1958) Experimentalle Untersuchungen zur Bionomie und Ethologie des Kaisermantels *Argynnis paphia L.* (Lep. Nymph.). *Z. Tierpsychol.*, **15**, 397–426.

Meinwald, J., Meinwald, Y.C. and Mazzocchi, P.H. (1969) Sex pheromone of the queen butterfly: chemistry. *Science*, **164**, 1174–1175.

Moglich, M., Maschwitz, U. and Holldobler, B. (1973) Tandem calling: a new kind of signal in ant communication. *Science*, **186**, 1046–1074.

Myers, J. and Brower, L.P. (1969) A behavioural analysis of the courtship pheromone receptors of the queen butterfly *Danaus plexippus berenice* Cr. *J. Insect Physiol.*, **15**, 2117–2130.

Norris, M. (1968) Some group effects on reproduction in locusts. *Coll. Intern. du. CNRS, Paris 1967*, **173**, 147–161.

Obara, Y. (1970) Studies on the mating behaviour of the cabbage white butterfly, *Pieris rapae crucivora. III.* Near-ultra-violet reflection as the signal of intraspecific communication. *Z. vergl. Physiol.*, **69**, 99–116.

Pliske, T.E. (1975) Courtship behaviour and use of chemical communication by males of certain species of ithomiine butterflies (Nymphalidae, Lepidoptera). *Ann Ent. Soc. Am.* **68**, 935–942.

Prokopy, R.J. (1982) Oviposition-deterring pheromone system of apple maggot flies. In *Management of Insect Pests with Semiochemicals* (ed. E.R.Mitchell), Plenum Press, New York, pp. 477–494.

Regnier, F.E. and Wilson, E.O. (1971) Chemical communication and 'propaganda' in slave-maker ants. *Science, N.Y.*, **172**, 267–269.

Rutowski, R.L. (1978) The courtship behaviour of the small sulphur butterfly, *Eurema lisa* (Lepidoptera: Pieridae). *Anim. Behav.*, **26**, 892–903.

Sappington, T.W. and Taylor, O.R. (1990) Developmental and environmental sources of pheromone variation in *Colias eurytheme* butterflies. *J. Chem. Ecol.*, **16**, 2771–2786.

Schaner, A.M., Brenner, A.M., Lev, R.D. and Jackson, C.C. (1989) Aggregation pheromone of *Drosophila mauritiana, Drosophila yakuba*, and *Drosophila rajasekari J. Chem. Ecol.*, **15**, 1249–1257.

Schildknecht, H. and Koob, K. (1970) Plant bioregulators in the metathoracic glands of myrmecine ants. *Agnew. Chem. Int. Ed. Eng.*, **9**, 173.

Schneider, D., Boppre, M., Zweig, J., *et al.* (1982) Scent organ development in *Creatonotus* moths: regulation by pyrrolizidine alkaloids. *Science*, **215**, 1264–1265.

Schneirla, T.C. (1971) *The Army Ants*, W.H. Freeman, San Francisco.

Scott, D.R. and Richmond, R.C. (1987) Evidence against an antiaphrodisiac role for *cis*-vaccenyl acetate in *Drosophila melanogaster. J. Insect. Physiol.*, **33**, 363–369.

Scott, D.R., Richmond, R.C. and Carlson, D.A. (1988) Pheromones exchanged during mating: a mechanism for mate assessment in *Drosophila. Anim. Behav.*, **36**, 1164–1173.

Silberglied, R.E. and Taylor, O.R. (1978) Ultraviolet reflection and its behavioural role in the courtship of sulfur butterflies, *Colias eurytheme* and *C. philodice* (Lepidoptera, Pieridae). *Behav. Ecol. Sociobiol.*, **3**, 203–243.

Stowe, M.K. (1988) Chemical mimicry. In *Chemical Mediation of Coevolution* (ed. K.C. Spencer), Academic Press, New York, pp. 513–580.

Tinbergen, N. (1932) Uber die Orientierung des Bienenwolfes. *Z. vergl. Physiol.*, **16**, 305–334.

Tinbergen, N., Meeuse, B.J.D., Boerma, L.K. and Varossisan, W.W. (1942) Die Balz des Samtfalters, *Eumenis* (= *Satyrus*) *semele (L.). Z. Tierpsychol.*, **5**, 182–226.

Tumlinson, J.H., Silverstein, R.M., Moser, J.C. *et al.* (1971) Identification of the trail pheromone of a leaf-cutting ant *Atta texana. Nature. Lond.*, **234**, 348–349.

Vale, G.A., Hall, D.R. and Gough, A.J.E. (1988) The olfactory responses of tsetse flies, *Glossina* spp. (Diptera: Glossinidae) to phenols and urine in the field. *Bull. Ent. Res.*, **78**, 293–300.

Vander Meer, R.K., Alvarez, F. and Lofgren, C.S. (1988) Isolation of the trail recruitment pheromone of *Solenopsis invicta. J. Chem. Ecol.*, **14**, 825–838.

Vander Meer, R.K., Lofgren, C.S. and Alvarez, F.M. (1989) The orientation inducer pheromone of the fire ant, *Solenopsis invicta. Physiol. Ent.*, **15**, 483–488.

Vilela, E.F. and Howse, P.E. (1984) Territoriality in leaf-cutting ants, *Atta* spp. In *Fire Ants and Leaf-cutting Ants, Biology and Management* (eds C.S. Lofgren and R.K. Vander Meer), Westview Press, Colorado, p. 157–171.

Whitten, W.M., Young, A.M. and Williams, N.H. (1989) Function of glandular secretions in fragrance collection by male euglossine bees (Apidae: Euglossini). *J. Chem. Ecol.*, **15**, 1285–1295.

Williams, N.H. and Dodson, C.H. (1972) Selective attraction of male euglossine bees to orchid floral fragrances and its importance in long distance pollen flow. *Evolution*, **26**, 84–95.

Willis, M.A. and Birch, M.C. (1982) Male lek formation and female calling in a population of the arctiid moth *Estigmene acrea. Science*, **218**, 168–170.

Wilson, E.O. (1962) Chemical communication among workers of the fire ant *Solenopsis saevissima* (Fr. Smith). I. The organization of mass-foraging. *Anim. Behav.*, **10**, 134–147.

Wittmann, D., Radtke, R., Zeil, J. *et al.* (1990) Robber bees (*Lestrimelitta limao*) and their host. Chemical and visual cues in nest defense by *Trigona (Tetragonisca) angustata* (Apidae: Meliponinae). *J. Chem. Ecol.*, **16**, 631–641.

# 3

# Factors controlling responses of insects to pheromones

This chapter will concentrate on the factors affecting the emitter and receiver of a pheromonal signal. Chapter 1 showed that the interpretation of the signal (i.e. for all practical purposes the response) depends on the context and internal state of the receiver. There are numerous factors that can affect the probability of a male insect approaching a source of female sex pheromone (Fig. 3.1), and it is vitally important to take these into account in both laboratory experimentation and field testing. Equally, there are many factors that affect the occurrence of calling behaviour of females.

## 3.1 PHEROMONE RELEASE

The amount of pheromone that insects release is generally extremely low. There are technical difficulties in estimating release rates accurately, because pheromone is often adsorbed on to adjacent surfaces, including the body of the insect itself. Pheromone of the oriental fruit moth, *Grapholitha molesta*, is adsorbed on to wing scales (Baker *et al.*, 1980). In *Antheraea polyphemus*, a scale esterase is present which degrades adsorbed pheromone, preventing persistent slow release that might be of advantage to searching parasitoids (Vogt *et al.*, 1985).

Lacey and Sanders (1992) estimate that the principal component of *Grapholitha molesta* is emitted at 8.5 ng $h^{-1}$ during the calling period. An average figure for moths is probably an order of magnitude higher. The cabbage looper moth, *Trichoplusia ni*, releases 880 ng $d^{-1}$ with peaks of up to 50 ng over 5 minute periods (Bjostad *et al.*, 1980). The highest release rates have been recorded from the arctiid moth *Holomelina lamae*, reaching 350 ng over 10 minutes (Schal *et al.*, 1987). Such bursts are

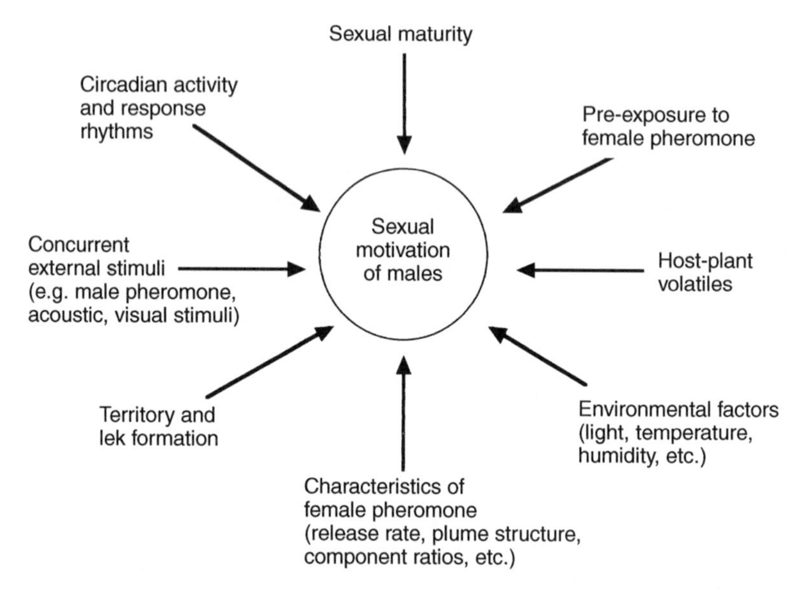


**Fig. 3.1** Factors potentially affecting sexual motivation of male moths (and other male insects) responding to female-produced pheromone.

achieved by extruding and retracting the gland rhythmically, which also produces a modulated signal. This occurs in the arctiids *Utetheisa ornatrix* (Conner *et al.*, 1980) and *Pyrrharctia isabella* (Krasnoff and Roelofs, 1989). In *U. ornatrix* the pheromone gland of the female is protruded rhythmically at a rate of approximately 1.5 per second. In *P. isabella* the rhythmic protrusion results in dispersal of pheromone as an aerosol. Rhythmic dispersal is likely to improve the signal transmission by increasing both the signal to noise ratio and the redundancy. It has been calculated that a sinusoidal modulation of release rate can double the range of attraction, and instantaneous puffing can increase the range by an order of magnitude if the detection threshold of a volatile is high (Dusenbery, 1989).

In some cases, pheromone is not released from what can be thought of as a point source. The effective dimensions of the source depend on where the calling insect is sitting (Aylor, 1976). In scolytid bark beetles, hundreds of calling insects can generate a massive plume the size of an entire tree trunk (Murlis *et al.*, 1992). The pheromone emission of calling pea moths (*Cydia nigricana*) is adsorbed on to the plant foliage, again broadening the source. Wheat leaves, taken from immediately downwind of pea moth pheromone traps, albeit baited with very high doses of pheromone, were still releasing pea moth pheromone after 4 hours when assayed by gas chromatography and electroantennography of male moths. Moths were still attracted to the vegetation in the field more than one hour after the removal of traps; they landed and began wing-fanning as in the initial stages of the courtship ritual (Wall *et al.*, 1981). If air containing the pheromone released by a single female cabbage moth. (*Mamestra brassicae*) is passed over leaves of brussels sprouts, sufficient is adsorbed to attract males over a distance of 0.5 m in a wind tunnel (Noldus *et al.*, 1991). This is also sufficient to stimulate *Trichogramma* egg parasitoids to search over leaves, which remain attractive to the wasps for between 4 and 24 hours. As the moths call at night and the parasitoids are active during daylight, leaf adsorption extends the window of opportunity of the parasitoids.

Species isolation in certain neotropical cockroaches is achieved partly by the females calling at different heights in the vegetation (Fig. 3.2). Schal (1982) found eight species that segregated themselves by height above ground. The males also perched in the vegetation at characteristic

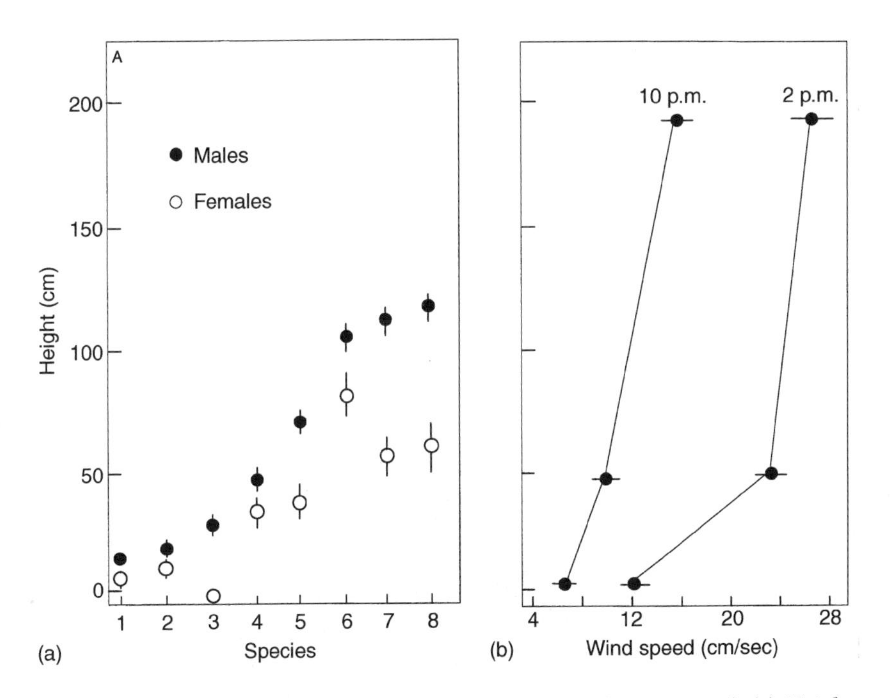


**Fig. 3.2** Communication channels separated by height above ground. (a) Heights in vegetation of Costa Rican forest from which eight species of female cockroach were observed calling, and heights at which males of the corresponding species were found perching in positions to intercept the female odour plume. (b) Changes in wind speed with height at two times of day in the same location, resulting in upward drift of air currents. (Modified from Schal, 1982.)

heights, always significantly higher than their females. The explanation appears to be that a thermal gradient develops at night in the air near the ground and pheromone plumes tend to rise as a result. The males are then in positions where the likelihood that they will encounter the pheromone plume of the females of their own species is increased.

Release rate of pheromone is affected by the posture of the insect concerned. Simply by raising its wings away from the abdominal tip and lifting its abdomen from the substrate, the arctiid moth *Spilosoma*

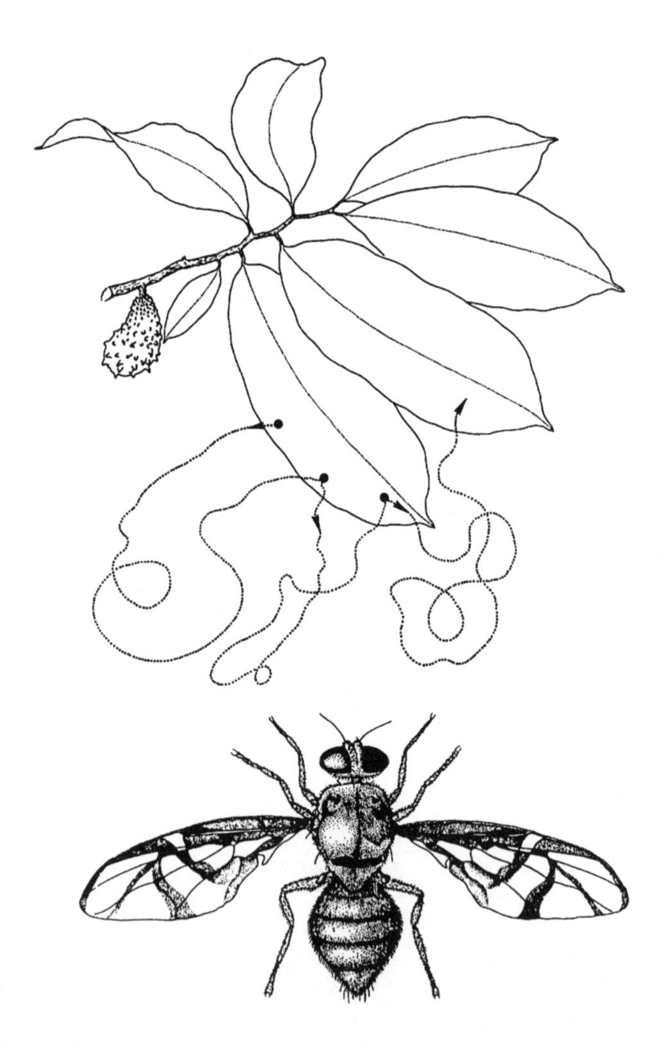

**Fig. 3.3** Looping flights by tephritid fruit fly *Anastrepha robusta* when calling from a leaf territory. (From Aluja, 1993, with permission.)

*congrua* can increase the velocity of the air over the tip of its abdomen sevenfold, resulting in an estimated 220% increase in the pheromone release rate (Conner and Best, 1988). The male of the fruit fly *Anastrepha robusta*, in common with many Tephritidae, establishes a leaf territory from which it calls. This species amplifies its calling by making frequent looping flights extending on average about 15 cm from the leaf (Aluja, 1993) (Fig. 3.3). The pheromone will thus be liberated from an extended zone into air currents away from the interior of the vegetation.

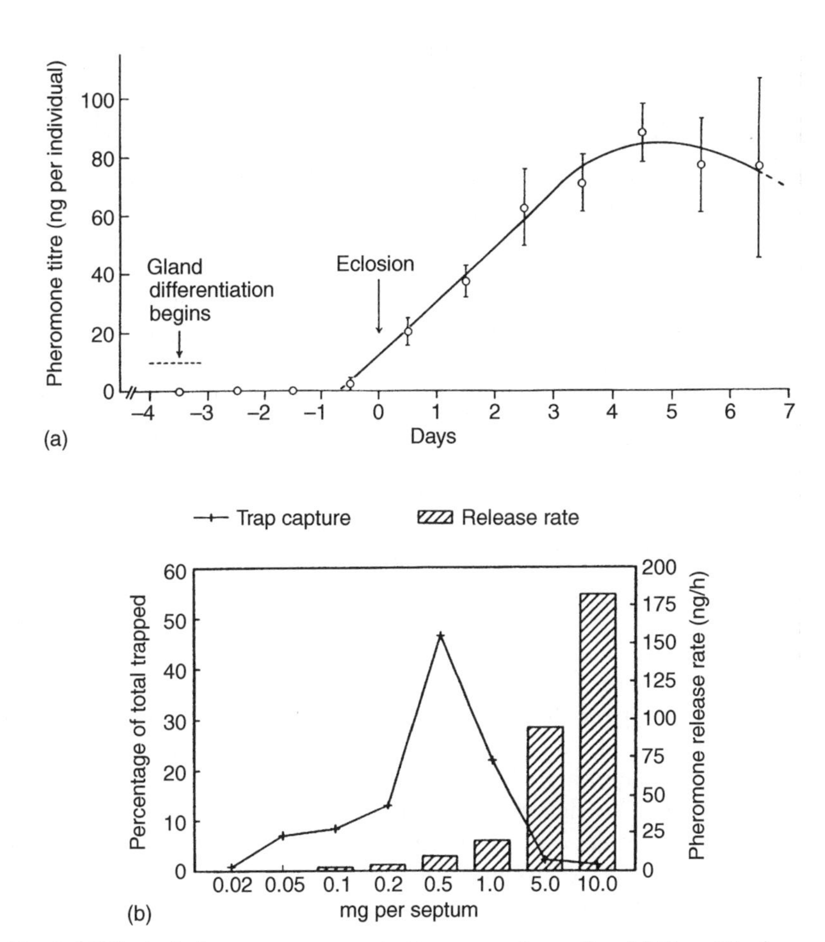


**Fig. 3.4** Effect of pheromone concentration on male moths. (a) Quantity (mean + standard deviation) of sexual pheromone produced by female redbanded leaf-roller moths, *Argyrotaenia velutinana*. (From Miller and Roelofs, 1977.) (b) Relationship between trap catches of male velvetbean caterpillar moths, *Mocis disseverans*, and release rates of an optimal ratio of its two main sex pheromone components on rubber septa. (From Tumlinson, 1988.)

Male Lepidoptera typically have both upper and lower thresholds, and responsiveness follows a bell-shaped curve (Fig. 3.4). The simple consequence of this is that it is not usually possible to trap large numbers of insects in the field by releasing pheromone at a high rate. Controlled release technology is therefore of prime importance for the use of pheromones in pest monitoring (Chapters 9 to 12).

## 3.2 CIRCADIAN CHANGES IN THRESHOLDS

Circadian changes in activity and responsiveness pervade every aspect of insect physiology and behaviour; they limit feeding, mating or even growth to certain more or less restricted periods of day or night. Outside these periods, the probability of insects responding (even to high levels of stimulation) may be so low as to make any behavioural test pointless. Attraction of blood-feeding Diptera to hosts is conspicuously under circadian control and often limited to a brief period. In Uganda, Haddow *et al.* (1968) found that some species of mosquito were active for a period as short as 20 minutes per day, and a succession of species exploited different parts of the dusk period. Circadian rhythm governs a variety of behavioural patterns in tsetse flies (*Glossina* spp.) including spontaneous flight activity, biting, optomotor responses, mating and responsiveness to human odour (Brady, 1972).

Pheromone release by the female cabbage looper moth, *Trichoplusia ni* (a noctuid), reaches a peak about eight hours after dark (Sower *et al.*, 1970). The responsiveness of the male undergoes a parallel change, slightly out of phase, reaching a peak 6–7 hours after dark so that males are highly sexually motivated by the time the females begin to call.

The extent of threshold shifts that can take place in responsiveness to female pheromone was demonstrated by the striking results obtained by Castrovillo and Carde (1979) on the codling moth, *Cydia* (= *Laspeyresia*) *pomonella*. Here the thresholds of the male for activation, upwind flight and hair-pencilling behaviour undergo enormous parallel threshold shifts (Fig. 3.5). Responsiveness of males increases about fourfold at the beginning of the scotophase (dark period). The implications of this result for laboratory bioassays are of profound importance: first, they must always be done at the same time of day, otherwise results will not be comparable, and secondly, for the greater part of the 24-hour period insects may, for behavioural bioassay purposes, be unresponsive.

Caaling behaviour of the female codling moth follows a similar pattern (Fig. 3.5), completing a narrow channel of communication that is defined in temporal as well as in chemical terms. Female calling is influenced by ambient temperature. The calling begins earlier and goes on for longer at lower temperatures, which may be an adaptation to

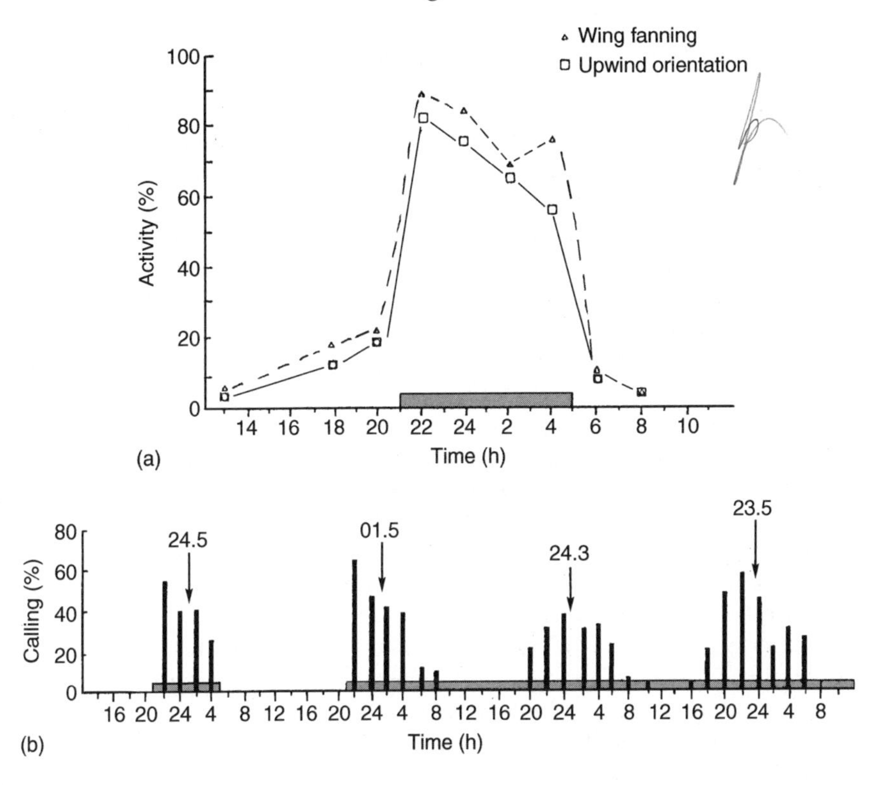


**Fig. 3.5** Rhythms of calling and response in codling moth, *Cydia pomonella*. (a) Changes during the light/dark cycle in percentages of male moths showing aspects of sexual behaviour at a fixed pheromone concentration. (b) Percentages of female moths calling during scotophase (dark period) when this is extended, showing persistence of a circadian rhythm under constant conditions. (Modified from Castrovillo and Carde, 1979.)

take advantage of the relatively higher temperatures in the early part of the night.

Calling of sympatric species of moths generally occurs during different night-time periods, but this is not the only means of maintaining species isolation. Published data on 10 species of tortricid moth in the eastern USA that produce similar sex pheromones shows that flight times and periods of adult emergence during the year tend to coincide (Carde and Baker, 1984). Furthermore, many of these species share a restricted range of five sex pheromone components (tetradecenyl acetates and dodecenyl acetate). The barriers to cross-mating appear to be differences in the relative proportions of these components in each species.

## 3.3 SEXUAL MATURITY AND MATURATION

There have been various attempts to show that the endocrine system of the corpus cardiacum and the corpora allata are implicated in the control of pheromonal communication; some have shown this to be so, others have not. For example, the corpus cardiacum does not affect calling or the onset of oviposition behaviour in the silk moth *Antheraea polyphemus* (Sasoki *et al.*, 1983) and these behaviour patterns appear to be entirely under nervous control. On the other hand, the corpora allata are important in the silver-Y (black cutworm) moth (*Agrotis ypsilon*). Allatectomy eliminates sexual behaviour, while reimplantation of corpora allata restores it (Gadenne *et al.*, 1993). Gating of responsiveness to pheromone is a central process: the response of the antennal sensilla (EAG) stays the same. This insect undergoes an autumn migration, and few are caught in pheromone traps at that time when the CA are reduced in size. Juvenile hormone also controls pheromone release in the German cockroach (*Blattella germanica*). Pheromone production by females is correlated with oocyte growth, and is slowest in those that are carrying an ootheca (Schal *et al.*, 1990). Application of juvenile hormone analogue to allatectomized insects stimulates pheromone production. Implantation of wax-filled oothecae in virgin females inhibits oocyte growth and lowers pheromone production.

Competence for mating varies considerably among different species and between males and females of the same species. Pheromone synthesis may begin in the pupal stage. Female *Heliconius charitonia* butterfly pupae attract males several days before eclosion (Gilbert, 1976) which begin mating as soon as the pupal case splits. The two main components of the sex pheromone of the red-banded leaf-roller moth (*Argyrotaenia velutinana*) are first detectable by chromatography one day before eclosion from the pupa (fig. 3.4a), and increase in concentration to a peak of about 80 ng per insect at four to six days, after which the titre diminishes (Miller and Roelofs, 1977). It is possible that sexual maturity is attained only after several days of adult life in most insects; for example, sexual activity in tephritid fruit flies such as *Anastrepha suspensa* (the Caribbean fruit fly) tends to begin after 7–10 days (Nation, 1990) but then continues for up to 35 days of age.

In the desert locust, *Schistocerca gregaria*, sexual maturity in the field is delayed in the dry season until the onset of rains, which may mean several weeks' delay. Extensive research in the laboratory on the effects of external factors failed to reveal the trigger for the onset of sexual maturity, and it was not until this phenomenon was studied in the field (Carlisle *et al.*, 1965) in Ethiopia that the environmental cue was found. Just before the rains, when the humidity rises, many desert trees and shrubs begin to bloom. These include *Boswellia* and *Commiphora*, sources of frankincense, myrrh and balm of Gilead, which release into the air fragrances rich in monoterpenes such as pinenes, limonene and eugenol. It was shown in the laboratory that these trigger the development of sexual maturity so that the females are ready to lay eggs in the moist sand, which is necessary for their development. Female *S. gregaria* also produce a pheromone that accelerates the development of sexual maturity in other females, so that broad synchrony is obtained, and males similarly produce a pheromone that accelerates sexual maturity in other males (Norris, 1968).

There is accumulating evidence that behavioural responsiveness of insects to semiochemicals can be influenced by changes in thresholds of sensory receptors, which, in turn, affect behavioural thresholds to potential mates, or hosts. Host-seeking behaviour of the yellow fever mosquito, *Aëdes aegypti*, begins at around 24–48 hours after eclosion, reaching a maximum at 100 hours (Davis, 1984). This is paralleled by changes in the response of the antennal receptors (EAG – see 4.3.1) to lactic acid, which is an important host attractant. After a blood feed, the threshold temporarily rises again by almost an order of magnitude (and behavioural responsiveness is reduced) as a result of a blood-borne factor. The magnitude of the EAG of honey bees also changes with maturity. In workers, potential changes of around 0.5 mV were recorded in response to a standard concentration of *iso*-pentyl acetate (an alarm pheromone component) which over the course of the first few days of adult life rises to about 1.8 mV, coinciding with the ability to respond to queen substance (Masson and Arnold, 1984). It was also found that, over about the first three weeks of adult life, bees kept in groups gave EAG responses roughly twice the size of bees kept isolated. Pheromone sensitivity in certain moth species is also affected by blood-borne factors (Davis, 1984) and titres of juvenile hormone appear to be associated with such changes. The EAG response of tsetse flies to host kairomones such as 4-heptanone, 3-nonanone, acetone and 1-octen-3-ol is influenced by the age, sex, and food-deprivation of the insects (Den Otter *et al.*, 1991).

## 3.4 PRE-EXPOSURE TO PHEROMONE

The majority of insect pheromones are odourless to the human nose, or nearly so, and they are often secreted in nanogram quantities. It is all too easy to assume that all experiments, whether in the laboratory or in the field, are carried out against the background of an olfactory *tabula rasa*. In fact, this is extremely unlikely: volatile chemicals may be absorbed and adsorbed by materials other than vegetation. These include glass, plastic materials, vegetation and even the human body. A standard method of entrapping volatiles for analysis involves adsorption on to glass (Chapter 6), and plastic materials are commonly used as slow-release substrates for pheromones.

Perhaps more startling is the tendency of pheromone researchers themselves to absorb and release pheromone over long periods. Cameron (1983) was exposed to the synthetic pheromone of the gypsy moth from 1971 to 1977. In 1981 he was still very attractive to the male gypsy moth, as shown in replicated tests with five other individuals. Curiously, the ability to attract was present in other members of his family, roughly in proportion to the closeness of the relationship. This phenomenon has its amusing side (with headlines such as 'Sex-crazed moths attack scientist') but should be taken extremely seriously by experimenters. It has been clearly shown that behavioural thresholds of insects undergo an upward shift on continuous exposure for relatively short times. This is known as **habituation**, a term which is conventionally used for a type of learning process. Pheromone researchers tend to use the term 'habituation' to describe any kind of waning of responsiveness, without always establishing whether it is a central nervous process. Habituation to pheromones, however, unlike sensory adaptation, is usually relatively persistent and may be detectable for hours after a period of exposure. The presence of persistent pheromone sources in a laboratory can easily be overlooked; indeed it may be impossible to detect some of them, but their influence on the thresholds of insects under test may be considerable.

Bartell and Lawrence (1977) showed that when males of the light brown apple moth, *Epiphyas postvittana*, were exposed to pulses of female pheromone at intervals of two minutes, during which time sensory adaptation can occur, the reduction in responsiveness was far greater that that to a continuous exposure of the same duration. The female pheromone has two major components which must both be present before the males will show a sexual response. There is no response if one component is presented immediately after the other, but if males are exposed to alternating pulses of the two components they show a reduction in responsiveness when tested with both components some minutes or hours later. Habituation in this insect must therefore be a central nervous phenomenon and not a form of sensory adaptation.

## 3.5 THE CONTEXT OF OTHER STIMULI

### 3.5.1 Host-plant volatiles

The situation in which most of the antennal receptors of an insect are specialists, tuned almost exclusively to the reception of pheromone components, is an extreme case. In many insects, host-plant volatiles have an additive or synergistic role in pheromone communication systems. There appear to be two advantages to this kind of dual attractant system: first, the females are more likely to encounter the males by orienting to host-plant odour; and second, when the females have mated they are already on or near the host plant.

The extent to which host-plant volatiles influence pheromone production and liberation is very variable and the degree of intervention, when it occurs, is also difficult to assess. Interest in pheromone research was reduced considerably at one point when it was hypothesized that lepidopteran sex pheromones were derived from secondary plant compounds of the host (Hendry *et al.*, 1975). Research has since shown that they are generally synthesized *de novo* (Roelofs and Wolf, 1988). A question-mark of another kind arose when Riddiford and Williams (1967) produced evidence to show that the female polyphemus moth (*Antheraea polyphemus*) would call only in the presence of volatiles emitted from the leaves of the foodplant, red oak (*Quercus rubra*). The active component was later identified as (Z)-2-hexenal (Riddiford, 1967). A repeat of this work by Carde and Taschenberg (1984), however, failed to confirm the earlier conclusion.

More cases are coming to light in which the importance of host-plant volatiles to sex pheromone communication in moths has been clearly established. Female small ermine moths (*Yponomeuta* spp.) were found to call more often in the presence of odours from their host plant (Hendrikse and Vos-Bunnemeyer, 1987). In the absence of the host-plant odour, the onset of calling was delayed for one or two days, or the number of females calling was lower than when the plant odours are present. Older virgin females showed a preference for calling in areas in which the host plant was present. The sunflower moth, *Homoeosoma electellum*, is attracted towards newly opened sunflowers for oviposition and the first instar larvae feed on the pollen. The presence of pollen stimulates the onset of calling in virgin females and there is some evidence that more males are attracted to females calling from sunflowers that are producing pollen (McNeil and DeLisle, 1989).

Raina *et al.* (1989) showed that the calling behaviour and sex pheromone production of the noctuid moth *Heliothis phloxiphaga* are markedly influenced by the presence of its host plant, the Texas paint brush tree (*Castilleja indivisa*). Females reared from larvae collected in the field and exposed to the plant are much more likely to call, and produce about 50 times as much pheromone as those that are kept without it. The synergy declines in insects reared on artificial diet in the laboratory and virtually disappears after 10 generations. The related species, *Helicoverpa zea* (the corn earworm), has about 70 different host plants, including maize (*Zea mais*). Raina *et al.* (1991) found no calling at all in wild-caught females unless host-plant material such as an ear of maize with 'silk' was present. They traced the effect to emission of ethylene from the plant tissue and showed that there was a direct relationship between the amount of pheromone produced by females and the concentration of ethylene (up to 1000 ppm).

The importance of host-plant volatiles as adjuncts to pheromones has been established not only in Lepidoptera, but also in certain Diptera and Coleoptera. In the field, male papaya fruit flies (*Toxotrypana curvicauda*) perch on green papaya fruit, establish a territory and call by inflating pleural glands on the abdomen (Landolt and Hendrichs, 1983). In laboratory tests they were more attractive to females in the presence of green papaya fruit peel or peel extract (Landolt *et al.*, 1992). The females were more likely to encounter males by orientating to host-plant odour, and when they had mated they were already on or near the larval food plant.

The importance of host-plant volatiles as adjuncts to pheromones has been established not only in *Dendroctonus brevicomis* (Chapter 2), but also in many other species of bark beetle (Scolytidae). The boll weevil, *Anthonomus grandis*, is weakly attracted by volatiles such as (*E*)-2-hexen-1-ol, but when this was added to the synthetic aggregation pheromone (grandlure) in field traps, the trap catch was more than twice that of grandlure alone and relatively more females were caught (Dickens, 1989).

Plant volatiles do not always potentiate the effects of pheromone. The so-called 'green leaf volatiles', hexan-1-ol and hexanol, inhibit the response of the southern pine beetle, *Dendroctonus frontalis*, to its aggregation pheromone and to the host-tree pheromone synergist, turpentine (Dickens *et al.*, 1992). This can be interpreted as a means of diverting the insects from areas in which non-host trees predominate. A similar phenomenon has been recorded in aphids which feed on hops, in which (−)-β-caryophyllene inhibits the response to aphid alarm pheromone, (*E*)-β-farnesene (Dawson *et al.*, 1984).

### 3.5.2 Acoustic stimuli

Acoustic signals are common among bark beetles (Rudinsky *et al.*, 1976), and *Dendroctonus pseudotsugae*, the Douglas fir beetle, produces several kinds. Females stridulate when they have established burrows in the tree. This repels other females that land nearby, and so spaces out burrows. In response to pheromone, males produce different stridulatory signals as they enter the female galleries: a single attractant chirp and an 'interrupted' chirp. The latter repels other males that arrive and synergizes the effect of inhibitory pheromone produced by both sexes. Males of *Ips* species, in contrast to *Dendroctonus*, enter the burrow first and the female stridulates before entry, but there is no evidence that stridulation in this species elicits pheromone release.

The lesser wax moth, *Achroia grisella*, is the only moth in which mate location has been found to depend upon acoustic orientation (Spangler, 1984). The males produce sounds of ultrasonic frequency (around 100 kHz) and, at the same time, release pheromone while fanning with their wings. Female moths are able to find their way to the males from a distance of about 1 m by running or flying. Although the pheromone may play some part in pair formation, there is no evidence that moths with both tympana destroyed can find females (Spangler and Hippenmeyer, 1988). Ultrasound clicks are produced by some arctiid moths, including the dogbane tiger moth, *Cycnia tenera*, in which male clicks, along with the presence of pheromone emitted from the male's coronemata, make the female receptive to mating (Conner, 1987).

In general, night-flying moths respond to ultrasound pulses with evasive action. This is one of their means of escaping from approaching insectivorous bats (Roeder, 1963). Baker and Carde (1978) showed that ultrasonic pulses induced sudden changes in the orientation of male gypsy moths (*Lymantria dispar*) flying upwind towards a pheromone source. They suddenly deviated from their course and flew away rapidly, which was attributed to the anti-predator behaviour elicited.

Acoustic stimuli play an important role in the courtship of many species of fruit fly, including *Drosophila* species. Locomotion and courtship activity of males increased when recordings of courtship song were played to them in the presence of females (Schilcher, 1976). The effect was relatively persistent: after 60 seconds of playback, male locomotor activity continued at a high level for more than 3 minutes. Male Caribbean fruit flies sing calling songs and precopulatory songs to females (Sivinski *et al.*, 1984). The males aggregate in leks and liberate sexual pheromone while generating pulsed calling song by wing-fanning. As in *Drosophila*, females are more active in the presence of the song. The precopulatory song, produced in the initial stages of copulation, is believed to demonstrate male vigour and thereby affect female choice.

In some social insects, there is also evidence that acoustic signals add to or synergize the effect of pheromone. The leaf-cutting ant *Atta cephalotes* produces ultrasonic sounds by stridulation when trapped in earth or grasped by an attacker. This lowers the threshold of response to alarm pheromone released from glands in the head, but headless insects can be located at close range by nest-mates following the gradient of acoustic intensity (Markl, 1965). The acoustic stimulus in such cases, which may be common among social insects, can thus be thought of as an alerting signal. In the ants *Novomessor albisetosus* and *N. cockerelli*, the first worker to attack prey will stridulate and this enhances the release of pheromone from the poison gland in the next ant to arrive (Holldobler *et al.*, 1978).

### 3.5.3 Visual signals

Many day-active insects (butterflies, dipteran flies, etc.) respond to a mixture of long-range visual signals and short-range chemical signals in courtship (Chapter 2). Male oriental fruit moths (*Grapholitha molesta*) increase the frequency of close-range interactions when a visual stimulus is present in addition to female pheromone (Baker and Carde, 1979). Visual cues are also used in orientation towards animal hosts or host plants, and in such cases may facilitate responses to pheromones. The best-studied examples are bark beetles, which tend to orientate towards vertical trunk-like silhouettes. Wyatt *et al.* (1993) examined the use of visual cues by the predatory beetle *Rhizophagus grandis* orientating to its bark beetle prey, *Dendroctonus miceus*. *Rhizophagus* is attracted by a kairomone in the frass of the bark beetle and it was found, rather surprisingly, that visual stimuli provided by three-dimensional cylindrical objects took second place to turbulence, changes in wind velocity and wind eddies.

### 3.5.4 Male repellents and anti-aphrodisiac pheromones

The males of danaid and ithomiine butterflies use their hair pencil secretions to repel conspecific males. Pursued males of *Hymenitis andromica* slow their flight and cant their hindwings backwards, displaying the hair pencils (Fig. 2.3) at the front margins of the hindwings (Pliske, 1975). This stops conspecific males from pursuing further, and also inhibits males of other species that have lactone hair pencil secretions. The female passion vine butterfly, *Heliconius erato*, exudes an odour resembling witch hazel. Gilbert (1976) found that this odour is produced only by mated females, and while it may repel bird predators it also repels conspecific males. In the mating process, a pheromone is transferred from the abdominal glands of the male that is odourless to the human nose but that appears to be a precursor for the anti-aphrodisiac secretion.

There is evidence for inhibitory pheromones also in some Coleoptera, Diptera and moths. Each sex of the mealworm beetle, *Tenebrio molitor*, secretes a pheromone that attracts the opposite sex, but the male also secretes an inhibitory pheromone after it has been stimulated by female scent (Happ, 1969) This remains on the female after mating and such marked females attract fewer males than virgin females.

Pheromone is dispensed by male moths from various sources but most commonly from hair pencils on the eighth abdominal segment and the margin of the forewings (Phelan, 1992; Birch *et al.*, 1990) The role of male moth secretions in inhibiting the approach of other conspecific males is not clearly understood and appears to vary from species to species. Hirai *et al.* (1978) showed that the presence of caged males of the true army worm (*Pseudaletia unipunctata*) upwind in wind tunnels inhibited the responsiveness of others downwind, and so would reduce male competition. However, Fitzpatrick *et al.* (1988) were unable to detect such inhibitory effects in wind tunnel experiments carried out with hair pencil extracts and excised male hair pencils placed upwind, and concluded that the results of Hirai *et al.* were due to impeded air flow in the tunnel. In the light brown apple moth, *Epiphyas postvittana*, far from there being any inhibition, males were more likely to land at a site where other males were present (Rumbo, 1993). A single resident male increased the attractancy of a source by a factor of 1.40, and two residents by 1.77, compared with an empty source. Male pheromone from abdominal hair pencils of *Grapholitha molesta* attracts females at close range, and also other males (Baker, 1983). The males of the velvet-bean caterpillar moth, *Anticarsia gemmatalis*, have hair pencils and scent brushes on the tip of the abdomen, which produce (Z,Z,Z)-3,6,9-heneicosatriene (Heath *et al.*, 1988), a compound that is also present in the female sexual pheromone blend. Its main role appears to be to increase the acceptance of females during courtship, but as a pure compound it is also attractive to other males.

In various mosquitoes, including *Aëdes aegypti*, sexual receptivity of the female is terminated and egg-laying stimulated by an accessory gland secretion of the male that is transferred in copulation (Craig, 1967; Hiss and Fuchs, 1972). The secretion of *A. aegypti* is known as **matrone** and has two proteinaceous fractions, one of which stimulates mating, though both are necessary for inhibition of mating. This pheromone has further priming effects in the mated female of *Anopheles gambiae*, which affect her blood-feeding behaviour, producing (among other things) a change in circadian flight activity (Jones and Gubbins, 1977). The activity of the virgin female is mainly confined to the first hour of the dark period, but mated females spread their activity more or less evenly over the hours of darkness.

## 3.6 PHEROMONE BLENDS

Distinctiveness of a chemical signal is easily achieved by adding to the components. In insects this is often done by biosynthesis of structural or optical isomers of a compound (Chapters 5–8). For a chemist, syntheses of such compounds are often fraught with difficulty because *in vitro* syntheses commonly yield mixtures (so called **racemic** mixtures) of two or more isomers, and isolation of any one in a pure form can be a very complex procedure. Living organisms, on the other hand, synthesize chemicals on enzymatic templates, so that it is difficult to avoid producing compounds that are 100% pure, and the production of isomers in a fixed ratio appears to pose little difficulty for an insect.

A beautiful example of the achievement of species specificity of response is provided by the European species of Yponomeutidae (small ermine moths). Six species of *Yponomeuta* share the (*E*) and (Z) isomers of 11-tetradecenyl acetate (11-TDA) (Löfstedt, 1993). Three species (*Y. padellus, Y. evonymellus and Y. viginipunctatus*), which produce approximately the same ratios of the (*E*) to (Z) compounds, show cross-attraction of males in wind tunnel assays (Fig. 3.6) but these are all found on different host plants. Three other species (*Y. cagnatellus, Y. rorellus and Y. plumbellus*), which produce very different ratios one from the other and show no (or very little) cross-attraction, share the same host plant.

Löfstedt *et al.* (1990) made the intriguing discovery that males of some *Yponomeuta* species have specialized antennal receptors that do not respond to the pheromone of their own species, but are sensitive to the principal component of the pheromone of their close relatives. They

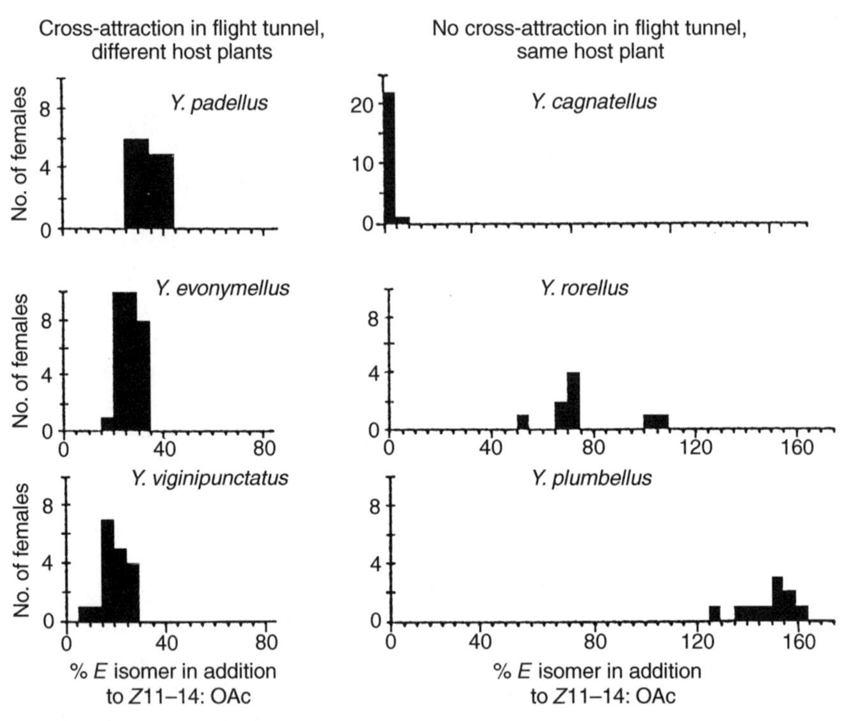


**Fig. 3.6** Pheromone component ratios as a species-isolating mechanism in small ermine moths (Yponomeutidae). Females of the six species produce various ratios of two main components: (*E*)- and (Z)-11-tetradecenyl acetate. Those that have different foodplants have similar ratios of the two components and cross-attract in wind tunnels. Those that share same foodplant have ratios radically different from each other. (From Lofstedt *et al.*, 1990.)

**Table 3.1** Factors in species isolation of four species of Sesiidae (clearwing moths) which share (*ZZ*)- and (*EZ*)-3,13-octodecadienyl acetate as female sex pheromone components (approximate information, taken from data in Greenfield and Karandinos, 1979)

| Species | Ratio *ZZ*:*EZ* isomer | Activity time (and peak) | Season |
|---|---|---|---|
| *Podosesia syringae* | 100:1 | 08.30–14.00 (09.00–10.00) | May |
| *Carmenta bassiformis* | 100:1 | 16.00–19.00 (17.00–18.00) | July |
| *Synanthedon scitula* | 100:1 | 19.00–21.00 (20.00–21.00) | July |
| *Synanthedon pictipes* | 1:100 | 09.00–14.00 (09.00–10.00) | May–June |

are tuned in to the communication channel of their relatives to avoid responding to them. Males of *Y. rorellus* have receptors for Z-11TDA though their females do not produce this compound. Adding only 1% of Z-11TDA to a sample of the female pheromone almost abolishes the attraction of males to pheromone-baited traps.

Similar studies have been carried out on North American species of tortricid moths (Carde and Baker, 1984), showing that many sympatric species have sex pheromones consisting of similar mixtures of components but that species specificity appears to depend upon the precise pheromone blend. Greenfield and Karandinos (1979) investigated species of Sesiidae (clearwing moths) in Wisconsin. As in the small ermine moths discussed above, some species have very similar ratios of the common sex pheromone components, (Z,Z)- and (*E*,Z)-3,13-octadecadien-1-ol acetate (ODDA). However, those species that do not differ markedly in their pheromone blend are active either at different times of day (Table 3.1) or during different months of the year. Separation by locality or parts of the biotope in which they are found does not appear to be an important means of species isolation.

## 3.7 POPULATION AND GENETICAL DIFFERENCES

In addition to interspecific variation, geographical variation in pheromone blends is commonly encountered. This can be seen as either quantitative differences (affecting ratios of compounds) or qualitative ones. The green stink bug, *Nezara viridula*, provides an example of both. This species has a world-wide distribution in the tropics and subtropics. Mature males release a pheromone which includes various sesquiter-

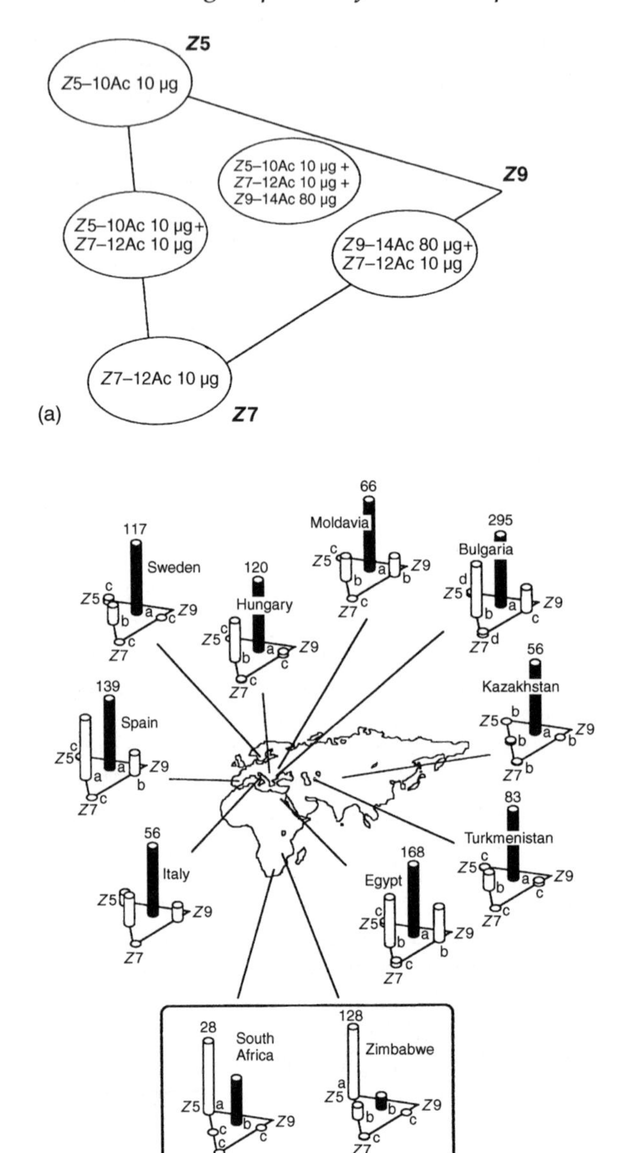


**Fig. 3.7** Results of trapping experiments with turnip moth, *Agrotis segetum*. (a) Compounds (Z)-5-decenyl acetate, (Z)-7-dodecenyl acetate and (Z)-9-tetradecenyl acetate, and ratios tested at each site below. (b) Percentage trap catches at 11 sites in Europe, Asia and Africa. Numbers above columns are total catches with best lure. Where letters at bottoms of columns are the same, there is no significant difference between effects of lures. (Modified from Tóth *et al.*, 1992.)

**Table 3.2** Pheromone 'dialects' in the turnip moth, *Agrotis segetum*: ratios* of the three main components of the female sex pheromone in moths collected from different regions of Europe, with (in boxes) relative numbers (%) of specialist sensilla on antennae of males from same populations (data from Lofstedt, 1993; Hansson *et al.*, 1990)

| Region | (Z)-5-decenyl acetate | | (Z)-7-dodecenyl acetate | | (Z)-9-tetradecenyl acetate | |
|---|---|---|---|---|---|---|
| Sweden | 4 | 66 | 52 | 33 | 44 | 1 |
| Britain | 9 | 40 | 45 | 56 | 46 | 4 |
| France | 47 | 87 | 40 | 12 | 13 | 1 |
| Armenia | 1 | 9 | 52 | 90 | 47 | 1 |
| Bulgaria | 1 | 6 | 42 | 92 | 57 | 2 |

*Note that figures are means and that the variation within each population is not given but may be high.

pene isomers. Bugs from Brazil lack a sesquiterpene component that is present in populations from the southern United States and southern France, and sesquiterpene isomer ratios differ between French and North American bugs (Aldrich, 1988). Hawaiian *N. viridula* were not parasitized by the parasitoid *Trichopoda pennipes*, imported from Florida, suggesting that their sex pheromone has yet another constitution.

One of the most detailed studies of pheromone variation in Lepidoptera is that of Hansson *et al.* (1990) on the turnip moth, *Agrotis segetum*. The three main components of the female sexual pheromone are (Z)-5-decenyl acetate, (Z)-9-dodecenyl acetate and (Z)-9-tetradecenyl acetate (Fig. 3.7a). The ratios of these compounds in populations from Sweden, Britain, France, Armenia and Bulgaria show marked differences (Table 3.2). By single cell recording it was possible to determine the numbers of specialized receptors for each compound, which show broadly similar but by no means well correlated differences.

Trapping experiments resulted in the Swedish type being attracted mainly to the three-component blend. The Armenian and Bulgarian types were attracted in greater numbers by mixtures of Z-7TDA, Z-9TDA and a third component, (Z)-7-dodecenol. What is perhaps most striking in this work is the weakness of the correlation between numbers of receptor types and pheromone component ratios. However, the results do suggest that the presence of only about 1% of one component in the female or, on the other hand, 1% of competent sensilla in the male can make a crucial difference to the communication channel and the numbers of males attracted.

Toth *et al.*, (1993) went on to assess the effects of five pheromone blends on turnip moth over two continents (Fig. 3.7) The arresting data are suggestive of a N-S cline in responsiveness to component ratios.

The blend ratio of the pink bollworm, *Pectinophora gossipiella*, shows little variation among insects from California, Argentina, Brazil, China, Egypt, Mexico and Pakistan (Haynes and Baker, 1988). Ratios of (Z,Z)-7,11-16 acetate to its (*Z,E*) isomer varied from a maximum of 63.1% (+2.1%) in Pakistan to a minimum of 57.5% (+4%) in China. The males remained receptive (though not equally so) to blends over a range of 40–70% of the (Z,Z) : (*Z,E*) compounds, and so are not tuned to a precise blend of female pheromone. Most variation was found in the emission rates, which varied from 0.057 ng $min^{-1}$ in a Chinese population to 0.119 nb $min^{-1}$ute in Californian moths. Furthermore, the emission rate of Californian moths increased by about 20% between 1982 and 1985. Haynes and Baker suggest that this may be incipient resistance evolving under the selection pressure of widespread attempts to control pink bollworm in California by mating disruption.

Dramatic changes in pheromone blend are produced by mutation of a single autosomal gene in the cabbage looper moth, *Trichoplusia ni* (Haynes and Hunt, 1990). Here, emission of the major component is reduced in quantity, a minor component, (Z)-5-dodecenyl acetate, is almost eliminated and the emission rate of a trace component, (Z)-9-tetradecenyl acetate, increases by about 20 times. The mixture released by such mutants does not attract others of the same species.

The North American population of the European corn borer, *Ostrina nubilalis*, has proved to be an ideal insect for the study of the genetical control of chemical communication (Roelofs *et al.*, 1987). This insect has two forms, one of which produces 97–99% (Z)-11-tetradecenyl acetate (Klun and Maini, 1979) and the other 97–99% of the (*E*) isomer. It is likely that the (*E*) strain was introduced originally into the United States from Italy and the (Z) isomer from western Europe. Hybrid females emit a 65 : 35% mixture of the two isomers. Contrary to what one might expect, it has been found that the genes controlling production and reception are largely independent (Baker, 1989). Furthermore, receptor types are controlled by autosomal genes, and the response specificity (which is mediated by the central nervous system) by a gene on the sex chromosomes.

## 3.8 ROLE OF PHEROMONE BLENDS IN COMMUNICATION

There are three main factors that influence the number of receptor types in insect sexual communication systems. These are: the requirement for high sensitivity; the need to detect components of the pheromone of other closely related species (which may then inhibit response to those species); and the requirement for more recognition cues in the pheromone blend of the species. These factors in turn will favour the production of more complex blends. The option of emitting greater quantities of pheromone to improve signal transmission does not seem to have been adopted, perhaps because it requires more metabolic energy or because it would aid host location by predators and parasites.

Reference to *Agrotis segetum* (Table 3.2) and experience with other species shows that high sensitivity to the precise female blend is not necessarily achieved by matching the relative numbers of receptors to the concentration of components within the blend; rather it would seem that the central nervous system plays an important role in analysis of afferent messages. To understand this role better, we need to examine the effects of incomplete blends and single compounds on behaviour.

In insects such as ants, components of a multicomponent alarm pheromone separate out to an extent by diffusion (Chapter 2) and particular roles can be assigned to individual components, with relatively little overlap between their different functions. Where pheromones are transmitted by air currents, such segregation by differing diffusion rates is negligible compared with that due to dispersion of the molecules by turbulent air movement (see below). In such cases, it is generally not appropriate to compare the active spaces of different compounds. Nevertheless, it is still possible to ascribe the control of different elements of behaviour sequence to different components (Howse *et al.*, 1986).

Bradshaw *et al.* (1983) studied the behaviour of a noctuid, the pine beauty moth (*Panolis flammea*). This insect has three main sex pheromone components: (Z)-9-tetradecenyl acetate (Z-9-TDA), (Z)-11-tetradecenyl acetate (Z-11-TDA) and (Z)-11-hexadecenyl acetate (Z-11-HDA), which are present in a ratio of 100:1:5 (Baker *et al.*, 1982). In the field, the Z-9-TDA is essential for trapping males, but when presented to male moths in a wind tunnel it was ineffective without the addition of Z-11-TDA or Z-11-HDA. The latter compound also increased the number of landings near the source, but the presence of Z-11-HDA was required for copulation attempts to take place after landing. Upwind zigzag flight occurred when Z-9-TDA was present as a plume in the presence of Z-11-TDA. On the basis of these results it is possible to define releasers for different parts of the behavioural sequence as follows:

| | | |
|---|---|---|
| Z-9-TDA | ⟶ | Orientation cue |
| Z-11-TDA | ⟶ | Activation, recruitment to plume of Z-11-TDA. Releaser for landing |
| Z-11-HDA + Z-9-TDA | ⟶ | Copulation attempts |

Early work on the oriental fruit moth, *Grapholitha molesta*, by Baker and Carde (1979) led to a similar view of the control of behaviour

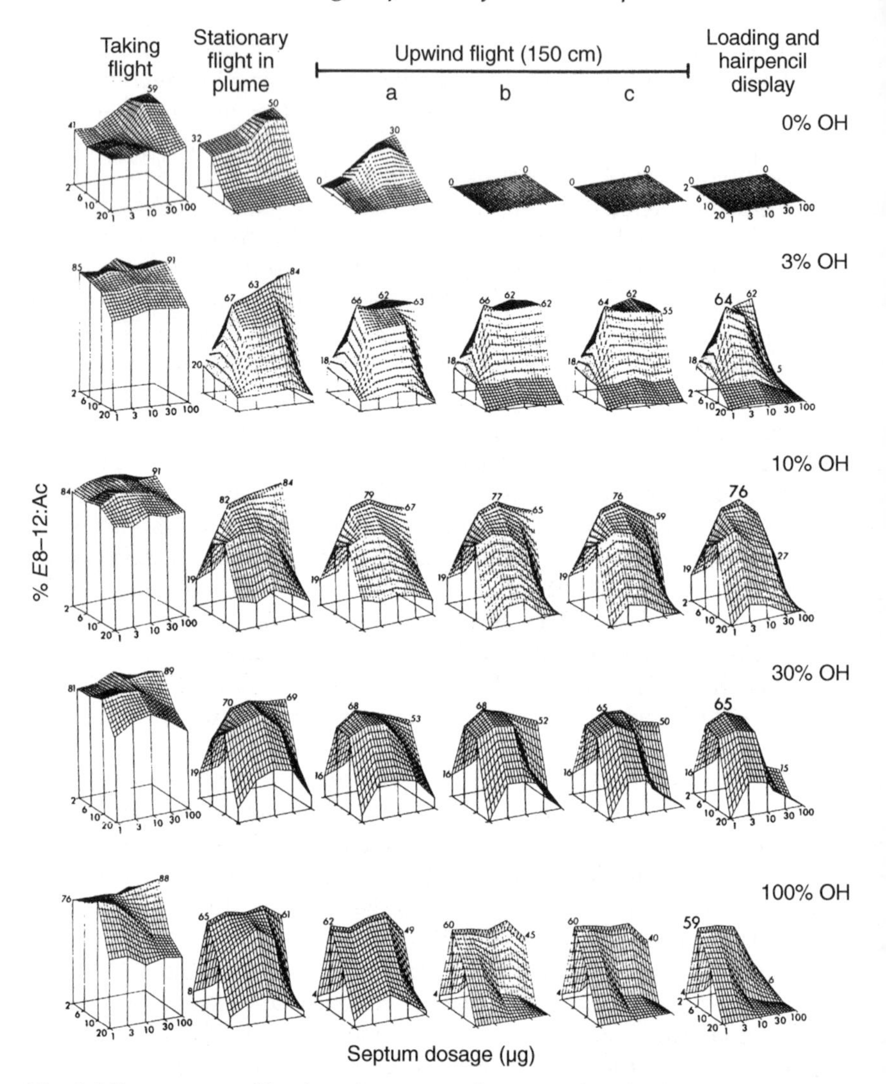


**Fig. 3.8** Response profiles (number responding out of 100) of male oriental fruit moths, *Grapholitha molesta*, to graded blends and doses of female pheromone components (*E*)- and (Z)-8-dodecenyl acetate, along with various percentages of (Z)-8-dodecenyl-1-ol* (rows). (From Linn and Roelofs, 1983.)

sequences of males. This insect has four main sex pheromone components: (Z)-8-dodecenyl acetate (Z-8-DDA), its (*E*) isomer (*E*-8-DDA), (Z)-8-dodecen-1-ol and dodecanol. They found that a sequence of behaviour from activation to landing near the female and wing-fanning could be elicited by the complete blend, but if elements were missing the sequence was less likely to go to completion. There was some evidence that the two acetates were long-range attractants, and that the dodecanol elicited a greater likelihood of landings and wing-fanning. Subsequently, Linn and Roelofs (1983) undertook the painstaking exercise of testing the flight responses of male *Grapholitha* to 100 different blend ratios of the two acetates and the (Z)-8-dodecen-1-ol. The results (Fig. 3.8) show that optimal responsiveness of male moths is achieved with particular ratios of the three compounds. Later, Linn *et al.* (1987) showed that 10 μl of the three-component blend elicited walking and wing-fanning at about 20 m from the source, while the same amounts of Z-8-DDA or of a bicomponent mixture were effective only at less than half that range. The conclusion – that sequential behaviour of male moths is controlled by exposure to a fixed blend of compounds, rather than consisting of a chain of successive responses to selected components or component mixtures – is now generally accepted. However, we must bear in mind that insect behaviour abounds with examples of behavioural sequences which continue once released, the orientation of which is guided by cues from the changing environment (e.g. Fig. 4.1). Thus degrading the available cues does not necessarily block the chain of actions, but just makes it less likely that it will continue to its natural conclusion. The finding that progressive degradation of the natural pheromone blend leads to a corresponding lessening of effectiveness does not conflict with the view that pheromones may have built-in redundancy in the form of overlap of function of the various components. Thus it is wise not to generalize: each case should be investigated using the most heuristic hypothesis.

## 3.9 THE NATURE OF THE ODOUR PLUME

The males of most species that have been studied fly continuously upwind until they are close to the female, when their behaviour changes to landing and walking. the male potato tuberworm moth *Phthorimaea operculella,* is an exception. This insect approaches from a distance in a chain of short flights, most of which are less than 1 m in length (Ono and Ito, 1989), and it may be reorienting itself into the odour plume each time that it lands. The majority of species approach a source of windborne odour in constant flight using complex zigzag flight patterns, which suggests that they are continuously assessing their trajectory. Such orientation is upwind (positive anemotaxis), chemically induced and visually guided. The mechanisms of orientation have been reviewed by Baker (1985, 1989), Murlis *et al.* (1992) and Kennedy (1983).

A simple concept for understanding such behaviour in night-flying moths was developed by Bossert and Wilson (1963), using mathematical models originally developed after World War I for studying the movements of clouds of gases in the atmosphere. This provides an analogy which, while it is not very exact, helps us to understand the basic concept of the orientation method.

Odour released into the wind from a point source is carried downwind in a corridor, producing an active space in which the concentration is above threshold for the release of upwind flight, and the boundaries of which can be defined approximately (Fig. 3.9a). The length of this corridor depends on the turbulence of the airflow: the odour molecules are dispersed by turbulence in increasing amounts as they are carried downwind. Turbulence is greater in higher wind velocities: the margins of the corridor are eroded away to a greater extent. The overall effect is that the active space of the odour is smaller than in low wind velocities.

This model provides an explanation of the way in which a male moth

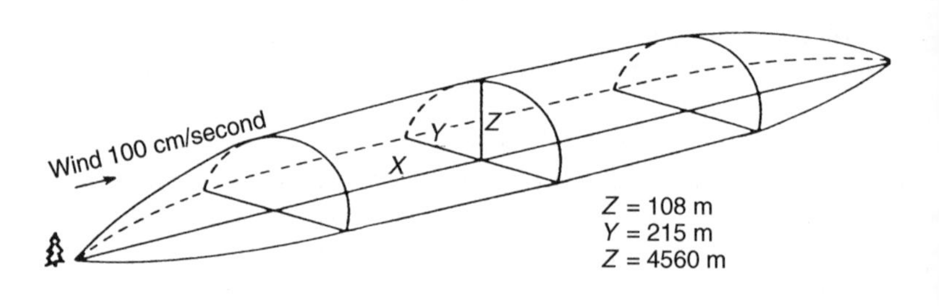


(a)

(b)

**Fig. 3.9** Two models of pheromone plume structure: (a) time-averaged model of Bossert and Wilson, 1963; (b) meandering filamentous structure of real plume. (From Murlis *et al.*, 1992.)

locates a calling female. A male that is downwind within the active space of pheromone released by a calling female will fly against the wind. If it flies out of the active space it begins turning movements across the wind which will enable it to relocate the pheromone plume. Thus by modulating its upwind flight as a result of experiencing variations in odour concentration it will arrive at the calling female.

The Bossert and Wilson model has been found to be inadequate, mainly because it describes the concentration of odour molecules averaged over a time period of three minutes. Insects do not average their sensory input over such a long time period. They react almost instantaneously: a male *Grapholitha molesta* that loses contact with the pheromone plume in a wind tunnel responds with a change in course within 0.15 seconds (Baker and Haynes, 1987). In fact, an odour plume is similar to the smoke plume from a factory chimney (Murlis *et al.*, 1992), appearing from a distance as a discrete undulating cloud. At close range a fine structure is evident, to which Wright (1958) first drew attention, in which there are no sharp boundaries to the active space and in which the odour is not in the form of a homogeneous cloud but is condensed into interlacing filamentous microstreams of air that twist together as they are carried downwind. It is believed that a moth experiences this fine structure when it is flying into the wind. Furthermore, comparison of the intensity of stimulation of the two antennae is not essential. Male *Heliothis virescens* will fly upwind along the axis of a plume even if one antenna has been ablated (Vickers and Baker, 1991). Sensory input must therefore converge in the central nervous system before it triggers turning mechanisms.

The physical characteristics of odour plumes have been described in detail by Murlis *et al.* (1992). As a cloud of molecules is carried away from the source, it expands, mainly as a result of turbulent diffusion. Molecular diffusion is insignificant in the time scales involved. However, the mechanics of turbulence dictate that the odour plume is initially affected only by molecular diffusion, and so the plume from a small source changes very little over an initial distance that is unlikely to exceed a few centimetres in the case of a calling insect. Beyond this distance (known as the Kolmogoroff length), eddies begin to break up the plume, smaller eddies stirring the filaments and larger ones of several metres in diameter producing undulations. In stable atmos-pheric conditions the plume meanders horizontally, but in unstable conditions (at night, for example) the plume will also undulate vertically.

The difference between a plume defined by a mathematical model of the type used by Bossert and Wilson (in which the concentration of molecules is averaged over a three-minute period) and the structure of a real plume is shown in Fig. 3.9b. The real plume meanders in a roughly sinusoidal manner. While mathematical models have been developed to describe the meandering structure, there is no method yet available of describing the fine structure. As soon as it was understood that odour plumes are not homogeneous and rectilinear, as the Bossert and Wilson model supposes, but have a filamentous nature, and also that they meander and are affected by large-scale eddies in many environments, attention focused on the method that a male moth uses in orientating towards the source of such a plume.

Most of the evidence suggests that insects use a mechanism known as upwind optomotor guided anemotaxis. That is to say, they steer into the wind, gauging their progress by comparing their heading with the apparent movement of the ground pattern beneath them. The optomotor response needs some explanation. Many animals adjust their position relative to their surroundings by purely visual mechanisms. When placed in an arena with a striped pattern on the wall they will turn as the arena rotates, in the direction of rotation. In a normal environment this optomotor response would enable the insect to regain its original position after having been blown out of position or (in the case of an aquatic insect) if it had drifted downstream.

Now, imagine you are an insect flying across this page from the bottom into a scented air stream from the top of the page. You cannot sense the direction from which the wind is coming because you are suspended in it, but you can monitor your progress to the top by the slippage of lines beneath you, and if you drift to one side you will detect that too, and alter your heading. You can also assess your rate of progress upwind from the slippage, but it follows, of course, that an insect cannot orientate in this way unless there is sufficient light to see the surroundings.

We have seen that the plume does not have clearly definable boundaries, so the simple explanation that a moth maintains its position in the plume by turning back as it traverses the boundary into clean air is inadequate. Experiments in wind tunnels have shown that under certain conditions moths are able to continue on an upwind trajectory after the wind has been stopped (Farkas and Shorey, 1972). Loss of contact with the pheromone, however, normally results in wide cross-wind zigzags known as 'casting movements' (Fig. 3.10). These manoeuvres increase the probability that the insect will encounter the plume again. Kennedy (1983) has argued that these excursions are part of a 'self-steered' or pre-programmed behaviour pattern of **counter-turning**, which produces a narrow zigzag flight pattern at one end of a continuum and casting flight manoeuvres at the other extreme. In zigzag flight, the turns are at around 90° to the wind, while in casting flight they are at around 180° to the wind (Fig. 3.10) so that the insect remains stationary in the wind and moves neither upwind nor

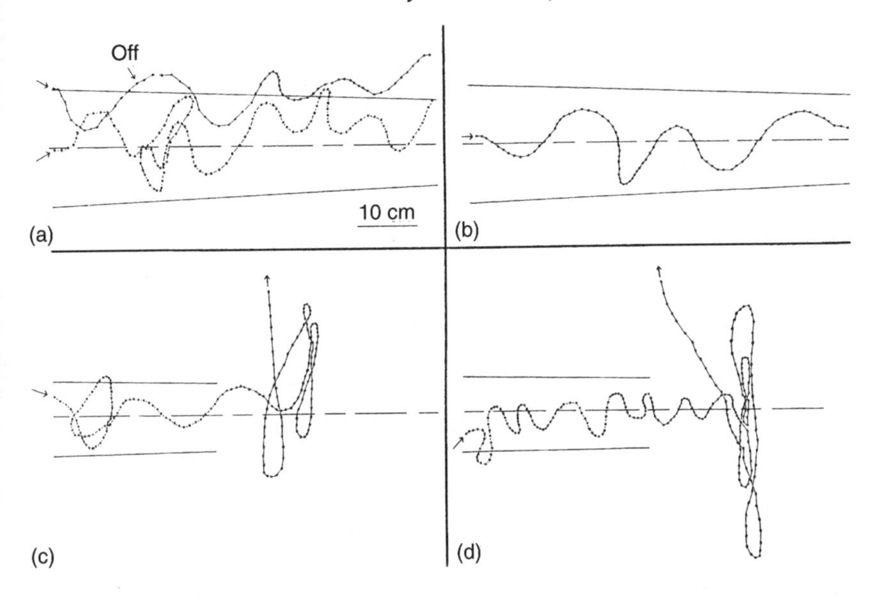


**Fig. 3.10** Tracks of male oriental fruit moth, *Grapholitha molesta* flying upwind in a wind tunnel towards a source of female pheromone. Outer lines are approximate boundaries of plume; dashed line is axis. Tracks were filmed from above; dots give moths' position at intervals of 1/60th of a second. (a) Lower track is of a moth flying upwind. Upper track shows continuation of upwind flight after the wind has been stopped at 'off'. (b) Track of moth continuing upwind flight towards the source after the wind had been switched off just before it came into view. (c), (d) Tracks of moths before and after pheromone source was removed, showing counter-turning without upwind progression. (From Kuenan and Baker, 1983, reproduced with permission from *Physiological Entomology*.)

downwind. Once released by a change in chemical stimulation (not necessarily loss of contact with the pheromone plume, but also increase in odour concentration) the programme will continue without further chemical stimulation.

The evidence for this interpretation comes from wind tunnel experiments with a variety of moths. Pink bollworm (*Pectinophora gossipiella*) males continued flying along the same trajectory for about a metre in a wind tunnel after the wind had been stopped (Farkas and Shorey, 1972). Studying *Grapholitha molesta,* Baker and Kuenan (1982) showed that removal of the pheromone and stoppage of the wind immediately resulted in casting movements without upwind progression (Fig. 3.10). Male apple moths (*Adoxophyes orana*) made counterturning sequences when inside a homogeneous cloud of odour, so contact with the edges of the plume is not essential to initiate turning (Kennedy *et al.*, 1981). It was also shown that male *Adoxophyes* would fly upwind in a cloud

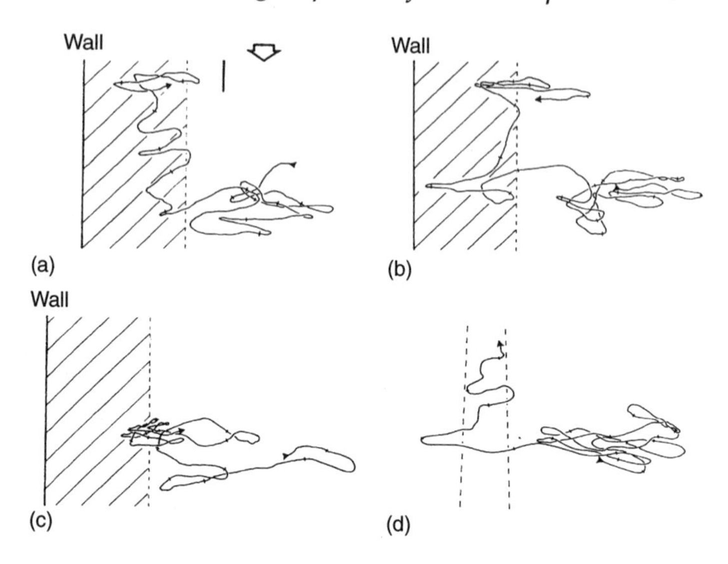


**Fig. 3.11** Tracks of the apple moth, *Adoxophyes orana*, in the wind tunnel. Diffuse cloud of pheromone was maintained on one side (hatched area) in (a), (b) and (c). When a pheromone plume was removed in (a), (b) and (c), counter-turning resulted which led to some upwind flight in the pheromone cloud. In (d), a narrow plume was superimposed on a cloud of pheromone filling the whole tunnel; the moth locked on to this and followed the plume as if it was in clean air. (Kennedy and Marsh, 1974; reproduced with permission from *Science*.)

of pheromone if a narrower plume was introduced into the cloud (Fig. 3.11).

Kennedy (1983) argues that there are two components to the chemical signals that mediate counterturning: odour concentration within the plume and odour 'pulse' frequency. Odour concentration is experienced by the moth as peaks and troughs, with the following effects:

- The higher the peaks and the lower the troughs, the narrower is the counterturning. The insect thus 'locks on' to filamentous plumes which have high concentrations.
- With low peaks, the counterturns become wider. This will occur downwind from the source, where concentration is low.

As the frequency of pulses or peaks decreases (i.e. the time between successive pulses increases), the counterturning will widen. With a high frequency of pulses (reaching a maximum in the case of a homogeneous cloud), the counterturning will narrow. While decrease in odour concentration results in casting movements and reduction results in displace-

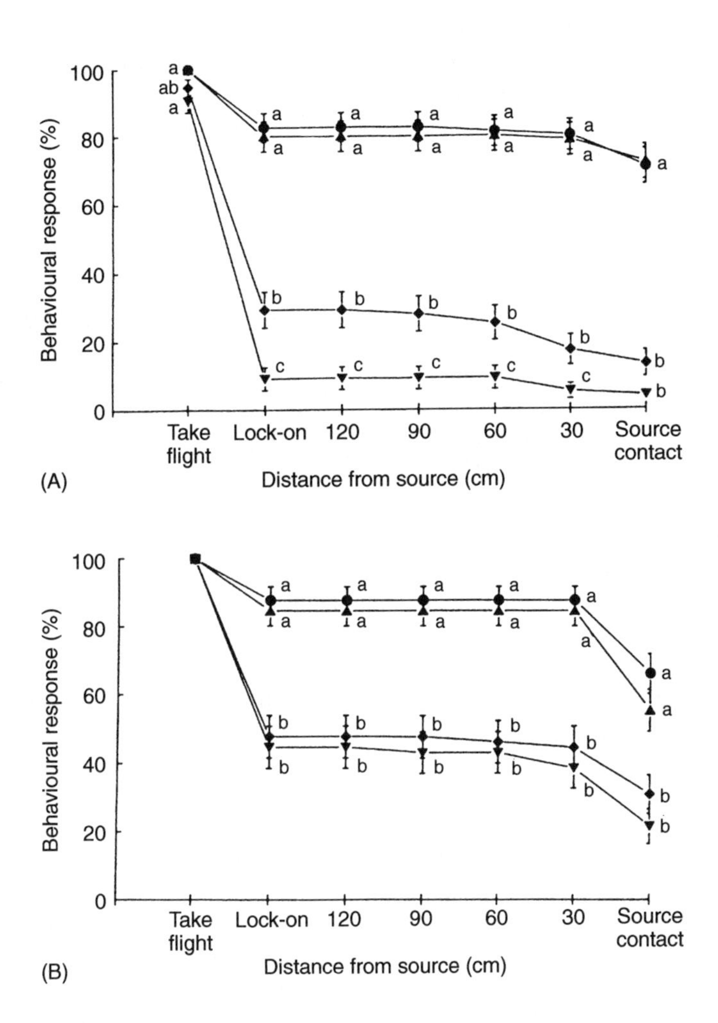


**Fig. 3.12** Behavioural responses of male cabbage looper moths, *Trichoplusia ni*, to components of female pheromone and inhibitor, (Z)-7-dodecanol, released from the same and different sources in wind tunnel. (A) Two sources of inhibitor were placed 5 cm either side of sources of (i) complete 6-component female pheromone blend (●), (ii) complete blend + inhibitor (▼), (iii) complete blend + inhibitor from separate abutting sources (◆), and (iv) complete blend + inhibitor on separate sources 5 cm apart (▲). (B) A source of inhibitor was placed 10 cm upwind of pheromone source; key as for (A) except (iv), in which the source (complete blend + inhibitor) was also placed 10 cm upwind. Values with same letter are not significantly different. (From Liu and Haynes, 1992, reproduced with permission from *J. Chem. Ecol.*)

ment upwind, increase in concentration (as an insect gets closer to the source, for example) has different effects. In response to increased concentration of pheromone in a wind tunnel, *Grapholitha molesta* increases the frequency of zigzag turns and decreases its flight (air)speed (Baker and Haynes, 1987).

This explanation takes into account the meandering filamentous nature of the odour plume and the behaviour often observed in the field. Zigzagging narrows as an insect approaches the pheromone source, and traps producing narrow plumes are generally more effective than those that produce diffuse ones (Lewis and Macaulay, 1976). It presupposes that the reaction times of moths are very short, and this has indeed been found to be so: loss of pheromone triggers a change from upwind to casting flight in around 0.5 seconds in *Antheraea polyphemus* (Baker and Vogt, 1988), corresponding with the ability of antennal receptors to resolve 20 ms pulses at 2–5 per second (Rumbo and Kaissling, 1989). *Grapholitha molesta* has the even faster behavioural reaction time of 0.15 seconds (Baker and Haynes, 1987).

It thus appears that male moths can detect a single filament of odour into which they fly. Further convincing evidence for this comes from wind tunnel experiments on the cabbage looper moth, *Trichoplusia ni*. Liu and Haynes (1992) tested the response of males to the full pheromone blend of the species on a rubber septum. This releases a sequence of events, including taking flight, upwind flight and contact with the source (Fig. 3.12). It is known that (Z)-7-dodecenyl acetate (Z 7-DDA) inhibits attraction of males and so should do this when mixed with the natural pheromone blend. In fact, strong inhibition occurs only if the two are released from the same septum. Inhibition occurs when the two are released separately from abutting septa, but is significantly less strong, and if the two septa are placed 5 cm apart each side of the pheromone blend septum, there is no significant inhibition. Furthermore, if the Z 7-DDA is released 10 cm upwind so that the pheromone plume with the full blend is enclosed within the inhibitor plume, there is no significant inhibition. These are very important results because they show, firstly, that it is highly likely that moths can detect individual pheromone plumes and lock on to them; secondly, that the inhibitor must be detected simultaneously from the same source with the attractant to be effective; and thirdly, that moths (by some unknown mechanism) are able to detect filaments of pure inhibitor. This has the effect of narrowing the communication channel so that the insect is less likely to respond to pheromone plumes of related species calling at the same time. The results may also explain the ability of moths in general to follow a narrow plume within a broader one of the same composition if, for example, the filaments of the narrower plume have a higher concentration of odour molecules.

## REFERENCES

Aldrich, J.R. (1988) Chemical ecology of the Heteroptera. *A. Rev. Ent.*, **33**, 211–238.

Aluja, M. (1993) Unusual calling behavior of *Anastrepha robusta* flies (Diptera: Tephritidae) in nature. Florida Entomologist **76**, 391–395.

Aylor, D.L. (1976) Estimating peak concentrations of pheromones in the forest. In *Perspectives in Forest Entomology* (eds J.F. Anderson and M.K. Kaya), Academic Press, New York, pp. 177–188.

Baker, T.C. (1983) Variations in male oriental fruit moth courtship patterns due to male competition. *Experientia*, **39**, 112–114.

Baker, T.C. (1985) Chemical control of behaviour, In *Comparative Insect Physiology, Biochemistry and Pharmacology*, Vol. 9 (eds G.A. Kerkut and L. Gilbert), Pergamon Press, Oxford, pp. 621–672.

Baker, T.C. (1989) Sex pheromone communication in the Lepidoptera: new research progress. *Experientia*, **45**, 248–262.

Baker, T.C. and Carde, R.T. (1978) Disruption of gypsy moth male sex pheromone behaviour by high frequency sound. *Environ. Ent.*, **7** 45–52.

Baker, T.C. and Carde, R.R.T. (1979) Analysis of pheromone-mediated behaviours in male *Grapholita molesta*, the oriental fruit moth (Lepidoptera: Tortricidae). *Environ. Ent.*, **8**, 956–968.

Baker, T.C. and Haynes, K.F. (1987) Manoeuvres used by flying male oriental fruit moths to relocate a sex-pheromone plume in an experimentally shifted wind-field. *Physiol. Ent.*, **12**, 263–279.

Baker, T.C. and Kuenan, L.P.S. (1982) Pheromone source location by flying moths: a supplementary non-anemotactic mechanism. *Science*, **216**, 424–427.

Baker, T.C. and Vogt, R.G. (1988) Measured behavioural latency in response to sex-pheromone loss in the large silk moth *Antheraea polyphemus. J. Exp. Biol.*, **137**, 29–38.

Baker, T.C., Carde, R.T. and Miller, J.R. (1980) Oriental fruit moth pheromone component emission rates measured after collection by glass-surface adsorbtion. *J. Chem. Ecol.*, **6**, 749–758.

Baker, R., Bradshaw, J.W.S. and Speed, W. (1982) Methoxymercuration–demercuration and mass spectrometry in the identification of the sex pheromones of *Panolis flammea*, the pine beauty moth. *Experientia*, **38**,233–234.

Bartell, R. J. and Lawrence, J.A (1977) Reduction in responsiveness of male apple moths, *Epiphyas postvittana*, to sex pheromone following pulsed pheromonal exposure. *Physiol. Ent.* **2**, 1–6.

Birch, M.C., Poppy, G.M. and Baker, T.C. (1990) Scents and eversible scent structures of male moths. *A. Rev. Ent*, **35**, 25–58.

Bjostad, L.B., Gaston, L.K. and Shorey, H.H. (1980) Temporal pattern of sex pheromone release by female *Trichoplusia ni. J. Insect Physiol.*, **26**, 493–498.

Bossert, W.H. and Wilson, E.O. (1963) The analysis of olfactory communication among animals. *J. Theor. Biol.*, **5**, 443–469.

Bradshaw, J.W.S., Baker, R., and Lisk, J.C. (1983) Separate orientation and releaser components in a sex pheromone. *Nature*, **304**, 265–267.

Brady, J. (1972) The visual responsiveness of tsetse fly *Glossina morsitans* West. (Glossinidae) to moving objects: the effects of hunger, sex, host odour and stimulus characteristics. *Bull. Ent. Res.*, **62**, 257–279.

Cameron, E.A. (1983) Apparent long-term bodily contamination by disparlure, the gypsy moth (*Lymantria dispar*) attractant. *J. Chem. Ecol.*, **9**, 33–38.

Carde, R.T. and Baker, T.C. (1984) Sexual communication with pheromones. In

*Chemical Ecology of Insects* (eds W.J. Bell and R.T. Carde), Chapman & Hall, London and New York, pp. 355–386.

Carde, R.T and Taschenberg, E.F. (1984) A reinvestigation of the role of (E)-2-hexenal in female calling behaviour of the polyphemus moth (*Antheraea poyphemus*). *J. Insect Physiol.*, **30**, 109–112.

Carlisle, D.B., Ellis, P.E. and Betts, E. (1965) The influence of aromatic shrubs on sexual maturation in the desert locust *Schistocerca gregaria. J. Insect Physiol.*, **11**, 1541–1558.

Castrovillo, P.J. and Carde, R.T. (1979) Environmental regulation of female calling and male pheromone response periodicities in the codling moth (*Laspeyresia pomonella*). *J. Insect Physiol.*, **25**, 659–667.

Conner, W.E. (1987) Ultrasound: its role in the courtship of the archiid moth *Cycnia tenera. Experientia*, **43**, 1029–1031.

Conner, W.E. and Best, B.A. (1988) Biomechanics of the release of sex pheromone in moths: effects of body posture and local airflow. *Physiol. Ent.*, **13**, 15–28.

Conner, W.E., Eisner, T., Vander Meer, R.K. *et al.* (1980) Sex attractant of an arctiid moth (*Utetheisa ornatrix*): a pulsed chemical signal. *Behav. Ecol. Sociobiol.*, **9**, 227–235.

Craig, G.B (1967) Mosquitoes: female monogamy induced by male accessory gland substance. *Science*, **156**, 1499–1501.

Davis, E.E. (1984) Peripheral chemoreceptors and regulation of insect behaviour. In *Mechanisms in Insect Olfaction* (eds T.L. Payne, M.C. Birch and C.E.J. Kennedy), Oxford University Press, Oxford, pp. 243–251.

Dawson, G.W., Griffiths, D.C., Pickett, J.A. *et al.* (1984) Natural inhibition of the aphid alarm pheromone. *Ent. Exp. Appl.*, **36**, 197–199.

Den Otter, C.J., Tchicaya, T. and Schulte, A.M. (1991) Effects of age, sex and hunger on the antennal olfactory sensitivity of tsetse flies. *Physiol. Ent.*, **16**, 173–182.

Dickens, J.C. (1989) Green leaf volatiles enhance aggregation pheromone of boll weevil, *Anthonomus grandis. Entomol. Exp. Appl.*, **52**, 191–203.

Dickens, J.C., Billings, R.F. and Payne, T.L. (1992) Green leaf volatiles interrupt aggregation pheromone response in bark beetles infesting southern pines. *Experientia*, **48**, 523–524.

Dusenbury, D.B. (1989) Calculated effect of pulsed pheromone release on range of attraction. *J. Chem. Ecol.*, **97**, 971–977.

Farkas, S.R. and Shorey, H.H. (1972) Chemical trail-following by flying insects: a mechanism for orientation to a distant odor source. *Science*, **178**, 67–68.

Fitzpatrick, S.M., McNeil, J.N. and Dumont, S. (1988) Does male pheromone effectively inhibit competition among courting true armyworm males (Lepidoptera: Noctuidae)? *Anim. Behav.*, **36**, 1831–1835.

Gadenne, C., Renou, M. and Streng, L. (1993) Hormonal control of responsiveness in the male black cutworm *Agrotis ipsilon. Experientia*, **49**, 721–724.

Gilbert, L.E. (1976) Postmating female odor in *Heliconius* butterflies: a male-contributed anti-aphrodisiac. *Science*, **193**, 419–420.

Greenfield, M.D. and Karandinos, M.G. (1979) Resource partitioning of the sex communication channel in clearwing moths (Lepidoptera:Sesiidae) of Wisconsin. *Ecol. Monogr.*, **49**, 403–426.

Haddow, A.J., Casley, J.P., O'Sullivan, P.M. *et al* (1968) Entomological studies from a high tower in Zika forest, Uganda. Part II. The biting activity of mosquitoes above the canopy in the hour after sunset. *Trans. R. Ent. Soc. Lond.*, **120**, 219–236.

Hansson, B.S., Toth, M., Lofstedt, C. *et al.* (1990) Pheromone variation among eastern European and a western Asian population of the turnip moth, *Agrotis segetum*. *J. Chem. Ecol.*, **16**, 1611–1622.

Happ, G.M. (1969) Multiple sex pheromones of the mealworm beetle, *Tenebrio molitor*, L. *Nature, London*, **222**, 180–181.

Haynes, K.F. and Baker, T.C. (1988) Potential for evolution of resistance to pheromones: worldwide and local variation in the chemical communication system of pink bollworm moth, *Pectinophora gossipiella*. *J. Chem. Ecol.*, **14**, 1547–1560.

Haynes, K.F. and Hunt, R.E. (1990) Interpopulational variations in emitted pheromone blend of the cabbage looper moth, *Trichoplusia ni*. *J. Chem. Ecol.*, **16**, 509–519.

Heath, R.R., Landolt, P.J., Leppla, N.C. and Dieben, B.D. (1988) Identification of a male-produced pheromone of *Anticarsia gemmatalis* (Hubner) (Lepidoptera: Noctuidae) attractive to conspecific males. *J. Chem. Ecol.*, **14**, 1121–1130.

Hendrikse, A. and Vos-Bunnemeyer, E. (1987) Role of host-plant stimuli in sexual behaviour of small ermine moths (*Yponomeuta*). *Ecol. Ent.*, **12**, 363–371.

Hendry, L. B., Wichmann, J.K., Hindenlang, D.M. *et al.* (1975) Evidence for origin of insect sex pheromones: presence in food plants. *Science*, **188**, 59–62.

Hirai, K., Shorey, H.H. and Gaston, L.K. (1978) Competition among courting moths: male-to-male inhibitory pheromone. *Science*, **202**, 644–645.

Hiss, W.B. and Fuchs, M.S. (1972) The effect of matrone on oviposition in the mosquito *Aedes aegypti*. *J. Insect Physiol.*, **18**, 2217–2227.

Holldobler, B., Stanton, R.C. and Markl, H. (1978) Recruitment and food-retrieving behaviour in *Novomessor* (Formicidae, Hymenoptera). *Behav. Ecol. Sociobiol.*, **4**, 163–181.

Howse, P.E., Lisk, J.C. and Bradshaw, J.W.S. (1986) The role of pheromones in the control of behavioural sequences in insects. In *Mechanisms in Insect Olfaction* (eds T.L. Payne, M.C. Birch and C.E. Kennedy), Clarendon Press, Oxford, pp. 157–162.

Jones, M.D.R. and Gubbins, S.J. (1977) Modification of circadian flight activity in the mosquito *Anopheles gambiae* after insemination. *Nature, London*, **268**, 731–732.

Kennedy, J.S. (1983) Zigzagging and casting as a programmed response to wind-borne odour. A review. *Physiol. Ent.* **8**, 109–120.

Kennedy, J.S. and Marsh, D. (1974) Pheromone-regulated anemotaxis in flying moths. *Science, NY*, **184**, 999–1001.

Kennedy, J.S., Ludlow, A.R. and Sanders, C.J. (1981) Guidance of flying male moths by wind-borne sex pheromone. *Physiol. Ent.*, **6**, 395–412.

Krasnoff, S.B. and Roelofs, W.L. (1989) Quantitative and qualitative effects of larval diet on the male scent secretions of *Estigmene acrea, Phragmatobia fuliginosa,* and *Pyrrharctia isabella* (Lepidoptera: Arctiidae). *J. Chem. Ecol.*, **15**, 1077–1093.

Klun, J.A. and Maini, S. (1979) Genetic basis of an insect communication system: the European corn borer. *Environ. Ent.*, **8**, 423–426.

Kuenan, L.P.S. and Baker, T.C. (1983) A non-anemotactic mechanism used in pheromone source location by flying moths. *Physiol. Ent.*, **7**, 193–202.

Lacey, M.J. and Saunders, C.J. (1992) Chemical composition of sex pheromone of oriental fruit moth and rates of release by individual female moths. *J. Chem. Ecol.*, **18**, 1421–1435.

Landolt, P.J. and Hendrichs, J. (1983) Reproductive behavior of the papaya fruit fly, *Toxotrypana curvicaudata* Gerstaeker (Diptera: Tephritidae). *Ann. Ent. Soc. America*, **76**, 413–417.

Landolt, P.J., Reed, H.C. and Heath, R.R. (1992) Attraction of female papaya fruit fly (Diptera: Tephritidae) to male pheromone and host fruit. *Environ. Ent.*, **21**, 1154–1159.

Lewis, T. and Macaulay, E.D.M (1976) Design and evaluation of sex attractant traps for pea moth *Cydia nigricana* (Steph.) and the effect of plume shape on catches. *Ecol. Ent.* **1**, 175–187.

Linn, C.E. Jr and Roelofs, W.L. (1983) The effect of varying proportions of the alcohol component on sex pheromone blend discrimination in male oriental fruit moths. *Physiol. Ent.*, **8**, 291–308.

Linn, C.E. Jr, Campbell, M.G. and Roelofs, W.L. (1987) Pheromone components and active spaces: what do moths smell and where do they smell it? *Science*, **237**, 650–652.

Liu, Y.-B. and Haynes, K.F. (1992) Filamentous nature of pheromone plumes protects integrity of signal from background noise in cabbage looper moth, *Trichoplusia ni*. *J. Chem. Ecol.*, **18**, 299–306.

Löfstedt, C. (1993) Moth pheromone genetics and evolution. *Phil. Trans. Roy. Soc. Lond.* B, **340**, 167–177.

Löfstedt, C., Hansson, B.B., Dijkerman, H.J. and Herrabout, W.M. (1990) Behavioural and electrophysiological activity of unsaturated analogues of the pheromone tetradecenyl acetate in the small ermine moth *Yponomeuta rorellus*. *Physiol. Ent.*, **15**, 47–54.

Markl, H. (1965) Stridulation in leaf-cutting ants. *Science N.Y.*, **149**, 1392–1393.

Masson, C. and Arnold, G. (1984) Ontogeny, maturation and plasticity of the olfactory system of the worker bee. *J. Insect Physiol.*, **30**, 7–14.

McNeil, J.N. and Delisle, J. (1989) Host plant pollen influences calling behavior and ovarian development of the sunflower moth, *Homoeosoma electellum*. *Oecologia*, **880**, 201–205.

Miller, J.R. and Roelofs, W.L. (1977) Sex pheromone titer correlated with pheromone gland development and age in the redbanded leafroller moth, *Argyrotaenia velutinana*. *Ann. Ent. Soc. America*, **70**, 136–139.

Murlis, J., Elkinton, J.S. and Carde, R.T. (1992) Odor plumes and how insects use them. *A. Rev. Ent.*, **37**, 505–532.

Nation, J.L. (1990) Biology of pheromone release by male Caribbean fruit flies, *Anastrepha suspensa* (Diptera: Tephritidae). *J. Chem. Ecol.*, **16**, 553–572.

Noldus, L.P.J.J., Potting, R.P.J. and Barendregt, H.E. (1991) Moth sex pheromone adsorption to leaf surfaces: bridge in time for chemical spies. *Physiol. Ent.*, **16**, 329–344.

Norris, M. (1968) Some group effects on reproduction in locusts. *Coll. Intern. du CRNS, Paris 1967*, **173**, 147–161.

Ono, T. and Ito, M. (1989) Pattern of pheromone-orientated flight in male potato tuberworm moths. *J. Chem. Ecol.* **15**, 2357–2368.

Phelan, P.L. (1992) Evolution of sex pheromones and the role of asymmetric tracking. In *Insect Chemical Ecology, An Evolutionary Approach* (eds B.D. Roitberg and M.B. Isman). Chapman & Hall, New York and London, pp. 265–314.

Pliske, T.E. (1975) Courtship behavior and use of chemical communication by males of certain species of ithomiine butterflies (Nymphalidae, Lepidoptera). *Ann. Ent. Soc. Am.*, **68**, 935–942.

Raina, A.K., Jaffe, H., Kempe, T.G. *et al* (1989) Identification of a neuropeptide hormone that regulates sex pheromone production in female moths. *Science*, **244**, 796–798.

Raina, A.K., Stadelbacher, E.A. and Warthenn, J.O. (1991) Role of host plants in the reproductive behavior of females of *Heliothis and Helicoverpa* species. *Proc. Conf. Insect. Chem. Ecol., Tabor 1990*. Academia Prague and SPB Acad. Publ., The Hague, pp. 261–276.

Riddiford, L.M. (1967) *Trans*-2-hexenal: mating stimulant for polyphemus moths. *Science*, **158**, 139–141.

Riddiford, L.M. and Williams, C.M. (1967) Volatile principle from oak leaves: role in sex life of the polyphemus moths. *Science*, **155**, 589–590.

Roeder, K.D. (1963) *Nerve Cells and Insect Behavior*, Harvard University Press, Cambridge, Mass., 238 pp.

Roelofs, W.L, and Wolf, W.A. (1988) Pheromone biosynthesis in Lepidoptera. *J. Chem. Ecol.*, **14**, 2019–2031.

Roelofs, W., Glover, T., Tang, X.-H. *et al.* (1987) Sex pheromone production and perception in European corn borer moths is determined by both autosomal and sex linked genes. *Proc. Natl. Acad. Sci.*, **84**, 7585–7589.

Rudinsky, J.A., Ryker, L.C., Michael, R.R. *et al.* (1976) Sound production in Scolytidae: female sonic stimulus of male pheromone release in the *Dendroctonus* beetles. *J. Insect Physiol.*, **22**, 1675–1681.

Rumbo, E.R. (1993) Interactions between male moths of the lightbrown apple moth *Epiphyas postvittana* (Walker) (Lep.: Tort.), landing on synthetic pheromone sources in a wind-tunnel. *Physiol. Ent.*, **18**, 79–86.

Rumbo, R. and Kaissling, K.-E. (1989) Temporal resolution of odour pulses by three types of pheromone receptor cells in *Antheraea polyphemus*. *J. Comp. Physiol. A.*, **165**, 281–289.

Sasoki, M., Riddiford, L.M., Truman, J.W. and Moore, J.K. (1983) Reevaluation of the role of the corpora cardiaca in calling and oviposition behavior of a giant silk moth. *J. Insect Physiol.*, **29**, 695–705.

Schal, C. (1982) Intraspecific vertical stratification as a mate-finding mechanism in tropical cockroaches. *Science*, **215**, 1405–1407.

Schal, C., Burns, E.G. and Blomquist, G.J. (1990) Endocrine regulation of female contact sex pheromone in the German cockroach, *Blattella germanica*. *Physiol. Ent.*, **15**, 81–91.

Schal, C., Charlton, R.E. and Carde, R.T. (1987) Temporal patterns of sex pheromone titers and release rates in *Holomelina lamae* (Lepidoptera: Arctiidae). *J. Chem. Ecol.*, **13**, 1115–1129.

Schilcher, F. von (1976) The role of auditory stimuli in the courtship of *Drosophila melanogaster*. *Anim. Behav.*, **24**, 18–26.

Sivinski, J., Burk, T. and Webb, J.C. (1984) Acoustic courtship signals in the Caribbean fruit fly, *Anastrepha suspensa* (Loew). *Anim. Behav.*, **32**, 1011–1016.

Sower, L.L., Shorey, H.H. and Gaston, L.K. (1970) Sex pheromones of noctuid moths. XXI. Light:dark cycle regulation and light inhibition of sex pheromone release by females of *Trichoplusia ni*. *Ann. Ent. Soc. Am.*, **63**, 1090–1092.

Spangler, H.G. (1984) Attraction of female lesser wax moths (Lepidoptera: Pyralidae) to male-produced and artificial sounds. *J. Econ. Ent.*, **77**, 346–349.

Spangler, H.G. and Hippenmeyer, C.L. (1988) Binaural phonotaxis in the lesser wax moth *Achroia grisella* (F.) (Lepidoptera: Pyralidae). *J.Insect Behav.*, **1**, 117–122.

Toth, M., Lofstedt, C., Blair, B.W. *et al.* (1992) Attraction of male turnip moths *Agrotis segetum* (Lepidoptera: Noctuidae) to sex pheromone components and their mixtures at 11 sites in Europe, Asia and Africa. *J. Chem. Ecol.*, **18**, 1337–1347.

Tumlinson, J.H. (1988) Contemporary research frontiers in insect semiochemical research. *J. Chem. Ecol.*, **14**, 2109–2129.

Vickers, N.J. and Baker, T.C. (1991) The effect of unilateral antennectomy on the flight behaviour of male *Heliothis virescens* in a pheromone plume. *Physiol. Ent.*, **16**, 497–506.

Vogt, R.G., Riddiford, L.M. and Prestwich, G.D. (1985) The kinetic properties of a pheromone degrading enzyme: the sensillar esterase of *Antheraea polyphemus*. *Proc. Nat. Acad. Sci. USA*, **82**, 8827–8831.

Wall, C., Sturgeon, D.M., Greenaway, A.R. and Perry, J.N. (1981) Contamination of vegetation with synthetic sex attractant released from traps for the pea moth, *Cydia nigricana*. *Ent. Expl. et Appl.*, **10**, 111–115.

Wright, R.H. (1958) The olfactory guidance of flying insects. *Can. Entomologist*, **90**, 81–89.

Wyatt, T.D., Phillips, A.D.G. and Gregoire, J.-C. (1993) Turbulence, trees and semiochemicals: wind-tunnel orientation of the predator, *Rhizophagus grandis*, to its bark beetle prey *Dendroctonus micans*. *Physiol. Ent.*, **18**, 204–210.

# 4

# Bioassay methods

## 4.1 THE PROBLEM OF DEFINITION

The mere fact that a given chemical compound known to occur in an insect elicits a behavioural response in a conspecific does not prove that it is a semiochemical. For example, if we wish to know whether a given compound is a pheromone, or a component of a pheromone, we must show the following.

1. That it is emitted by the animal. Extracts of a gland can yield pheromone precursors rather than the compounds that are released into the air. This has been clearly demonstrated in the tobacco budworm, *Heliothis virescens*, and the corn earworm moth, *H. zea*, by Tumlinson and his colleagues (Tumlinson, 1988). Gland extracts of the female moths are dominated by alcohols, such as Z11-16:OH. These are oxidized enzymatically to the corresponding aldehydes as they pass through the cuticle, and so air collected from around the insects contains compounds that are not present in the glands.
2. That the receiver responds to the synthesized compound in the same way that it responds to the emitted extract.
3. That the behavioural response to the compound is limited to the species in question (possibly extending to a small number of congeneric species). We have to remember here that related species often share pheromone components, but that specificity of response is ensured by other methods – for example, the use of blends of compounds in sex pheromones – and that pheromones may function as kairomones, in which case a sexual pheromone may attract both the opposite sex of the same species and adult parasitoids.

## 4.2 BIOASSAYS AND PHEROMONE IDENTIFICATION

Investigation of semiochemicals normally begins with observations on the insect in the field, and preferably in its natural habitat. The second step (Fig. 4.1) is the construction of a method for reproducing the behaviour in the laboratory and finding appropriate aspects to measure. Analysis of chemical communication systems normally involves the following steps.

*1. Observation of the field mating system*
A good knowledge of the normal behaviour of the insect in the field is essential if various pitfalls are to be avoided. We need to ask several questions. Under what conditions does the insect appear to liberate pheromone? In the case of sexual pheromones, does the male or the female engage in calling, or do both? Can insects be attracted to caged virgin males or females in the field? Answers to questions of this nature can greatly simplify the design of laboratory experiments.

*2. Establishment of laboratory cultures*
Establishment of a continuous culture of insects is not always necessary but can speed up the progress of research considerably. One good reason for this is that the time-scales of chemists undertaking analytical work can be highly unpredictable at the outset (depending, for example, on the complexity of compounds and amounts that can be extracted) and can be many months out of synchrony with the appearance and availability of adult insects in the field.

*3. Design of a bioassay*
Ideally, the bioassay should be designed to measure the type of behaviour that has been observed in the field, and carried out in appropriate conditions (see below). It should be as simple as possible, capable of easy replication, and yet highly predictive of how the insect would respond in the field.

*4. Extraction of semiochemical*
Extraction methods are considered in Chapter 6. Biological information must be used to decide when and under what conditions insects liberate pheromone, and chemical expertise used to select the most appropriate extraction technique.

*5. Comparison of the responses to extracts with responses to live insects*
One means of checking whether the pheromone has been altered in some way by the extraction procedure is to test the extract against captive insects and make sure that they respond to it in the same way as they do to live conspecifics.

*6. Analysis of the extract*
Gas chromatography and fractionization can be used to determine the variety of compounds present and to separate them. These techniques are outlined in Chapter 6.

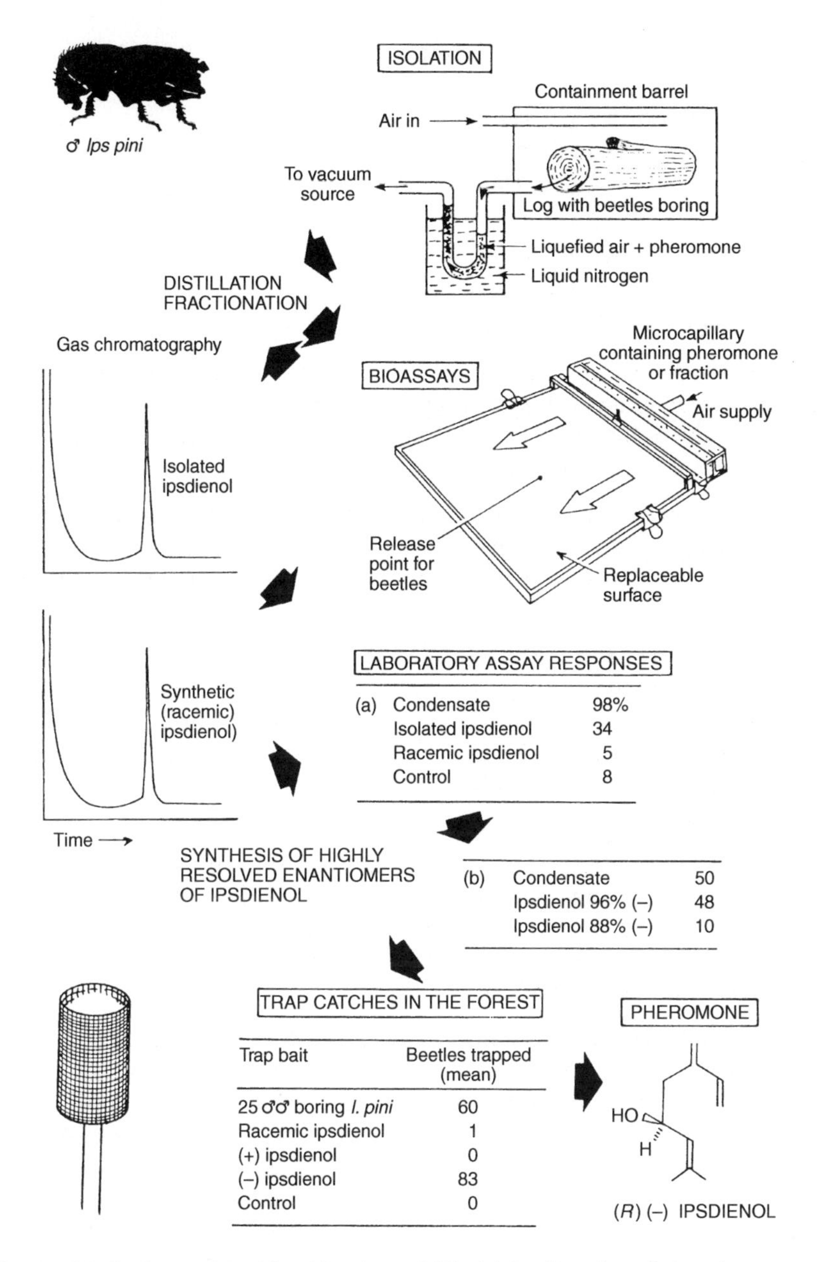


**Fig. 4.1** Methods used in identification of (*R*), (–)-ipsdienol as the main component of aggregation pheromone of the scolytid beetle *Ips pini*. After extraction of air from beetle-infested log, the condensate was analysed – starting with the use of gas chromatography and bioassay of fractions in two-dimensional assay chamber with airflow. Isolation of the active peak led first to synthesis and testing of racemic ipsdienol, and then of (+) and (–) optical isomers. Field-trapping experiments confirmed that the (–) isomer was the active one. (From Birch, 1984, in *Chemical Ecology of Insects*, eds Bell and Carde, Chapman & Hall.)

*7. Selection of the fraction*
Fractions of the extract are made and tested to determine which provokes the appropriate behavioural response in the chosen bioassay.

*8. Analysis of the major components*
Components of the active fraction are investigated using mass spectrometry, NMR, etc. (Chapter 7)

*9. Formulation of hypotheses about the chemical structure of the detected compound, followed by synthesis*
This may be an involved and complicated process, especially if the compound is new to nature.

*10. Comparison of the behavioural effects of the synthetic compound with those of the active fractions*
This may involve both laboratory assays and evaluation in the field using traps.

*11. Synthesis of enantiomers (if any)*
Insects sometimes respond to racemic mixtures, but this does not preclude the possibility that they respond optimally to one enantiomer or optical isomer. Compounds for test must therefore be synthesized in an extremely pure form.

*12. Field testing in slow-release formulations*
At this point, we may be confident of being able to make a good copy of the natural pheromone which is effective for a practical use in insect pest management. We also need to be able to reproduce the optimal release rate and develop dispensers that will maintain that release rate over periods of weeks or months.

In this scheme, the backbone of the procedure is the bioassay: unless this is properly designed, and highly discriminating, mistakes can be made. Any search through the literature reveals publications on the identity of this or that pheromone, and subsequent publications that report a totally different conclusion. This is a common experience in reviewing early pioneering studies in which analytical laboratory equipment was less sophisticated than it is now and when the importance of a really discriminating bioassay was not realized. In the study of Lepidopteran sex pheromones, one bioassay which led to errors was the activation and fluttering response induced when pheromone-laden air was passed over male moths. It was some time before it was found that many chemical compounds that were not natural pheromones could provoke the same response.

In the study of animal communication, as we have explained, the concept of the meaning of the message is inapplicable. We can measure only the responses of an animal and determine the effect of the context of the message on those responses. The responses of the sensory receptors can tell us what parameters of the signal have been lost in the transduction process, and which are transmitted to the central nervous system. Within the CNS, the specific sensory input is merged with input from other receptors, and with stored information from other parts of the CNS. The CNS is a dynamic system, in a constant state of activity and flux, and it is extremely difficult to follow the fate of the transformed signal through it, but very straightforward to map and chronicle the outputs in terms of movement patterns or the emission of answering signals of various kinds. Measurement of the total output, however, is too cumbersome a process to answer most of the questions we need to pose. Instead, it is more convenient to measure an easily recognizable part, or, in other words, to develop a bioassay.

The art of the bioassay is to select a response that is easily recorded and which is a good **predictor** of the final outcome. Not all elements of a behavioural sequence are equally good predictors and, in general, the closer one gets to the sensory receptor level, the less predictive value events have. Most methods involve recording of behavioural changes and utilize equipment that can be set up in standard laboratory conditions. Some bioassays do not involve the measurement of behavioural responses, but record changes in physiological activity. The best known and most commonly used of such methods, in which only part of the animal may be used, involve neurophysiological recording from sensory nerves.

## 4.3 ELECTROPHYSIOLOGICAL BIOASSAYS

Electrophysiology is a very powerful and direct method of measuring the aspects of the signal to which the animal is responsive. In the study of pheromone communication, its value lies in identifying the communication channel (and thereby directing the focus away from parts of the signal that the animal cannot detect). The most common assays of this kind involve recording from the whole antenna, or from individual sensory cells, though recording from muscles has occasionally been used.

### 4.3.1 The EAG

The most practical from of electrophysiology is electroantennography (EAG). Developed by Schneider (1957), this technique was first used on silk moth antennae. EAG can be practised either with an excised antenna or with the antenna still in place on the head. Basically, the change in potential between the tip and the base of the antenna is recorded as a

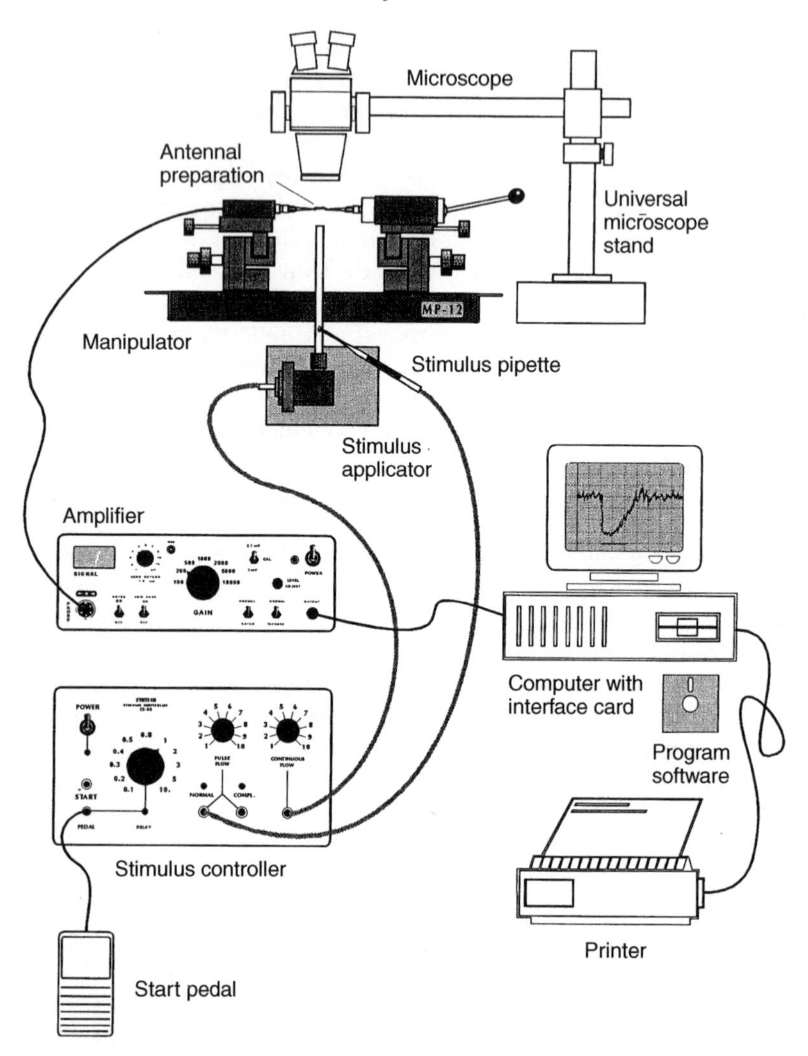


**Fig. 4.2** Apparatus for recording electroantennograms (EAGs). Antennal preparation is positioned between two electrodes mounted on micromanipulators. Odour stimulus is delivered in a controlled air stream produced by an applicator containing a solenoid. The potential change between electrodes is amplified by a DC amplifier and the output is stored in a computer or fed into a suitable printer. (Reproduced courtesy of Dr J.N.C. van der Pers, SYNTECH, The Netherlands.)

puff of odour passed over the antenna (Fig. 4.2). If antennal receptors are stimulated, their summed electrical activity is a slow DC depolarization of around 1–20 millivolts. What is recorded is not the summed action potentials, but the summed changes in potential of numerous receptor neurones. To understand this, it is necessary to appreciate the layout of chemoreceptors on the antenna. Each is innervated by one or a few sensory neurones, the axons of which project at an angle to the antennal nerve, of which they all form part. In the transduction process, which takes place on the dendritic membrane, a DC depolarization or receptor potential is generated. When this is of sufficient magnitude it spreads to the cell body and initiates the production of action potentials. Because of the time delay, the dendrite goes momentarily negative with respect to the axon; and as the axon lies proximally to the tip of the dendrite, and as many axons are likely to be involved, the whole antenna goes negative with respect to the base for a few seconds.

The EAG potential is therefore a relative measurement of the number of chemoreceptors stimulated by the presence of the odour molecules. It can, however, be a poor indicator of behavioural responsiveness for two main reasons. Firstly, the depolarization depends upon the number of neurones stimulated and does not distinguish between those that may have inhibitory or synergistic effects at various integrative levels. Secondly, a small number of neurones may control a key response. Löfstedt *et al.* (1982) found poor correlations between the magnitude of EAG responses and the relative amounts of pheromone in female pheromone glands in the noctuid moth *Agrotis segetum*. While two compounds, (Z)-5-decenyl acetate and (Z)-7-dodecenyl acetate, evoke high levels of response, (Z)-9-tetradecenyl acetate does not, although it is produced in large quantities by the female and is essential for good attraction in the field.

In the cockroach, *Periplaneta americana*, it has been found that, as well as the normal negative peak, positive peaks can be obtained (Contreras *et al.*, 1989). The positive peak was always associated with compounds which were repellent in behavioural tests, and negative peaks were associated with attractants (depending on the concentration). In general, however, the value of the EAG technique lies in its use for rapidly screening large numbers of compounds to eliminate those that apparently cannot be detected by the insect.

Techniques exist for coupling an EAG recording set-up with the output from a gas chromatograph (GC). The gas eluting from the column is divided into two streams by a splitter (see Fig. 6.8). One stream goes to the detector of the GC and the other is introduced into a clean air stream which is passed over the antenna. In this way, GC peaks can be correlated with EAG responses. Extracted mixtures can thus be screened rapidly without the necessity for fractionization and separate testing of fractions. Peaks that provoke no electrical response can therefore be ignored. Alternatively, the GC can be coupled to a single-cell recording apparatus (see below), or even to an actograph for recording changes in the activity of insects (Hummel and Miller, 1984).

In the latter case, part of the effluent is passed into a chamber containing live insects and their activity is continuously recorded either by an observer or by an automated method. The main problem with such coupled methods is that peaks may issue very close to each other in time, and it then becomes difficult or impossible to assign biological activity to a precise component.

EAG recording can be used to derive a hypothetical structure for a pheromone, a technique that has been used very successfully by Roelofs (1984) and his colleagues. A library of compounds is required, which is used to establish the 'response profile'. This usually begins with establishing the chain length that gives the optimal EAG response. Subsequently the optimal functional group, double bond positions, etc. are established. To take an example, Roelofs and Carde (1974) used the following series of clues to deduce the structure of the sexual pheromone of the lesser apple worm moth, *Grapholitha prunivora*.

- The highest EAG responses were given to 12-carbon atom compounds.
- A higher response was obtained to acetates than to other functional groups.
- *Z* isomers were more effective than the corresponding *E* isomers.
- (Z)-8-dodecenyl acetate gave the largest EAG response.

In the pea moth, *Cydia nigricana*, a study of response profiles led to the identification of (*E*)-10-dodecenyl acetate as one candidate pheromone component (Wall *et al.*, 1976). This was effective in field trials and used commercially as an attractant in monitoring, but the true pheromone was later identified, using conventional spectrometric and spectroscopic methods, as the less stable compound (*E,E*)-8,10-dodecadien-1-yl acetate (Greenway, 1984). The first compound is an example of a parapheromone.

Mobile EAG apparatus has been designed for use in the open air or in wind tunnels as a sensitive means of detecting pheromone. Baker and Haynes (1989) used such an apparatus to monitor the detection of odour filaments in a pheromone plume as a mounted male oriental fruit moth was moved upwind towards the pheromone source. They found that the peak–trough EAG response diminished in height close to the source, probably because of more constant stimulation of the antennal receptors, and this led to flight arrestment.

### 4.3.2 Single-cell recording

While EAGs provide a measure of the quantity of nerve impulses transmitted to the brain, recording from single sensory cells (SCR) tells us precisely which chemical compounds are detected and which are not. This technique requires penetration of the neuron with a microelectrode and makes possible the recording of action potentials. It has been used to study the physiology of chemoreception in saturniid moths (Chapter 1) and the response spectra of entire antennae (response characteristics of sensilla and numbers of sensilla of different types). In general, a good correlation is found between EAG and SCR results. It is always possible, however, that only a very few sensilla respond to an essential component of the pheromone blend and that they do not generate a detectable EAG response.

Van de Pers and Minks (1993) have designed a mobile single-cell recording unit. The preparation is mounted on a stage with micromanipulators and an air velocity sensor, and responses have been recorded in field conditions to sources of the sexual pheromone of two moth species. This method makes it possible to discriminate between various pheromone components and plot their position in an odour plume.

### 4.3.3 Electromyograms

Recording of muscle potentials is a simple way of registering the onset of a behavioural response and quantifying the response. It has been used, for example, by Obrecht and Hanson (1989) to measure parameters of the wing-fanning response in males of the gypsy moth, *Lymantria dispar*. A moth was attached by the underside of the thorax to the centre of a loudspeaker cone, and impaled on recording electrodes which projected into the thoracic musculature. The insect was then exposed to female pheromone and the loudspeaker was used as a transducer to pick up the thoracic vibration, and the electrodes to register simultaneously the muscle potentials.

## 4.4 BEHAVIOURAL BIOASSAYS

Chapter 3 showed that there are many factors that can affect the threshold responsiveness of animals. The environmental conditions, age and physiological state of the insects must therefore be standardized as far as possible. It is necessary to establish first the period of the day or night during which the insect is most likely to respond: attempting to experiment with insects outside those times may be totally unproductive. The factor that is most difficult to control is contamination. We have seen that pheromones are generally odourless to the human nose, active in extremely low concentrations, and readily adsorbed on to materials such as glass, plastic and cellulose, i.e. the materials from which apparatus is usually constructed. We can never be sure that the air in the laboratory is totally clean: pheromone may be carried on clothing, or introduced from insects brought in from time to time, or from extracts kept in refrigerators. Furthermore, we cannot predict the extent to which apparatus becomes contaminated.

It is of great importance to determine the number of insects used in a trial. Part of the normal behaviour of gregarious insects, such as locusts, or of social insects, such as ants, involves coordination with other members of the group: an isolated individual is not in its normal environment. Thus it makes sense to test responses of groups of insects in some cases, but then it is important to decide what to measure. For example, a group of locust hoppers will move upwind in an air stream carrying the odour of freshly cut grass. If 75 of a group of 100 do this, we can use the percentage as a measure of the strength of the response, but we cannot say that we have replicated the test 75 times, because all the individuals except the first one may have been influenced by what the others were doing. Likewise in a bioassay investigating the orientation of leaf-cutting ants in mazes (Fig. 4.3), only the choice made by the first ant of a group could be recorded (Vilela *et al.*, 1987).

Baker and Carde (1984) have emphasized the importance of controlling the emission rate of the chemical used in bioassays, both during the course of each assay and during successive replications of the assay. They also point to the importance of standardizing the internal state of the insect in each test: this can be affected by the kinds of factors listed in Fig. 3.1, and by habituation through persistent or repeated exposure to the pheromone. They classified behavioural bioassay techniques according to whether the insect is in moving air or still air, and whether the displacement of the insect is monitored or not. Most bioassays are set up for the laboratory, but to these we can add bioassays carried out in restricted field conditions – for example, in cages with the host plant set up in the kind of environment in which the insect is normally found.

Behavioural assays generally depend on displacement of the insect in space. The types of displacement that occur have been classified by Fraenkel and Gunn (1940). Their most far-reaching contribution to understanding behaviour was to distinguish between **taxes** and **kineses** as forms of orientation. In kinetic orientation, the animal changes its rate of locomotion or rate of turning, or time spent on locomotion. In tactic orientation, the path of the animal is directed towards, away from or at a fixed angle to a stimulus source.

Orientation movements are further classified according to the stimulus involved. We can, for example, speak of phototaxis, geotaxis or anemotaxis as responses to light, gravity and air currents, respectively. We can specify the angle to the source as in line with the source and towards it (e.g. positive phototaxis), away from it (negative phototaxis), or at a fixed angle to it (menotaxis). Yet a further refinement to the classification refers to the tendency of the animal to orientate according to the balance of stimulation received from more than one source. Presented with two light sources, a honey bee will usually orientate to one or the other, and is said to show a photo**telo**taxis, while a flour moth larva will orientate towards a midpoint between them if they are of equal intensity, or to one side of the midline if one light is stronger than the other. This is said to be a photo**tropo**taxis.

It is important to remember that these terms are purely descriptive, but the great advantage in using them is that we avoid describing behaviour by its end results. If a male gypsy moth arrives in a pheromone trap it is merely tautologous to say it is 'attracted': that tells us nothing about how it got there. The notion of 'attraction' is not helpful because it suggests that it is pulled in along some invisible lines of force. Clearly this is not so, and Carde (1979) has shown that the behaviour of a male gypsy moth approaching a calling female is in fact a sequence of steering and locomotory movements that begins with activation (see Fig. 4.10). The mechanism of upwind flight is categorized as an odour-modulated anemotaxis, although the mechanism of this is still incompletely understood.

Bell and Tobin (1982) have added other criticisms to the classification of orientation movements, emphasizing that insects may integrate different mechanisms and use them at the same time, or switch rapidly between them. Not only that, but orientation may well depend upon the input from a variety of different sensory receptors, responding to signals in different modalities. An example of this is provided by the work of Vilela *et al.* (1987) who found that leaf-cutting ants orientated with respect to their own odour trails in most circumstances, but if these were absent, or if the information the ants had was ambiguous, they used other cues which, in order of descending importance, were visual, spatial layout of the trail, odour differences on the trail, and gravitational cues. As is so often the case, the kind of answer that an experiment gives depends on the question that is being asked.

Kennedy (1977) has also stressed that many simple bioassays can give ambiguous results because they do not discriminate between different kinds of behavioural response that lead to an insect changing its position in space. For example, one component of a pheromone may inhibit approach at a close range, though it may be essential for long-range approach upwind. Or a compound may have some kind of arrestant effect that may affect choice of position or direction in an olfactometer, leading to the interpretation that it is an 'attractant', when in the field there may be no approach at all.

Kennedy argues in favour of what he calls 'behaviourally discriminating assays', which make it possible to separate the various components of approach or withdrawal from a stimulus, listing the basic design principles of such assays as follows.

1. The odour stimulus should be presented in a form that elicits only one kind of response.
2. The way in which the stimulus is presented should be carefully controlled by the experimenter so that it is not affected by the movements of the insect itself.
3. The effects of decreasing and stopping the stimulation should be evaluated as well as the effects of introducing and increasing it.
4. The insects must be in a state in which they are all equally and highly sensitive to the odour. This means ensuring that their motivational and physiological state is similar and also that external conditions (amount of space in which to manoeuvre, etc.) are adequate.

Given that insects normally respond to pheromone with a sequence of behaviour and that pheromone distribution is not at all homogeneous in the way that a light beam is, it is important that any bioassay should be a good predictor of the likelihood that the whole sequence of behaviour will result. Carde (1979) showed, for example, that activation is not a suitable bioassay response: it is one that can be induced by a wide variety of stimuli, most of which have nothing to do with pheromonal communication.

Chemical stimuli are extremely difficult to define in terms of concentration, distribution, and change of concentration with time. In contrast, an acoustic, electrical or visual field is relatively easily defined in quantitative and qualitative terms. The rate of diffusion from a point source in still air can be predicted, but only with difficulty (Bossert and Wilson, 1963) and on the basis of assumptions (that may be incorrect) about diffusion coefficients. Furthermore, air is relatively still only when it is close to a surface, within the so-called boundary layer. The depth of the boundary layer depends on the velocity of air movement above the surface, and it is unlikely to be more than a few millimetres thick. A small insect moving within the boundary layer will create turbulence by its own movements: an aggregation of rapidly moving ants would then virtually destroy an odour gradient around a point source within a very short time.

## 4.5 TRAIL BIOASSAYS

A chemical applied to the substrate – for example, in the form of an odour trail – offers the best possibility of a stable and discrete signal. Bioassays for trail substances are therefore set up with relative ease (Fig. 4.3). Flying insects, and probably many others, do not have recourse to the stability of the boundary layer; they are moving in turbulent air streams in which the laws of diffusion cannot be applied. Barriers between micro-streams containing a high concentration of volatile chemicals and those with very small amounts are impossible to define and are changing continuously with time on a very small time scale. This makes the design of adequate bioassays extremely difficult. One of the most conceptually simple bioassays is a Y-maze or tube, in which air is passed through both arms and a test volatile is introduced into the airflow of one arm. It is assumed that the two air streams remain discrete at the junction, so that, for example, an insect would have one antenna in the charged air stream and the other in the control stream. This is, however, very unlikely, as turbulent mixing will occur at the junction.

## 4.6 ACTIVITY METHODS

The use of activity measurements alone can be misleading: increase in activity can be brought about by a compound with an irritant effect, or, for example, a component of food odour which may have nothing to do with pheromonal communication. Notwithstanding such problems, activity measurements can be used to determine the optimal concentrations of known pheromones. A method of extreme simplicity has been developed by Block and Bell (1974) for assaying synthetic cockroach pheromone. The effect of the two pheromone components of the female *Periplaneta americana*, periplanone A and periplanone B, is to induce running, upwind orientation and wing-raising. Insects are put into a closed box with a straight line drawn across the middle of the floor from one side to the other. The frequency of line crossings made by male cockroaches is then monitored before and after the introduction of pheromone (referred to as the male activity count, MAC). This method has been used by Seelinger (1985) to compare the ability of *P. americana* and *P. australasiae* to detect periplanone A and periplanone B (Fig. 4.4).

## 4.7 FORCED TURNS: MEASUREMENT OF CHOICE

### 4.7.1 Orientation in mazes

Many researchers have used Y-maze bioassays. An insect orientates into an air stream until it encounters the junction, at which point the odour under test is in only one of the two odour streams – that from one arm carrying a test odour and the other not – and the insect makes a choice.

The scientific literature is redolent with references to 'attractants' which have been elucidated for house flies and other insects with the use of Y-maze bioassays (e.g. Dethier, 1947). The simplicity and replicability of this methods belie serious pitfalls, which are as follows.

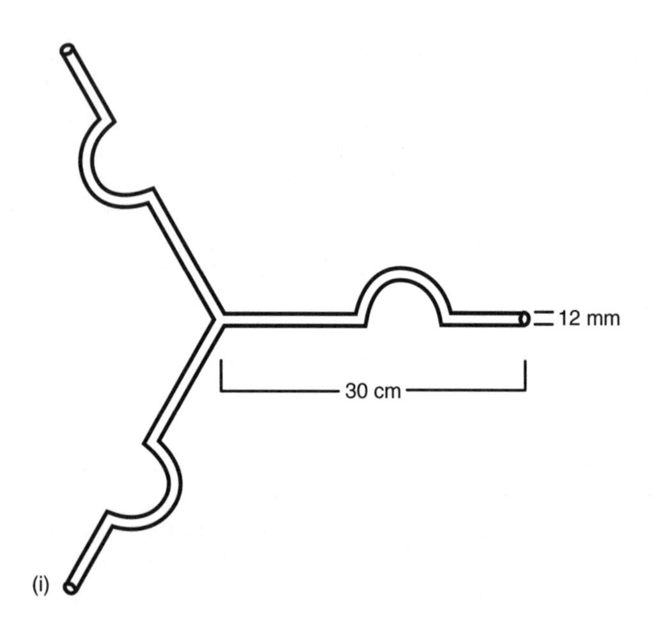
12 mm
30 cm
(i)

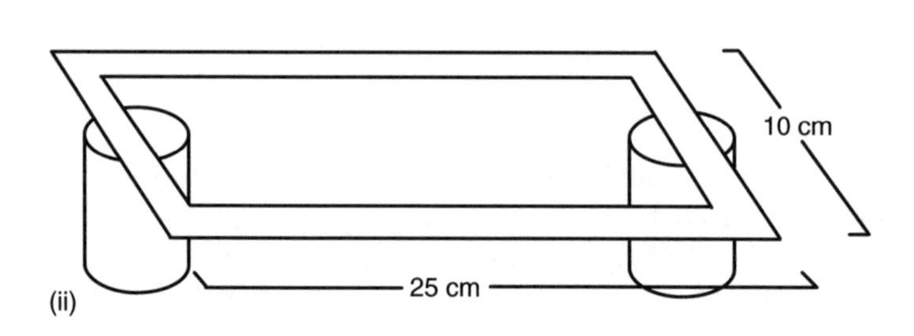
10 cm
25 cm
(ii)

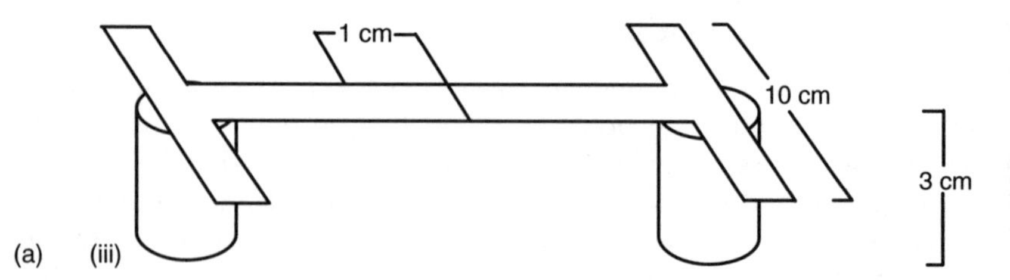
1 cm
10 cm
3 cm
(a)
(iii)

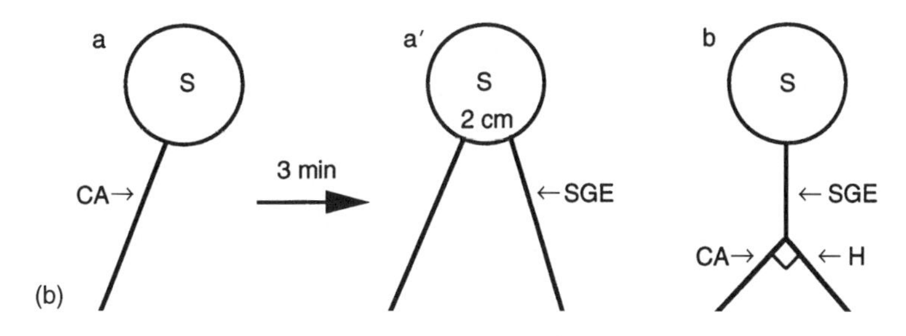


**Fig. 4.3** Some bioassay methods for trail pheromone. (a) Bridge mazes used for leaf-cutting ants and placed between the nest and a foraging table: (i) glass tube maze in which response to light cues can be minimized by covering with foil and by ox-bow bends in centre of each arm; (ii) bridge providing two symmetrical routes; (iii) bridge with single runway. (From Vilela *et al.*, 1987.) (b) Simple bioassay for separable effects of trail-following and recruitment to trails in termite *Nasutitermes costalis*. *CA*=trail made by synthetic trail component cembrene-A; SGE=trail made with sternal gland extract; H=hexane (solvent) trail. (From Traniello and Busher, 1985.)

- Volatiles inevitably adsorb on to the surfaces over which the first insect walks and can therefore affect the path of subsequent insects unless they are first removed.
- The insect is exposed to the odour before it reaches the choice point and may habituate, or adaptation of its sensory receptors may occur.
- Contact with the sides of the maze may determine the arm chosen.
- The airflow at the junction is very likely to be turbulent and so the insect may not have the simple choice of odour on one antenna and not on the other.
- Insects such as house flies respond to novel odours and may select them repeatedly on one day but appear to be insensitive to them thereafter.
- Some insects (Dingle, 1964), and perhaps many, show the phenomenon of turn alternation in which, having made a turn in one direction, they tend to make a compensatory turn at the next point at which they are free to do so. This can be difficult to control for in Y-mazes.
- Some insects may be able to orientate by methods other than the simultaneous comparison of the intensity of stimulation on each antenna.

### 4.7.2 Olfactometers

Most olfactometers represent a compromise between a Y-maze and a wind tunnel. They are usually designed on a pragmatic basis, i.e. they

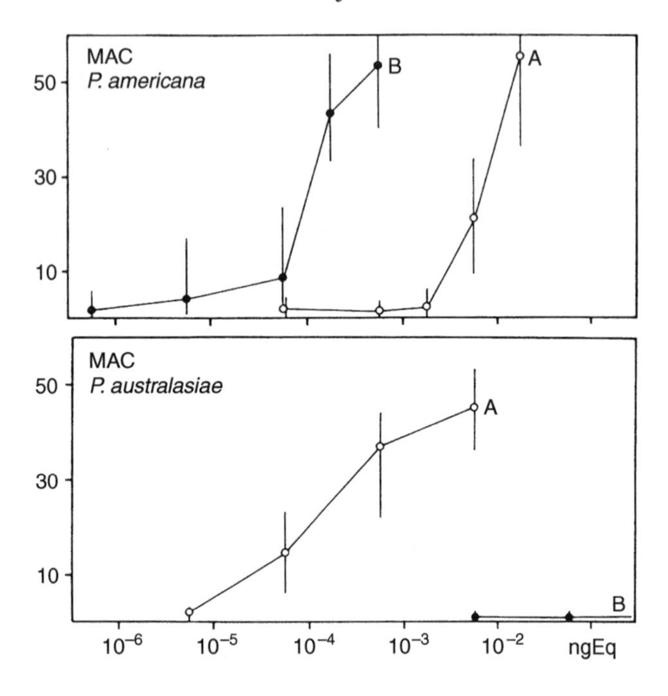


**Fig. 4.4** Bioassay based on activity. (a) Results of semi-automated method of tracking walking insects in two dimensions. Five male cockroaches in a container were exposed to air stream container containing two main components of sex pheromone of cockroach *Periplaneta americana*. MAC (male activity count) measured by recording the number of crossings made by insects over a line drawn down middle of the container. Results show that pheromone component Periplanone A is attractive to both *P. americana* and *P. australasiae*, but that Periplanone B inhibits activity of *P. australasiae*, providing a possible sexual isolating mechanism: 1ng Eq produces the same EAG response as 1ng of racemic Periplanone B. (From Seelinger, 1985.)

happen to work, but what works for one species may not do so for another. An example is the olfactometer designed by Katsoyannos *et al.* (1980), shown in Fig. 4.5 and designed for use with tephritid fruit flies and small moths. A fan is used to drive air through an antechamber, connected in series with a bait chamber, an intermediary chamber and a catch chamber. The catch chamber has a funnel-shaped entrance designed to prevent flies that enter the chamber from leaving it. The catch chamber opens into a large test chamber which contains the insects under test, and air passes through this to the outside of the cage.

The bait chamber may contain calling insects, pheromone extract or other candidate attractants. Flies in the test cage responding to the odour current orientate towards the catch chamber, and those that have

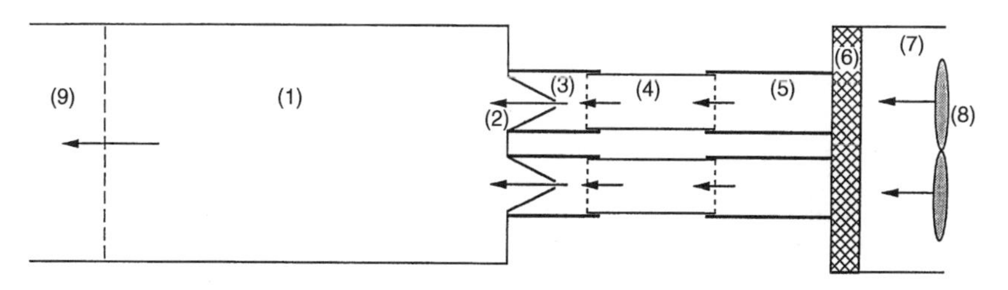


**Fig. 4.5** Dual-choice olfactometer, based on that described by Katsoyannos *et al.* (1980). Insects are put into test cage (1). By moving upwind they can pass through one of two funnels (2) into catch chamber (3). One of bait cage (4) contains control source, the other a test odour source (which could be calling insects). Air enters test cage through duct (5) abutting on to charcoal filter (6) at end of larger duct (7), which contains fan (8) driving air through apparatus to exhaust duct (9) leading to outside of room. (Courtesy of Ivanildo de Lima.)

entered are counted at the end of a certain time period. The olfactometer may be set up with two or more identical series of chambers, one of which acts as a control. The advantage of this method is that the insects are able to fly freely before making a 'choice', and quantification is achieved simply by counting the number of flies trapped in a given time period. However, once again, as soon as each fly is trapped the stimulus situation changes, and the effects of this on other flies can be neither easily assessed nor circumvented completely by removing each fly as it arrives.

## 4.8 FREE TURNS: TRACKING INSECT MOVEMENTS

### 4.8.1 Assays in still air

In still air, odour molecules disperse through diffusion, and phenomena such as turbulence and convection may play a negligible role. But the concept of still air is an abstraction: air in a closed, draught-free room may appear to be still but movement can be detected by watching dust particles or a fine hair under a magnifying lens or microscope. Even insects in confined spaces, such as social insects in their nest galleries, will generate turbulence by their own movements. In open field conditions, any concentration gradient may be broken up very rapidly; we can surmise that it will persist longer in the boundary layer, but we cannot say with any accuracy how deep that boundary layer might be at any one place and how it might change from moment to moment.

Laboratory bioassays generally assume that the air is still and that the odour from a point source spreads out according to the classical laws of

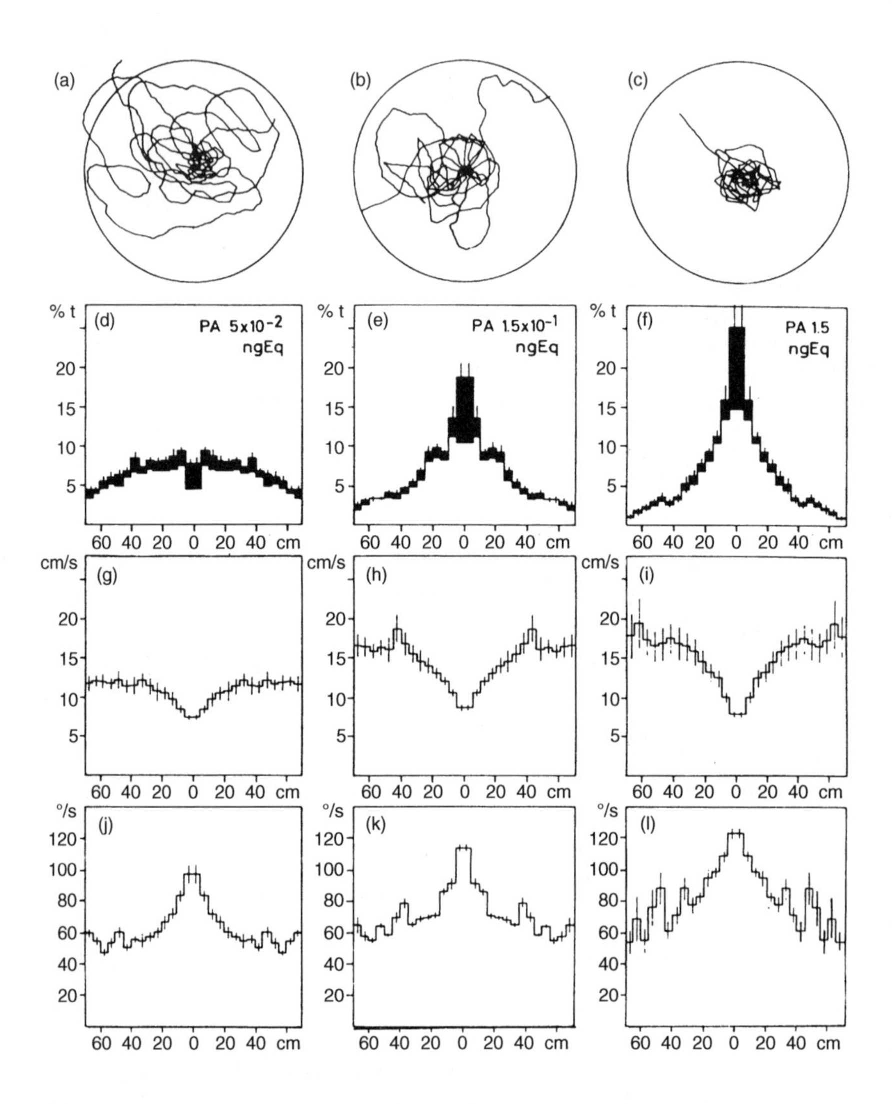


**Fig. 4.6** Bioassay based on activity. (b) Results obtained from videotapes of cockroaches (*P. americana*) in a 1.8 m diameter circular arena with a pheromone source suspended above the central point. Upper figure (a,b,c) shows typical runs of males exposed to different concentrations of pheromone Periplanone A (in ng equivalents: 1ng Eq produces the same EAG response as 1ng of racemic Periplanone B.); (d,e,f) relative times spent by males at different distances (cm) from source; (g,h,i) speed of movement (cm/s) at different distances from source; (j,k,l) angular velocity. (From Seelinger and Gagel, 1985.)

diffusion. 'Instantaneous' release will then give rise to a concentration gradient that changes rapidly with time. Initially the gradient will be very steep, but will become progressively shallower with time. In a closed container, all the molecules will be equally distributed after a certain time. The practical sense of this is that still-air bioassays should be done in containers that are as large as possible (but this means the air will be less still), and some estimate must be made of the rate of expansion of the active space. Bossert and Wilson (1963) concluded that the active space of a chemical released on a plane surface from a point source would be approximately hemispherical, with the radius expanding with time. They used the following equation to predict the radius ($R$) at time ($t$) of the active space of a volatile chemical released as an instantaneous puff:

$$R_t = \sqrt{4Dt} \log \left( \frac{2Q}{K(4\pi Dt)^{3/2}} \right)$$

where $D$ is the diffusion coefficient, $Q$, is the number of molecules released, and $K$ is the behavioural threshold of the animal.

From this we see that the radius will expand most rapidly when $Q$ is high and $K$ is low, but then the concentration gradient will be lost rapidly. Bossert and Wilson point out that alarm communication systems, in which rapid transmission of the signal and rapid fading are normally required, typically have high $Q/K$ ratios. Conversely, sex pheromone communication normally involves persistence of a signal with a low threshold sensitivity of the receiver, i.e low $Q/K$ values.

In fact, because the active space is defined by both the concentration of a chemical and the sensory threshold of the insect, it is most easily measured by changes in behaviour of an insect at its boundary. Bradshaw (1981) made such measurements on the African weaver ant, *Oecophylla longinoda*, and compared them with the boundaries calculated using Bossert and Wilson's formula (above), into which he substituted independent behavioural measurements of threshold sensitivity of the ants, and found a reasonable degree of coincidence.

A variety of automated methods has been developed to record and analyse the tracks of insects. Nowadays video recorders are in common use, and computer programs are available that will analyse tracks of single insects. A simple automated method for walking insects that can be used with a microcomputer has been devised by White *et al.* (1984). The movements of flies over a glass plate were followed in an arena with 30 cm high walls. A vertical mirror or glass plate was used to reflect the image of the flies on to a bit-pad digitizer, and the output was fed into a microcomputer. Computer programs could then be used to calculate selected parameters of a fly's path. This method has been further developed by Seelinger & Gagel (1985) and others to quantify responses of cockroaches to pheromone components. Figure 4.6 shows results derived from analysis of videotaped movements of cockroaches made by transferring pathways from frame-by-frame video recordings on to an X–Y digitizer and analyzing selected parameters with a computer program.

### 4.8.2 Turning movements: insects fixed in space.

One easily automated means of measuring the turning movements of walking insects is to hold the insect fixed in position so that as it walks the substratum beneath it moves rather than the insect itself. A simple form of such a bioassay was developed by Rust *et al.* (1976) to measure the response of the American cockroach, *Periplaneta americana,* to pheromone. A Y-maze globe (Fig. 4.7) was constructed from light styrofoam, sufficiently light for a suspended insect to grasp with its legs. At

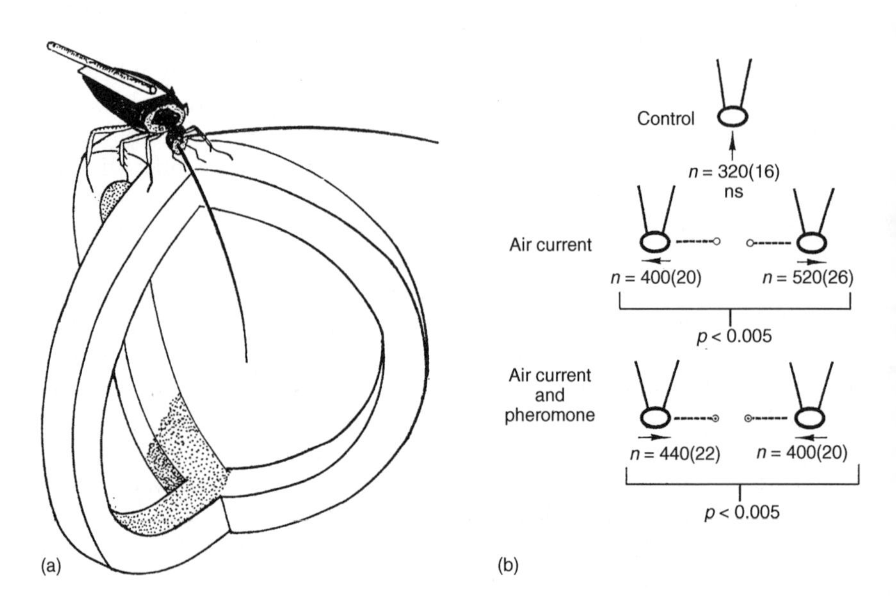


**Fig. 4.7** (a) Male cockroach, *Periplaneta americana,* fixed in position with Y-maze globe suspended by its feet. As the insect walks, it rotates the maze with its feet, coming at regular intervals to Y-junction at which it must 'choose' a new direction of orientation. (b) Typical results of assay for aggregation pheromone, with head and antennae shown schematically. n = no. of turns in direction of arrows (no. of insects tested in parentheses, each tested 20 times). Air current source indicated by small circles. Significance of turning tendency given by probability (p), ns = non-significant.

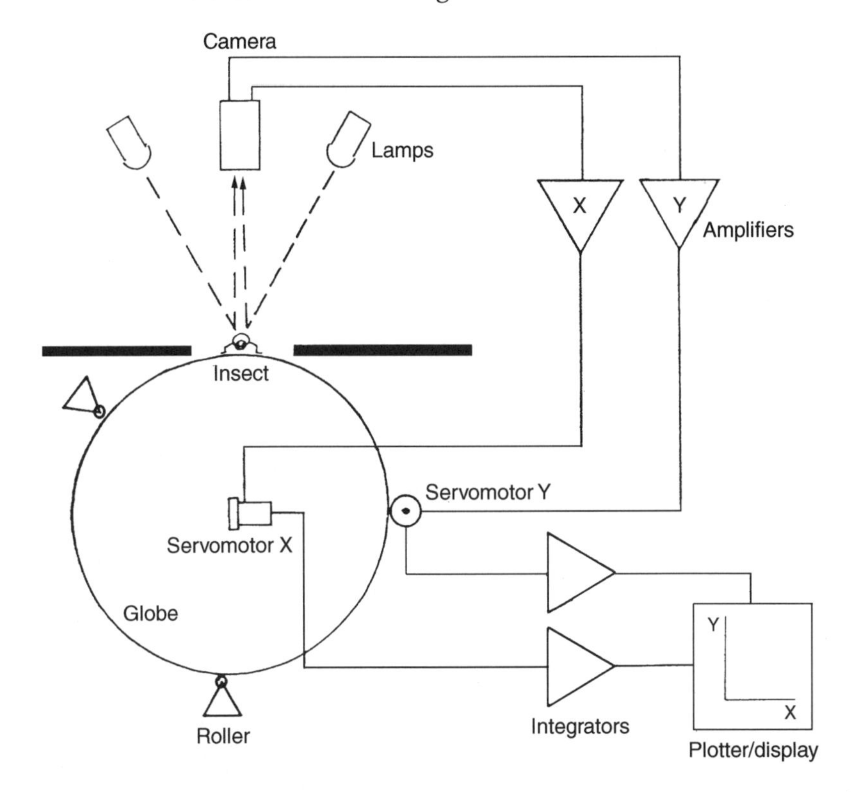


**Fig. 4.8** Principle of the Kramer servosphere method of recording orientation of a walking insect (modified from Erber, 1975). The insect sits on a globe which can be rotated in the horizontal axis by servomotor X and in the vertical axis by servomotor Y. Deviations of the insect from the camera axis are detected and signals amplified by two amplifiers which control the motors and restore the insect to a position in the camera axis. Output of motors is integrated and fed to the plotter or VDU.

regular intervals, the insect must make a left or right turn, and it is then possible to measure the turning tendency over a fixed period.

Completely automated methods using a servosphere were developed by Kramer (1975) from the Y-maze concept. Here, the insect with a reflective disc on its thorax is placed on the surface of a sphere (Fig. 4.8) exactly below a narrow infra-red light beam. The movements of the sphere may be controlled so that the insect stays in the same position in space beneath the beam and the movements of the sphere are then registered automatically by optical sensors which, connected to a computer, record indirectly the movements of the animal. This method has been used for cockroaches (Bell and Kramer, 1980) and for insect parasitoids (Papaj, 1993).

For moths, Priess and Kramer (1984) developed a 'flight simulator'. This consisted of a star-shaped plate maintained initially in a horizontal plane but capable of being tilted in any direction. The moth is mounted on the end of a thin rod, which is suspended magnetically within a tube. The displacement of this rod in a vertical plane is registered by a sensor and used to measure lift, while tilting of the star-shaped plate, acting against the force of retaining springs, measures thrust and flight direction. This method proved to be very useful for continuously registering changes in the flight tracks of a moth in response to controlled optical and olfactory stimulation.

### 4.8.3 Wind tunnels

The advantages of using wind tunnels are that flying insects are able to move freely in three dimensions, and that the conditions inside the wind tunnel can be closely regulated, including the air speed and (within limits) the structure of the pheromone plume. The difficulties lie in the quantification of the response: in the first place, which part of the behaviour sequence or flight parameter to measure, and secondly how to measure it.

A typical wind tunnel is a glass chamber with airflow controlled by one or more fans. Air is cleaned through filters and passed through a honeycomb structure of some kind in order to reduce the turbulence. The versatility of a wind tunnel can be increased if there is a belt with black stripes moving below the floor. The stripes provide visual referents for the insect, and so can be used to induce an optomotor response: if they travel in the same direction that the insect is flying the insect will tend to compensate by increasing its air speed, and vice versa. If the belt is moved at an appropriate speed contrary to the direction of flight of the insect (for example, when it is flying upwind towards a pheromone source), the insect can be made to adopt stationary flight for long periods.

Wind tunnels of this kind have been used with great effect by many researchers. They have the disadvantage that they are expensive to make and difficult to clean. They must be decontaminated regularly because chemicals adsorb on to the glass and other surfaces. In an attempt to overcome some of these problems, Jones *et al.* (1981) designed a simply constructed tunnel without glass, using plastic tubing for the walls (Fig. 4.9). The tubing is inflated by air pressure and can be replaced at any time, minimizing contamination problems.

Even within the standardized environment of a wind tunnel, there are a number of factors that can influence the behaviour of an insect. Carde (1979) has summarized these (Fig. 4.10) for the male gypsy moth, *Lymantria dispar*. The insect shows a sequence of behaviour, beginning

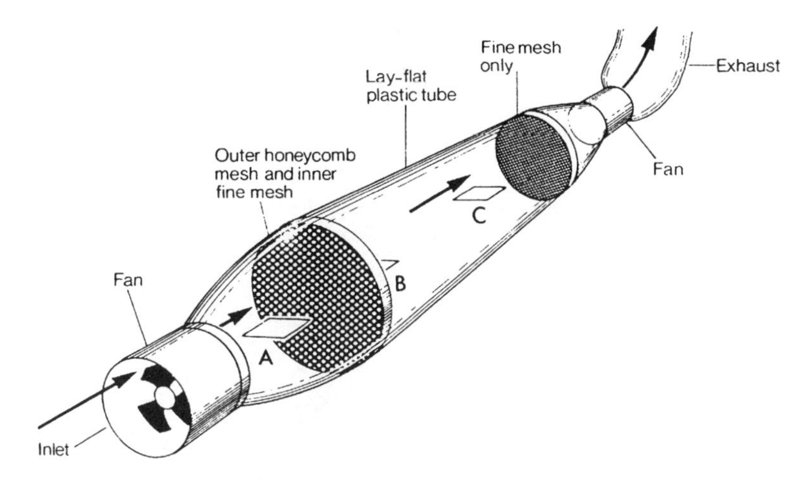


**Fig. 4.9** Simply constructed demountable wind tunnel in which walls are provided by inflatable plastic tubing. A and B are alternative release points for pheromone, depending on the nature of the plume required, and C the release point for insects. (From Jones *et al.*, 1981.)

with an 'activation' response to low concentrations of the pheromone carried to it in the air stream. The threshold for activation can be raised by increasing the light intensity or lowering the temperature. The male starts to walk and then takes off into the wind. It steers into the wind, but requires visual input from the surroundings (mainly from the floor of the wind tunnel) in order to do so and maintain its flight speed. If the concentration of pheromone suddenly falls, the insect is prompted into casting flight at right angles to its path. As it approaches the

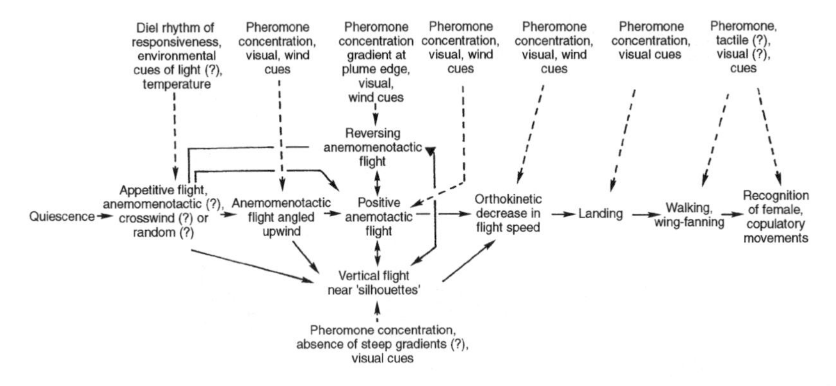


**Fig. 4.10** Factors affecting response of male gypsy moths, *Lymantria dispar*, to a calling female. (Modified from Carde, 1979.)

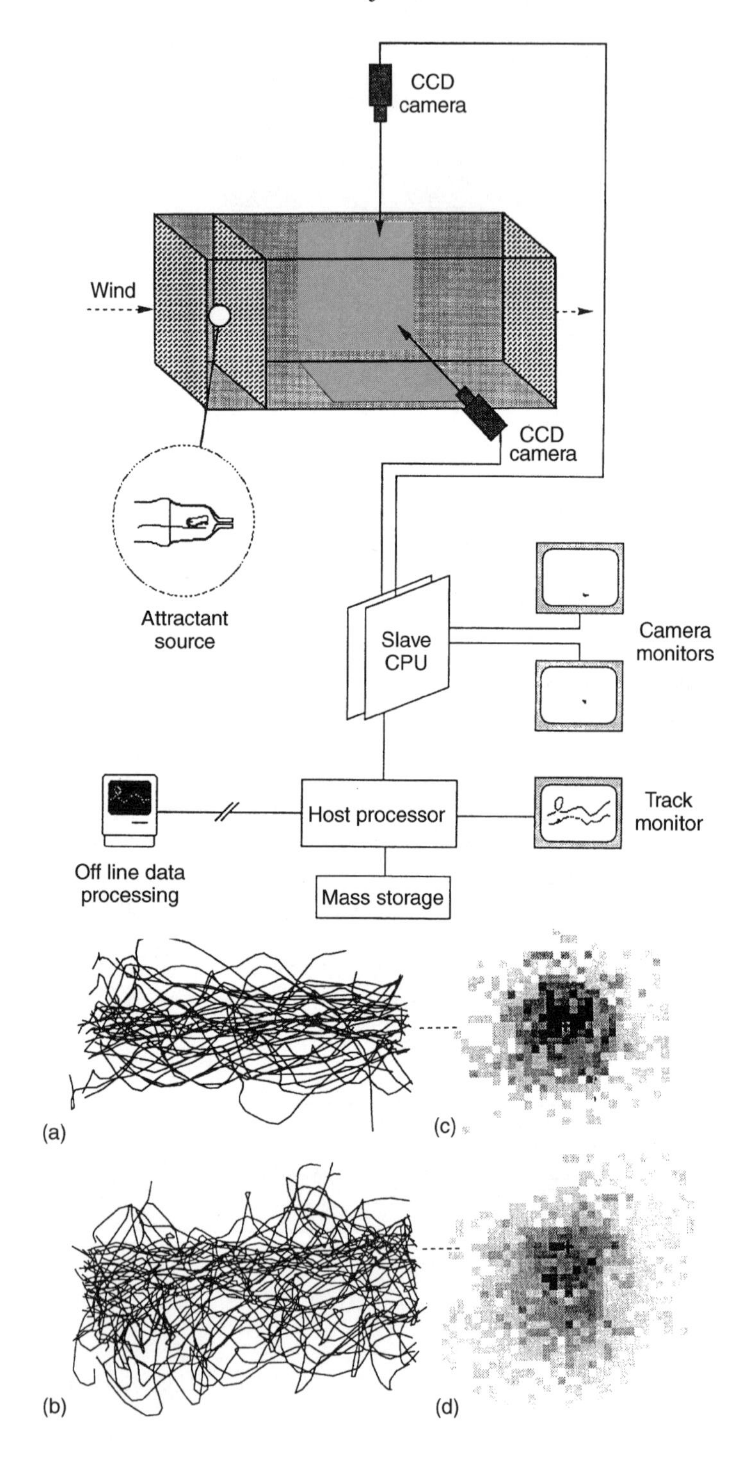
CCD
camera
Wind
CCD
camera
Attractant
source
Slave
CPU
Camera
monitors
Host processor
Track
monitor
Off line data
processing
Mass storage
(a)
(c)
(b)
(d)

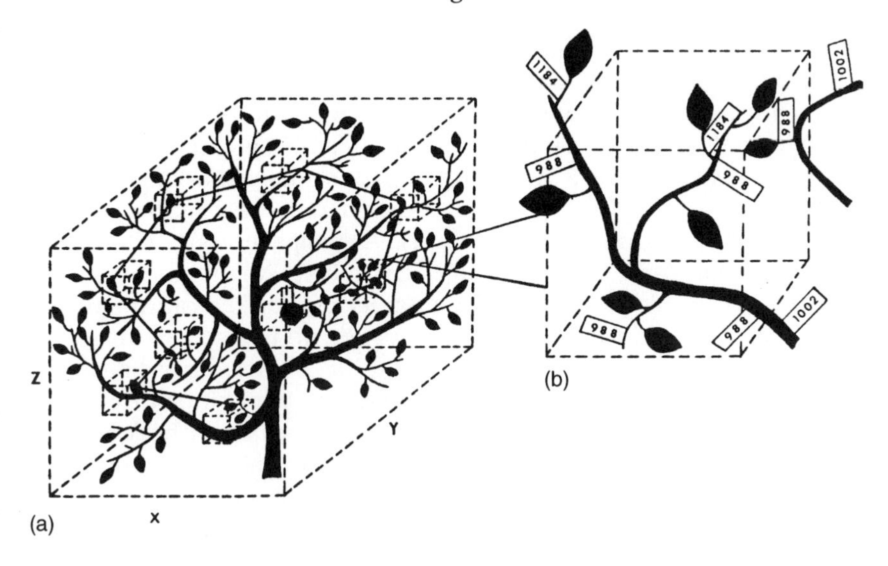


**Fig. 4.12** (a) Apple tree tagged to facilitate plotting of tracks of fruit flies (*Rhagoletis pomonella*). (b) Imaginary cube of space in which tags are numbered according to 3-D coordinates. (From Aluja *et al.*, 1989.)

source, the higher pheromone concentration elicits a reduction in flight speed, and then landing.

This description points to a number of different measurable behaviour patterns. Baker and Carde (1984) have listed 17 types that can be measured with the use of a stop-watch or, better, by analysis of video recordings. These include measurements of flight speed, position of landing, time spent on the source and number of copulatory attempts.

Again, using computers, much information can be gained from video recordings of flight paths in a wind tunnel. Fig. 4.11 shows a system for recording the flight tracks of insects in two axes in a wind tunnel. The information is digitised and fed into a computer where the tracks can be converted to 3-dimentional coordinates for analysis of trajectories and flight velocities. In this way, it was possible to compare the flight

---

**Fig. 4.11** Wind tunnel set up to record flight tracks in three dimensions. Below: flight tracks of male grape berry moths, *Lobesia botrana,* (a) towards calling females; (b) to three-component blend of synthetic pheromone components. (c,d) Respective probabilities of males being present at any one position: intensity of shading represents level of probability. (From Witzgall and Arn, 1990.)

speeds and deviations from the axis of an odour plume of male grape berry moths exposed to pheromone from calling female and blends of synthetic pheromone components (Witzgall and Arn, 1980).

Aluja *et al.* (1989) developed a method for tracking and quantifying the movements of insects in a large flight chamber. Working with the apple maggot, *Rhagoletis pomonella*, they marked every leaf of a caged tree with an individual tag (Fig. 4.12). A complete apple tree 3 m in height was enclosed in a mesh cage. The foliage was arranged so that all the leaves could be seen by an observer outside the cage, and then the *x*, *y* and *z* coordinates were used to divide up foliage into imaginary cubes of space. This allowed production of a three-dimensional tree map, in which individual twigs were identified with strips of masking tape with numbers on them and individual leaves could be identified. Female flies were then released singly into the cage and their movements recorded using a portable tape recorder and a stop-watch. The data were then entered into a computer which had a program capable of analysing the changes in position of flies with time. By this means it was possible to compare factors such as number of flies visiting different kinds of fruit, directness of flight towards a fruit, time spent on the tree before visiting a fruit, etc. This kind of quantification can be used to study many aspects of orientation in semi-natural conditions where automated direct recording methods cannot be used.

Bioassays in field cages can also be used with insects such as fruit flies. Although climatic conditions cannot be standardized, many other factors can be: if the cage is put around vegetation, other insects can be first removed with a non-persistent insecticide, and then known numbers of test insects of which the age and reproductive status are known can be introduced and their behaviour observed.

A method for tracking free-flying insects in the field in 3-D has been described by Riley *et al.* (1990). They followed the tracks of the noctuid moth *Helicoverpa (= Heliothis armigera)* over crops in India by illuminating the area with infra-red light and recording tracks stereoscopically with two video cameras.

## REFERENCES

Aluja, M.S., Prokopy, R.J., Elkinton, J.S. and Laurence, F. (1989) Novel approach for tracking and quantifying the movement patterns of insects in three dimensions under seminatural conditions. *Environ. Ent.*,**18**, 1–7.

Baker, T.C. and Carde, R.T. (1984) Techniques for behavioural bioassays. In *Techniques in Pheromone Research* (eds H.E. Hummel and T.A. Miller), Springer-Verlag, New York, pp. 45–73.

Baker, T.C. and Haynes, K.F. (1989) Field and laboratory electroantennographic measurements of pheromone plume structure correlated with oriental fruit moth behaviour. *Physiol. Ent.*, **14**, 1–12.

Bell, W.E. and Kramer, E. (1980) Sex pheromone-stimulated orientation of the American cockroach on a servosphere apparatus. *J. Chem. Ecol.*, **6**, 287–295.
Bell, W.E. and Tobin, T.R. (1982) Chemo-orientation. *Biol. Rev.*, **57**, 219–260.
Birch, M.C. (1984) Aggregation in bark beetles. In *Chemical Ecology of Insects* (eds W.J. Bell and R.T. Carde), Chapman & Hall, London and New York, pp. 331–354.
Block, E.F. and Bell, W.J. (1974) Ethometric analysis of pheromone receptor function in cockroaches. *J. Insect Physiol.*, **20**, 993–1003.
Bossert, W.H. and Wilson, E.O (1963) The analysis of olfactory communication among animals. *J. Theor. Biol.*, **5**, 443–469.
Bradshaw, J.S. (1981) The physiochemical transmission of two components of a multiple chemical signal in African weaver ant *(Oecophylla longinoda). Anim. Behav.*, **29**, 581–585.
Carde, R.T. (1979) Behavioural responses of moths to female produced pheromones and the utilization of attractant-baited traps for population monitoring. In *Movement of Highly Mobile Insects: Concepts and Methology in Research* (eds R.L. Rabb and G.G. Kennedy), North Carolina State University Press, Raleigh, pp. 286–315.
Contreras, M.L., Perez, D. and Rozaz, R. (1989) Empirical correlations between electroantennograms and bioassays for *Periplaneta americana. J. Chem. Ecol.*, **15**, 2539–2548.
Dethier, V.G. (1947) *Chemical Insect Attractants and Repellents*, Blakiston, Philadelphia.
Dingle, H. (1964) Further observations on correcting behaviour in boxelder bugs. *Anim. Behav.*, **12**, 116–112.
Erber, J. (1975) Turning behaviour related to optical stimuli in *Tenebrio molitor. J. Insect Physiol.*, **21**, 1575–1580.
Fraenkel, G.S. and Gunn, D.L. (1940) *The Orientation of Animals*, Oxford University Press, Oxford.
Greenway, A.R. (1984) Sex pheromone of the pea moth, *Cydia nigricana* (F), (Lepidoptera: Olethreutidae). *J. Chem. Ecol.*, **10**, 973–982.
Hummel, H.E. and Miller, T.A. (eds) (1984) *Techniques in Pheromone Research*, Springer-Verlag, New York.
Jones, O.T., Lomer, R.A and Howse, P.E. (1981) Response of male Mediterranean fruit flies, *Ceratitis capitata*, to trimedlure in a wind tunnel of novel design. *Physiol. Ent.*, **6**, 175–181.
Katsoyannos, B.L., Boller, E.F. and Remund, U. (1980) A simple olfactometer for the investigation of sex pheromones and other olfactory attractants in fruit flies and moths. *Z. ang. Ent.*, **90**, 105–112.
Kennedy, J.S. (1977) Behaviourally discriminating assays of attractants and repellents. In *Chemical Control of Insect Behaviour* (eds H.H. Shorey and J.J. McKelvey, Jr), Wiley, New York, pp. 215–259.
Kramer, E. (1975) Orientation of the male silkmoth to the sex attractant bombykol. In *Orientation and Taste, V* (eds D.A. Denton and J.P. Coghlan) Academic Press, New York, pp. 329–335.
Lofstedt, C., Van der Pers, J.N.C. and Lofquist, J. (1982) Sex pheromone components of the turnip moth, *Agrotis segetum*: chemical identification, electrophysiological evaluation and behavioural activity. *J. Chem. Ecol.*, **8**, 1305–1321.
Obrecht, C. and Hanson, F.E. (1989) Instrumental measurement of the gypsy moth pre-flight behavioural response to pheromone. *Physiol. Ent.*, **14**, 187–193.
Papaj, D.R. (1993) Automatic behaviour and the evolution of instinct: lessons

from learning in parasitoids. In *Insect Learning* (eds D.R. Papaj and A.C Lewis), Chapman & Hall, New York and London, pp. 243–272.

Priess, R. and Kramer, E. (1984) Pheromone-induced anemotaxis in simulated free flight. In *Mechanisms of Insect Olfaction* (eds T.L. Payne, M.C. Birch and C.E.J. Kennedy), Clarendon Press, Oxford, pp. 69–80.

Riley, J.R., Smith, A.D. and Bettany, B.W. (1990) The use of video equipment to record in three dimensions the flight trajectories of *Heliothis armigera* and other moths at night. *Physiol. Ent.*, **15**, 73–80.

Roelofs, W.L. (1984) Electroantennogram assays: rapid and convenient screening procedures for pheromones. In *Tehniques in Pheromone Research* (eds H.E. Hummel and T.A. Miller). Springer-Verlag, New York, pp. 131–139.

Roelofs, W.L. and Carde, R.T. (1974) Oriental fruit moth and lesser appleworm attractant mixtures redefined. *Environ. Ent.*, **3**, 586–588.

Rust, M.K., Burk, T. and Bell, W.J. (1976) Pheromone stimulated locomotory and orientation responses in the American cockroach, *Periplaneta americana. Anim. Behav.*, **24**, 52–67.

Schneider, D. (1957) Elektrophysiologische Untersuchung von Chem- und Mechanorezeptoren der Antenne des Seidenspinners *Bombyx mori* L. *Z. vergl. Physiol.*, **40**, 8–41.

Seelinger, G. (1985) Interspecific attractivity of female sex pheromone components of *Periplaneta americana. J. Chem. Ecol.*, **11**, 137–148.

Seelinger, G. and Gagel, S. (1985) On the function of sex pheromone components in *Periplaneta americana*: improved odour source localization with periplanone-A. *Physiol. Ent.*, **10**, 221–234.

Traniello, J.F.A. and Busher, C. (1985) Chemical regulation of polyethism during foraging in the noetropical termite *Nasutitermes costalis. J. Chem. Ecol.*, **11**, 319–332.

Tumlinson, J.H. (1988) Contemporary research frontiers in insect semiochemical research. *J. Chem. Ecol.*, **14**, 2109–2129.

Van de Pers, J.N.C. and Minks, A.K. (1993) Pheromone monitoring in the field using single sensillum recording. *Entomol. Exp. Appl.*, **68**, 237–245.

Vilela, E.F., Jaffe, K. and Howse, P.E. (1987) Orientation in leaf-cutting ants (Formicidae: Attini). *Anim, Behav.* **35**, 1443–1453.

Wall, C., Greenway, A.R. and Burt, P.E. (1976) Electroantennographic and field responses of the pea moth, *Cydia nigricana,* to sex attractants and related compounds. *Physiol. Ent.*, **1**, 151–157.

White, J., Tobin, T.R. and Bell, W.J. (1984) Local search in the housefly *Musca domestica* after feeding on sucrose. *J. Insect Physiol.*, **30**, 477–487.

Witzgall, P. and Arn, H. (1991) Recording flight tracks of *Lobesia botrana* in the wind tunnel. In *Insect Chemical Ecology* (ed. I. Hrdý), SPB Academic Publishing bv. Czechoslovakia, pp. 187–196.

# *Part Two*

# Chemical Aspects of Pheromones

*I.D.R. Stevens*

## INTRODUCTION

A major problem with all interdisciplinary research areas is that of vocabulary: each separate discipline has built up a system in which words are used in a specialized and specific sense, which is often not that of normal speech. The result of this is that it is important that the specialist from each discipline should acquire a working knowledge of the specialized vocabulary that is used to describe aspects that are of common interest to them both. Hence, within each of the sections of this chapter, the early part is devoted to a brief explanation of the nomenclature that the chemist employs in order to define the exact structure of a compound and also to the conventions that are used to abbreviate the rather lengthy systematic names for chemical structures.

The assumption is made that the reader is familiar with the concept of valency, the nature of co-valent bonding and the fact that carbon is 4-valent, nitrogen 3-valent, oxygen di-valent and hydrogen uni-valent.

# 5

# Chemical structures and diversity of pheromones

## 5.1 SIMPLE (STRAIGHT CHAIN) STRUCTURES AND FUNCTIONAL GROUPS

The basis for all structures of organic compounds is the hydrocarbon chain, in which carbon atoms are linked to each other in a linear fashion and the remaining valences of the carbons are satisfied by being bound to hydrogen – structure (**5.1.1**).

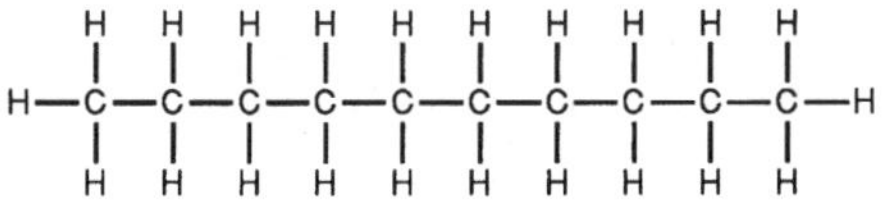

(**5.1.1**)

The bonds attached at each carbon are not, however, arranged in a planar manner, but are disposed tetrahedrally. This means that if one looks only at the carbon skeleton, the atoms lie in a zigzag arrangement as shown in structure (**5.1.2**). The hydrogen atoms then lie in two planes either above or below that defined by the central carbon chain, as in structure (**5.1.3**); the bonds to atoms lying above the plane of the paper are shown by solid wedges and those below by dashed wedges.

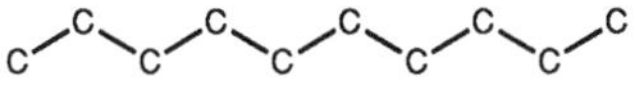

(**5.1.2**)

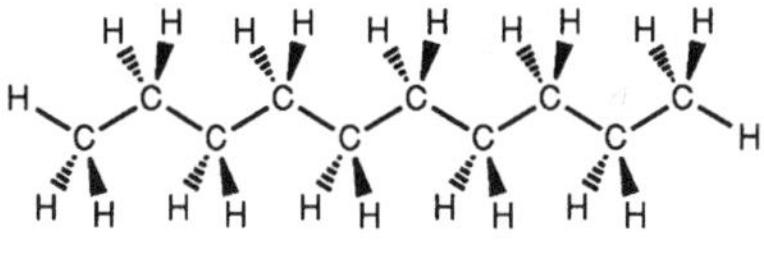

(**5.1.3**)

This leads to a simplified representation of such a molecule in which only the bonds to carbon are shown and those to the hydrogens are implied, thus structure (**5.1.4**) represents the same molecule as (**5.1.3**).

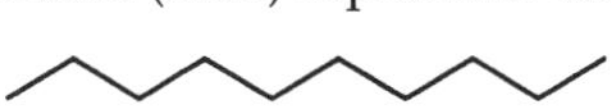

(**5.1.4**)

Skeletal structures like this will be used throughout this book.

The systematic nomenclature for hydrocarbon structures depends on the number of carbon atoms found in the straight part of the carbon chain, using the Greek-based numbering system, and is that devised by the International Union of Pure and Applied Chemistry (IUPAC).

Using this system, the compound shown in (**5.1.1**) is **decane**. The name is built from the prefix **dec**, indicating 10 (carbons), and from the ending **-ane**, showing that it is a hydrocarbon with no multiple bonds. The chain lengths most commonly found in pheromones are 10, 12 (**dodeca-**), 14 (**tetradeca-**), 16 (**hexadeca-**) and 18 (**octadeca-**). Appendix 5A at the end of the chapter gives the prefixes up to C 30.

### 5.1.1 Alcohols, esters and aldehydes

When one (or more) of the hydrogens in a hydrocarbon is replaced by the atom of another chemical element, this changes the chemistry associated with the compound and, as a consequence of this change, we talk of this introduction of the other element as creating a **functional group**.

A very widely found simple change is the introduction of one atom of oxygen into the structure to create the **alcohol** functional group, as in (**5.1.5**). The structure is shown in both full (**5.1.5a**) and skeletal form (**5.1.5b**). In the latter the alcohol group stands out clearly.

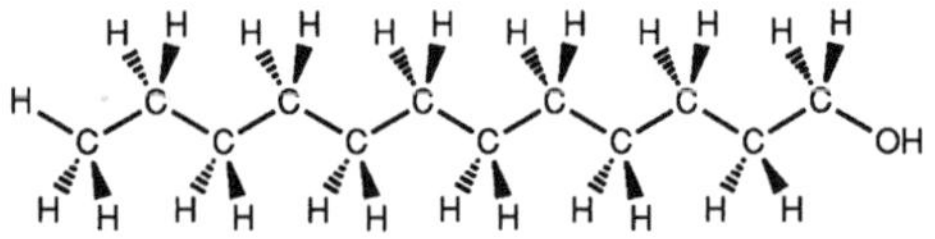

(**5.1.5a**)

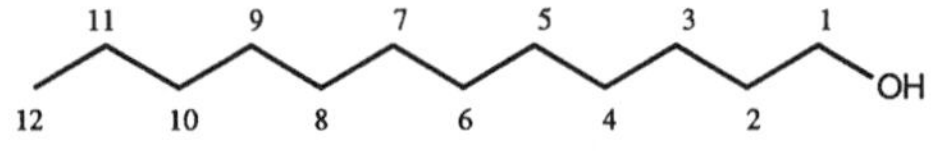

(**5.1.5b**)

In the majority of lepidopteran pheromones this alcohol functional group is found at the end of the chain and takes precedence in the name of the compound, appearing as the ending **-ol**; thus (**5.1.5**) is **dodecan-1-ol.** Each carbon atom along the chain is given a number, starting from the end of the chain nearest to the functional group (as in **5.1.5b**). hence the designator 1, showing that the alcohol functionality is at the end of the chain.

Compound (**5.1.5**) is an attractant for the males of the tea tortrix moth, *Homona coffearia*.

Other simple alcohols which act as attractants, or are found in female emissions, are **tetradecan-1-ol** (**5.1.6**) (from the sunflower moth, *Homoe-*

*soma electellum*) and **hexadecan-1-ol (5.1.7)** (from the cotton bollworm, *Heliothis armigera*).

OH (5.1.6)

OH (5.1.7)

Two chemical modifications of the alcohol group are also commonly found. In the first of these the alcohol is combined with an organic acid (with loss of water) to give an **ester**, as in (**5.1.8**). here we may recognize the alcohol part, which is (**5.1.5**), and the acid part, which is derived from acetic acid ($CH_3COOH$). The name is derived from the two parts, thus (**5.1.8**) is **dodecan-1-ol acetate**, which is found to act as a synergist for (**5.1.5**) in the attraction of *Homona coffearia* males and is the major component in the pheromone of the potato stem borer, *Hydraecia micacaea*. (Strict IUPAC nomenclature requires the replacement of the alcohol ending by **-yl** and the movement of the position designator to the beginning of the name. Thus (**5.1.8**) should be 1-dodecyl acetate; however, both names are recognized.)

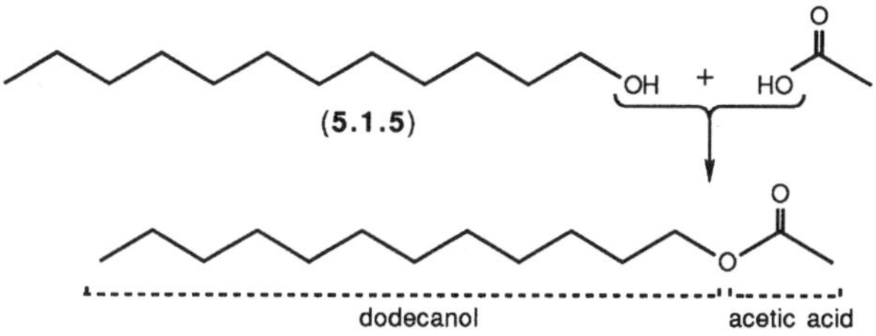

OH + HO O
(5.1.5)
O
O (5.1.8)
dodecanol acetic acid

Acetates are by far the most commonly occurring esters, but others that are found are the esters of formic acid (HCOOH), which gives rise to **formates**, and those of propanoic acid ($CH_3CH_2COOH$), which forms **propanoates**. Esters of longer chain or more complex acids are also found and these will be discussed later.

The second modification is the oxidation of the alcohol to an **aldehyde**, such as (**5.1.9**), which is one component of the pheromone mixture used by the tobacco budworm, *Heliothis virescens*. Chemically aldehydes are recognized by the —CH=O group, always found at the end of the chain, and in their names by the ending **-al**, which follows the chain length designator. Thus compound (**5.1.9**) is **tetradecanal**. (No position number is required, because the aldehyde group can only be at the end of the chain.)

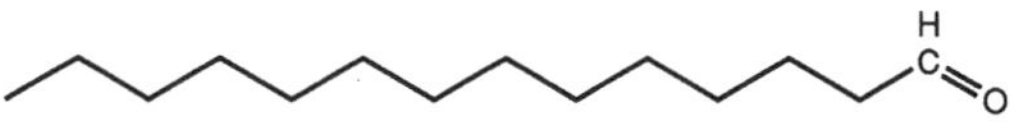

H
C
O (5.1.9)

The aldehyde group is chemically quite reactive in air and is readily oxidized further to the **carboxylic acid** group, —COOH. This change is of relevance both to the investigator, who may thus easily be misled by its occurrence during isolation of a pheromonal extract, and to the insect, where it serves to ensure that a 'message' is not persistent.

Acids are named using the chain length stem and the ending **-oic acid**. Thus hexanoic acid (**5.1.10**) is the female-produced sex pheromone from the Pacific coast wireworm, *Limonius canus*.

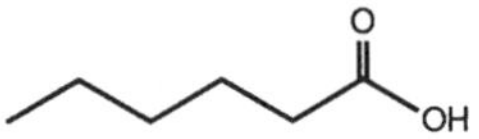

O OH (5.1.10)

Esters derived from these acids are named by placing the alcohol name first and adding the acid name with the ending changed to **-oate**.

This change of oxidation level from alcohol to aldehyde and from aldehyde to carboxylic acid is very often used to alter a semiochemical message from one of attraction to one of repulsion. Thus in many plant species an alcohol will act as an attractant for pollinators and the corresponding aldehyde as a repellent, the change happening once pollination has occurred; for example, (*E*)-2-hexen-1-ol (**5.1.11**) and (*E*)-2-hexenal (**5.1.12**).

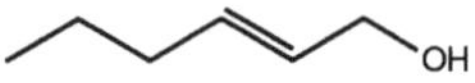

OH

(5.1.11)

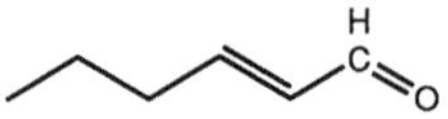

H C O

(5.1.12)

The reverse of this, where an aldehyde is an attractant and the corresponding alcohol acts as an inhibitor, has also been observed as with the astern spruce budworm, *Choristoneura fumiferana*, which uses (*E*)-1–tetradecenal (**5.1.13**) as an attractant and in which trap catches are strongly inhibited by the corresponding alcohol, (*E*)-11-tetradecen-11-ol (**5.1.14**).

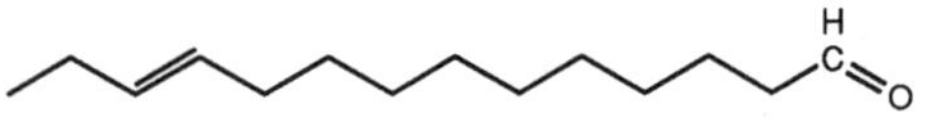

H C O (5.1.13)

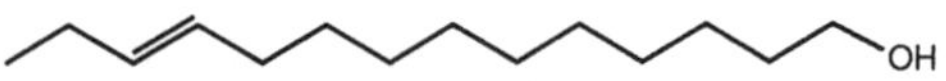

OH (5.1.14)

### 5.1.2 Double-bonds, positional and geometric isomerism

Another change which is commonly seen in the structures of pheromones is when two hydrogens are removed from adjacent carbon atoms in the chain, thus creating a **double-bond** between these two atoms: C=C. The double-bond is also a functional group in the molecule, but is one which ranks lower in precedence than an oxygen or nitrogen containing group.

Such a double-bond may occur anywhere along the hydrocarbon chain and, as a result, introduces the possibility of **positional isomerism**. This may be illustrated by compounds (**5.1.15**) and (**5.1.16**), which are minor constituents of the emissions of the oriental fruit moth, *Grapholita molesta*, and the Douglas fir cone moth, *Barbara colfaxiana*, respectively. The position of the double-bond is an important factor in species differentiation, allowing each species to recognize signals from members of its own kind and to distinguish them from those of other species. The method of distinction is almost certainly related to the number of carbons between the double-bond and the alcohol group at the end of the chain. In naming these compounds, we indicate the presence of the double-bond by changing the hydrocarbon ending from -**ane** to -**ene** and, to show the position of the double-bond, we use the number of the carbon atom at which the double-bond begins. Remembering that the numbering of the molecule starts from the alcohol end (section 5.1.1), we may thus name compound (**5.1.15**) as 8-dodecen-1-ol and (**5.1.16**) as 9-dodecen-1-ol.

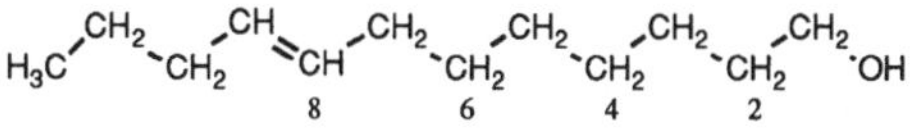

$H_3C-CH_2-CH_2-CH=CH-CH_2-CH_2-CH_2-CH_2-CH_2-CH_2-CH_2-OH$ (numbered 8, 6, 4, 2) **(5.1.15)**

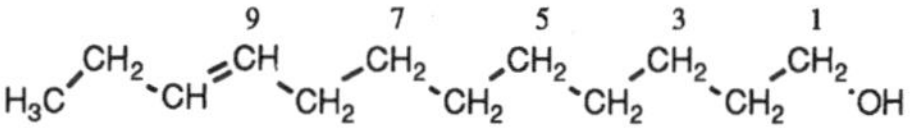

$H_3C-CH_2-CH=CH-CH_2-CH_2-CH_2-CH_2-CH_2-CH_2-CH_2-CH_2-OH$ (numbered 9, 7, 5, 3, 1) **(5.1.16)**

A second factor which is introduced as a result of the double-bond flows from the fact that the relative geometrical positions of the groups attached at either end of the double-bond are fixed. In contrast to a molecule with only single bonds, where the chains are free to rotate about the single bonds, in a molecule with a double-bond no rotation of the two ends of the bond relative to each other is possible, Thus,

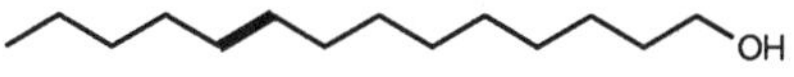

OH **(5.1.17)**

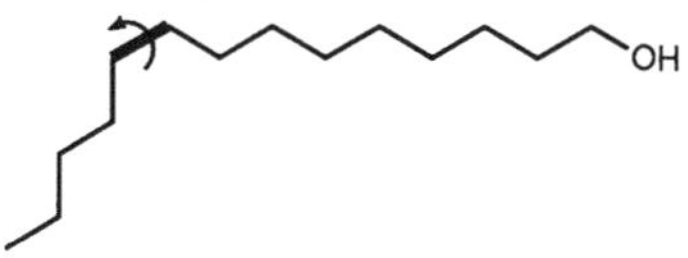

OH **(5.1.18)**

whereas (**5.1.17**) and (**5.1.18**) are the same molecule and may be twisted one into the other by rotation about the bond indicated, (**5.1.19**) and (**5.1.20**) are different molecules and cannot be transformed one into the other (except at very elevated temperatures). These two are **geometric isomers** of each other and this difference in geometry is readily perceived by an insect, which will normally respond only to one isomer and not to the other.

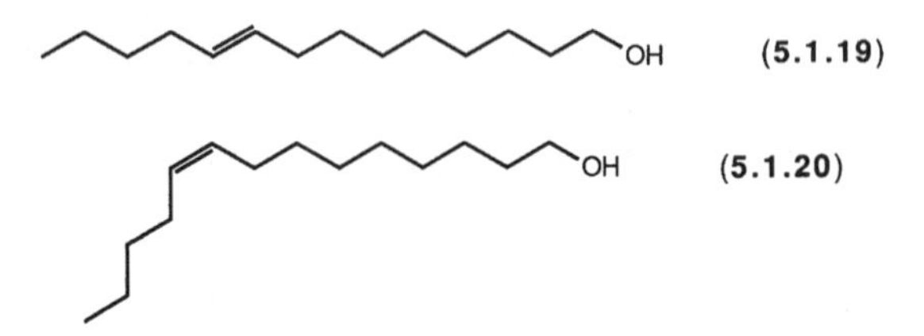


In naming geometric isomers, we use the designations *E* and *Z* to show whether the carbon chains that are attached are on opposite (*E*) or the same (Z) side of the double-bond. The designators are taken from the German words *Entgegen* (opposite) and *Zusammen* (together). Thus (**5.1.19**) with the chains on opposite sides is (*E*)-9-tetradecen-1-ol and (**5.1.20**) is (Z)-9-tetradecen-1-ol. The former is an attractant for *Epiplema plagifera,* while the latter is a major component from the sunflower moth, *H. electellum*. For tri- and tetra-substituted double-bonds, a precedence rule is used to decide which chains shall be used (section 5.2.1a).

When there is more than one double-bond in the molecule, then each of them needs to be defined in the same manner as for an isolated double-bond and we now include a multiplier in the name to show the number of double-bonds present. Hence (**5.1.21**), a minor component of the volatile effluvium from *H. electellum,* is named (Z,*E*)-9,12-tetradeca-dien-1-ol and the two compounds (**5.1.22**) and (**5.1.23**), used as a mixture by the potato tuberworm, *Phthorimaea operculella*, are (*E*,*Z*)-4,7-tridecadien-1-ol acetate and (*E*,*Z*,*Z*)-4,7,10-tridecatrien-1-ol acetate, respectively, the endings **diene** and **triene** showing that there are either two (di-) or three (tri-) double-bonds.

OH (5.1.21)

O
O (5.1.22)

O
O (5.1.23)

Note that the stereochemical designators are grouped together and placed in the same order as the position numbers of the double-bonds to which they refer. Hence (**5.1.21**) is not the same as (*E*,*Z*)-9,12-tetradecadien-1-ol, which is (**5.1.24**) and is a compound that has not yet been found in any pheromonal mixture.

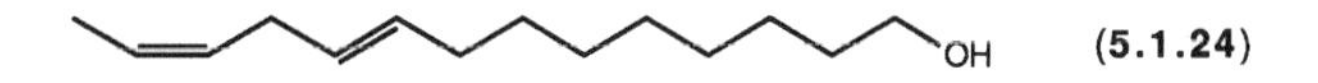


### 5.1.3 Abbreviations, acronyms and trivial names

As with all groups of specialists, workers in the field of insect semiochemicals have tended to build up a system of abbreviating the strict chemical names of the compounds that occur frequently and particularly so when dealing with these straight chain compounds. For the most part this has been done in a fairly methodical manner and two systems are in common use.

In the first, the IUPAC chemical name is shortened by taking the initial letters of the two numbers of the chain length designator and adding either the functional group ending or A if the functional group is an acetate. Hence compound (**5.1.5**), **d**o**d**ecanol would become DDOl with the bold letters capitalized and the ending retained. The corresponding acetate (**5.1.8**) would be abbreviated to DDA and the aldehyde as DDAl.

The presence of one double-bond is shown solely by the stereochemical and positional indicators discussed in section 5.1.2 and hence (**5.1.15**) and (**5.1.16**) become 8-DDOl and 9-DDOl, respectively. In fact *G. molesta* and *B. colfaxiana* each produce the Z-isomer of the relevant alcohol, i.e. the isolated components from these species are (Z)-8-DDOl and (Z)-9-DDOl. When several double-bonds are present, then an additional letter is added to the abbreviation to show the number of double-bonds present, with D for di- and T for tri- taken from the endings diene or triene. This may be illustrated with compounds (**5.1.21**), (*Z,E*)-9,12-tetradecadien-1-ol, and (**5.1.23**), (*E,Z,Z*)-4,7,10-tridecatrien-1-ol acetate, which are abbreviated to (*Z,E*)-9,12-TDDOl and (*E,Z,Z*)-4,7,10-TriDTA. This last example shows up the problems inherent in this method of shortening a name, for both tri- and tetra- have the same initial letter. Luckily the tridecane compounds form the only examples so far known. Straight chains with odd numbers of carbon atoms are very uncommon, otherwise confusion between hexadeca- (C-16) and heptadeca- (C-17) would also cause problems.

The second method uses a simple number to show the chain length and abbreviates the functional group; thus alcohol becomes -OH, aldehyde -Al and acetate -Ac. The position and stereochemistry of the double-bonds are shown as in the first method and the multiplier for the number of double-bonds is omitted. Hence, using the same set of examples as above, the saturated compounds (**5.1.5**) and (**5.1.8**) would be written as 12:OH and 12:Ac. The unsaturated compounds (**5.1.15**) and (**5.1.16**) become (Z)8–12:OH and (Z)9–12:OH, while (**5.1.21**) and (**5.1.23**) are written as (Z)9(*E*)12–14:OH and (*E*)4(Z)7(Z)10–13:Ac. This method has the advantage that the chain length is immediately clear, but one does need to be careful not to confuse the double-bond position indicator and the number showing the chain length.

Whereas the first method has been restricted to the three functional groups of alcohol, acetate and aldehyde, this second technique for abbreviating has been extended to cope with the functional groups of **ketone** (shown as Kt) and **epoxide** (as epo) and also to show hydrocarbons, where there are no oxygen functionalities, as the ending Hy. See section 5.1.4, for the nature of these groups.

This second style of shortening names has borrowed the abbreviation 'Ac' for acetate from the same one that is used much more widely in the chemical literature but uses it with a very significant change. All chemists employ the abbreviation 'Ac' to represent the grouping $H_3C{-}C{=}O$, which is acetyl. Hence, in a chemical structural formula, an acetate such as *n*-pentyl acetate would be written as $H_3C{-}(CH_2)_3{-}CH_2{-}O{-}Ac$, the O being essential to represent the **ester** group, for the formula $H_3C{-}(CH_2)_3{-}CH_2{-}Ac$ indicates a **ketone** – 2-heptanone (see section 5.1.4 for definition). This means that workers in the pheromone area need to be particularly careful both where and in which context they use these abbreviations.

Unfortunately, there is a growing tendency to refer to compounds by a trivial name, usually derived from the Latin name for the insect in question, shortened and with the addition of the ending **-lure**. Thus the mixture of (**5.1.25**) and (**5.1.26**), (Z,*E*)-7,11-hexadecadien-1-ol acetate and (Z,Z)-7,11-hexadecadien-1-ol acetate, used by the pink bollworm moth, *Pectinophora gossypiella*, is commonly referred to as **Gossyplure**.

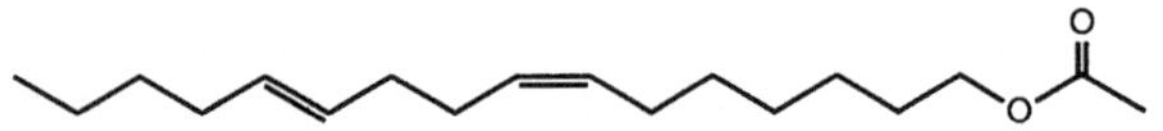

(5.1.25)

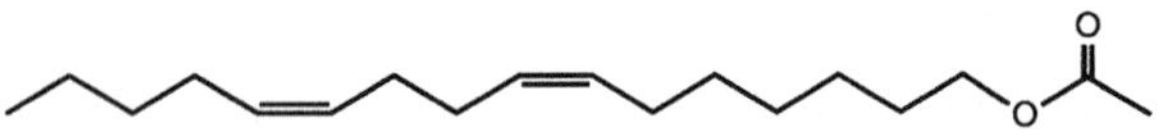

(5.1.26)

This obviously gives no clue as to the chemical structure and should be avoided. A major reason for this practice has been the pressure from commercial suppliers of pheromone-based insect control systems, where the need for a simple name for a product gives a marketing advantage.

### 5.1.4 Pheromones from females of lepidopteran species

In the vast majority of cases it has been observed that the pheromonal secretions from the females of lepidopteran species consist of compounds with straight carbon chains. The diversity amongst them is created by the interplay of five factors: the chain length; the functional group; the position of the double-bonds; the number of the double-bonds; and their stereochemistry.

Figure 5.1 gives a selection of compounds that have been identified from these secretions and the species from which they have been isolated. The table is not exhaustive and the compounds have been selected only to illustrate the range of structural complexity that has been identified. It may be noted that the five factors mentioned are quite sufficient to provide a formidable array of compounds and also to allow each species to define itself. In a substantial number of cases, certain genera are found to use the same major component and then the species differentiation is achieved by the addition of one or more other straight chain compounds at a specific ratio (see Chapter 8 for examples).

In addition, it has been found that certain chain lengths and double-bond positions are used by many families. Particularly common are: (Z)-7-dodecen-1-ol acetate, identified in 44 species (28 genera); (Z)-8-dodecen-1-ol acetate, in 47 species (22 genera); (Z)-9-tetradecen-1-ol acetate in 52 species (36 genera); (Z)-11-tetradecen-1-ol acetate in 65 species (37 genera); (*E*)-11-tetradecen-1-ol acetate in 58 species (38 genera); and (Z)-11-hexadecen-1-ol acetate, an attractant for 39 species (31 genera).

For all of these insects, the use of these straight chain compounds is highly likely to have its origin in their ready biosynthesis either directly from fatty acids or by using the same biochemical pathways by which the latter are formed (Baker, 1989; Tumlinson, 1988).

The compounds in Fig. 5.1 are classified by the functional group present and the degree of unsaturation. Within each class, they are ordered by chain length. For each compound, the name of one species for which the compound is a major pheromonal component is given. The majority of the compounds contain chains of even numbers of carbon atoms because fatty acids are built up from the two-carbon unit of acetic acid ($CH_3CO_2H$) and hence, as indicated in section 5.1.2, chains of odd numbers of carbons are very uncommon. However, it may be noted that those compounds which do not have an oxygen at C-1 (Fig. 5.1.3 to 5.1.5) are almost exclusively of odd chain length and a plausible explanation for this is that these compounds are formed by the decarboxylation (loss of $CO_2$) of the fatty acid containing one more carbon than the pheromone.

The first part of Fig 5.1 (i.e. 5.1.1a–c) consists of unsaturated acetates, alcohols and aldehydes. Although a number of saturated compounds with these functionalities have been found, the majority are rather minor components and have therefore not been illustrated (the most

**Fig. 5.1 Female sex pheromones of Lepidoptera**

***5.1.1 Acetates, alcohols and aldehydes***

*(a) Mono-unsaturated*

| Entry | Compound | Insect |
|---|---|---|
| 5.1.1 |  $OCOCH_3$ | Turnip moth (*Agrotis segetum*) (France) |
| 5.1.2 | 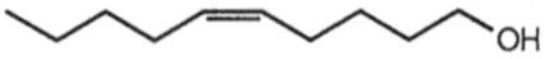 OH | |
| 5.1.3 |  $OCOCH_3$ | Peach twigborer (*Anarsia lineatella*) |
| 5.1.4 | 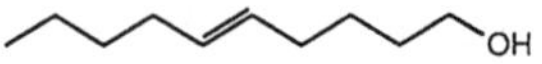 OH | |
| 5.1.5 |  $OCOCH_3$ | Turnip moth (*Agrotis fucosa*) |
| 5.1.6 |  $OCOCH_3$ | Potato tuber moth (*Scrobipalpopsis solanivora*) |
| 5.1.7 | 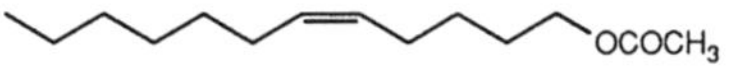 $OCOCH_3$ | European goat moth (*Cossus cossus*) |
| 5.1.8 |  $OCOCH_3$ | Cabbage looper (*Trichoplusia ni*) |
| 5.1.9 |  $OCOCH_3$ | *Grapholita* species |
| 5.1.10 | 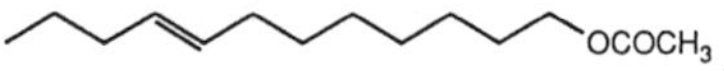 $OCOCH_3$ | |
| 5.1.11 | 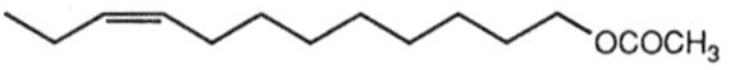 $OCOCH_3$ | Grape berry moth (*Endopiza viteana*) |
| 5.1.12 |  OH | Pine tip miner (*Rhyacionia zozana*) |
| 5.1.13 |  $OCOCH_3$ | *Agrotis (Scotia) exclamationis* |
| 5.1.14 | 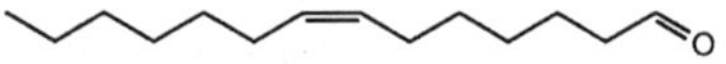 O | Citrus flower moth (*Prays citri*) |
| 5.1.15 |  $OCOCH_3$ | Summer fruit tortrix (*Adoxophyes orana*) |
| 5.1.16 |  $OCOCH_3$ | |
| 5.1.17 | 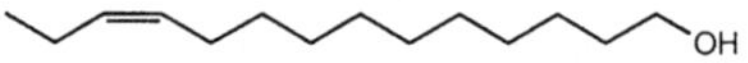 OH | Potato stem borer (*Hydraecia micacaea*) |
| 5.1.18 | 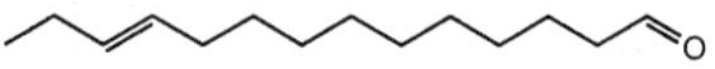 O | Spruce budworm (*Choristoneura fumiferana*) |
| 5.1.19 |  $OCOCH_3$ | Asian corn borer (*Ostrinia furnacalis*) |
| 5.1.20 | 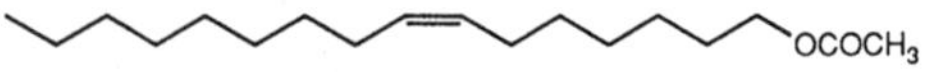 $OCOCH_3$ | Darksided cutworm (*Euxoa messoria*) |
| 5.1.21 |  $OCOCH_3$ | Rice green caterpillar (*Naranga aenescens*) |
| 5.1.22 |  $OCOCH_3$ | |
| 5.1.23 |  O | Corn earworm (*Heliothis zea*) |
| 5.1.24 |  O | |
| 5.1.25 |  OH | Pickleworm (*Diaphania nitidalis*) |
| 5.1.26 |  O | |

| **Entry** | **Compound** | **Insect** |
|---|---|---|
| 5.1.27 | 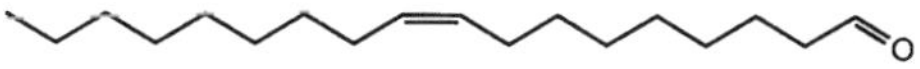 O | Yellow stem borer (*Scirpophaga incertulas*) |
| 5.1.28 |  O | Spotted bollworm (*Earias vitella*) |
| 5.1.29 |  $OCOCH_3$ | Sugar-cane borer (*Chilo sacchariphagus*) |
| 5.1.30 | 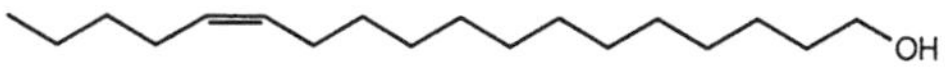 OH | |
| 5.1.31 |  O | South western corn borer (*Diatrea grandiosella*) |

*(b) Di-unsaturated*

(i) Conjugated

| **Entry** | **Compound** | **Insect** |
|---|---|---|
| 5.1.32 | 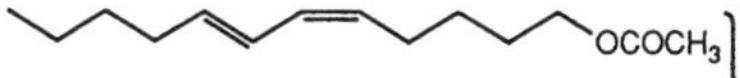 $OCOCH_3$ | Pine caterpillar (*Dendrolimus punctatus*) |
| 5.1.33 |  OH | |
| 5.1.34 |  O | Western tent caterpillar (*Malacosoma californicum*) |
| 5.1.35 |  $OCOCH_3$ | Grape berry moth (*Lobesia botrana*) |
| 5.1.36 | 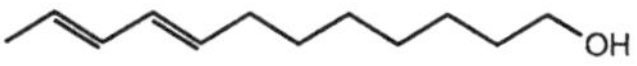 OH | Codling moth (*Cydia pomonella*) |
| 5.1.37 | 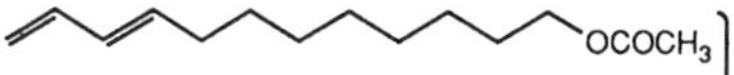 $OCOCH_3$ | Red bollworm (*Diparopsis castanea*) |
| 5.1.38 | 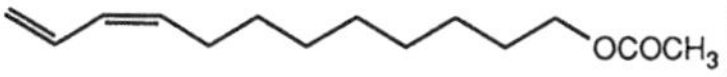 $OCOCH_3$ | |
| 5.1.39 | 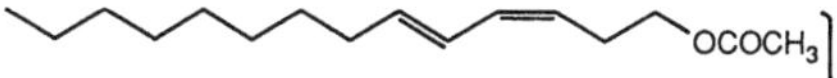 $OCOCH_3$ | Carpenter worm (*Prionoxystus robiniae*) |
| 5.1.40 | 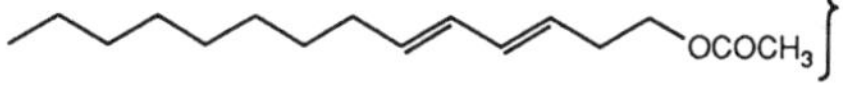 $OCOCH_3$ | |
| 5.1.41 |  $OCOCH_3$ | Egyptian cotton leafworm (*Spodoptera littoralis*) |
| 5.1.42 | 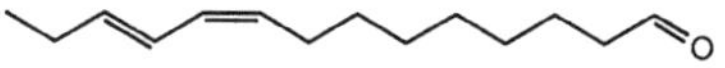 O | *Ectomyelois ceratoniae* |
| 5.1.43 |  $OCOCH_3$ | Light brown apple moth (*Epiphyas postvittana*) |
| 5.1.44 | 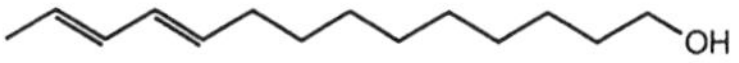 OH | *Maliarpha separatella separatella* |
| 5.1.45 |  $OCOCH_3$ | *Notocelia udmanniana* |
| 5.1.46 |  O | Sugar-cane borer (*Diatrea saccharalis*) |
| 5.1.47 |  OH | Silkworm moth (*Bombyx mori*) |
| 5.1.48 |  O | Tobacco hornworm (*Manduca sexta*) |
| 5.1.49 |  O | Spiny bollworm (*Earias insulana*) |
| 5.1.50 |  O | Navel orangeworm (*Amyelois transitella*) |

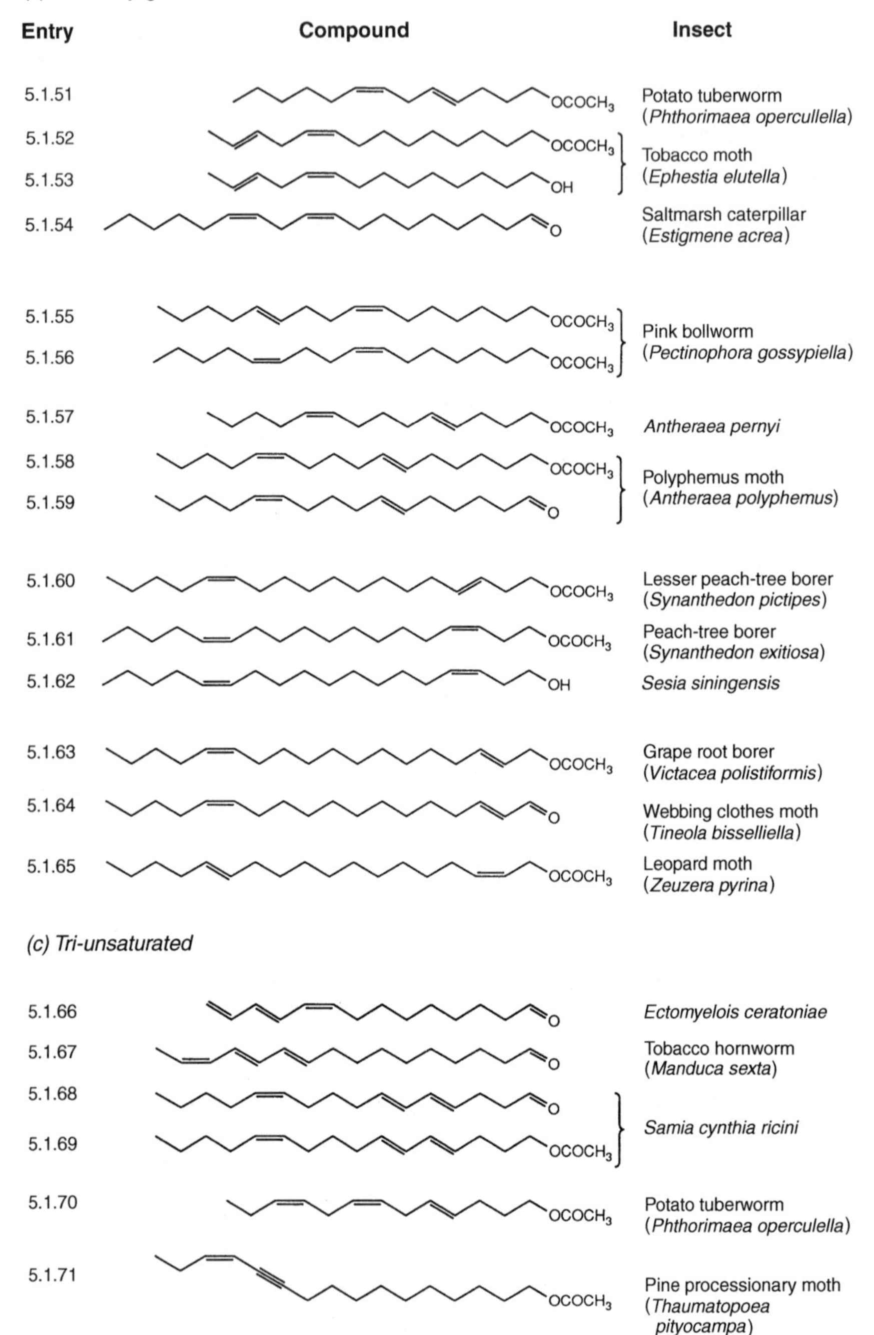
(ii) Non-conjugated
Entry
Compound
Insect
5.1.51
OCOCH3
Potato tuberworm (Phthorimaea opercullella)
5.1.52
OCOCH3
5.1.53
OH
Tobacco moth (Ephestia elutella)
5.1.54
O
Saltmarsh caterpillar (Estigmene acrea)
5.1.55
OCOCH3
5.1.56
OCOCH3
Pink bollworm (Pectinophora gossypiella)
5.1.57
OCOCH3
Antheraea pernyi
5.1.58
OCOCH3
5.1.59
O
Polyphemus moth (Antheraea polyphemus)
5.1.60
OCOCH3
Lesser peach-tree borer (Synanthedon pictipes)
5.1.61
OCOCH3
Peach-tree borer (Synanthedon exitiosa)
5.1.62
OH
Sesia siningensis
5.1.63
OCOCH3
Grape root borer (Victacea polistiformis)
5.1.64
O
Webbing clothes moth (Tineola bisselliella)
5.1.65
OCOCH3
Leopard moth (Zeuzera pyrina)
(c) Tri-unsaturated
5.1.66
O
Ectomyelois ceratoniae
5.1.67
O
Tobacco hornworm (Manduca sexta)
5.1.68
O
5.1.69
OCOCH3
Samia cynthia ricini
5.1.70
OCOCH3
Potato tuberworm (Phthorimaea operculella)
5.1.71
OCOCH3
Pine processionary moth (Thaumatopoea pityocampa)

### *5.1.2 Esters of other acids*

| Entry | Compound | Insect |
|---|---|---|
| 5.1.72 | | Soybean looper (*Pseudoplusia includens*) |
| 5.1.73 | | Soybean looper (*Pseudoplusia includens*) |
| 5.1.74 | | Bagworm moth (*Thyridopterix ephemeraeformis*) |
| 5.1.75 | | Western grapeleaf skeletonizer (*Harrisina brillans*) |
| 5.1.76 | | Pine Emperor moth (*Nudaurelia cytherea cytherea*) |
| 5.1.77 | | *Euproctis similis xanthocampa* |
| 5.1.78 | | *Euproctis chrysorrhea* |

### *5.1.3 Ketones*

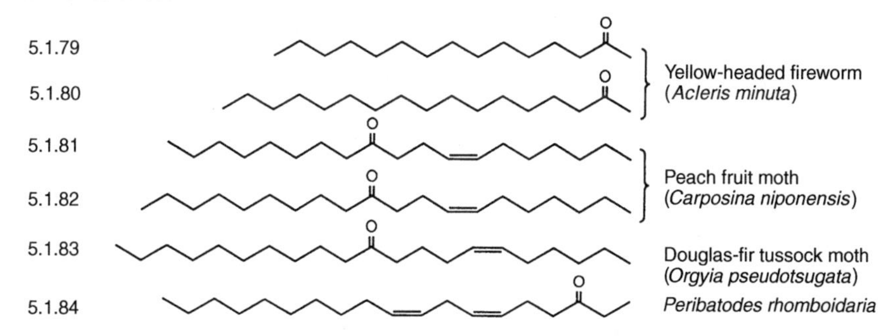

| Entry | Compound | Insect |
|---|---|---|
| 5.1.79 | | Yellow-headed fireworm (*Acleris minuta*) |
| 5.1.80 | | Yellow-headed fireworm (*Acleris minuta*) |
| 5.1.81 | | Peach fruit moth (*Carposina niponensis*) |
| 5.1.82 | | Peach fruit moth (*Carposina niponensis*) |
| 5.1.83 | | Douglas-fir tussock moth (*Orgyia pseudotsugata*) |
| 5.1.84 | | *Peribatodes rhomboidaria* |

### *5.1.4 Epoxides*

| Entry | Compound | Insect |
|---|---|---|
| 5.1.85 | | Gypsy moth (*Lymantria dispar*) |
| 5.1.86 | | *Xanthotype sospeta* |
| 5.1.87 | | *Abraxis grossulariata* |

| Entry | Compound | Insect |
|---|---|---|
| 5.1.88 | 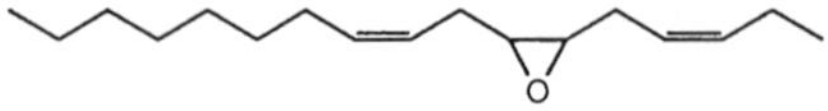 | *Semithesia ulsterata* |
| 5.1.89 | 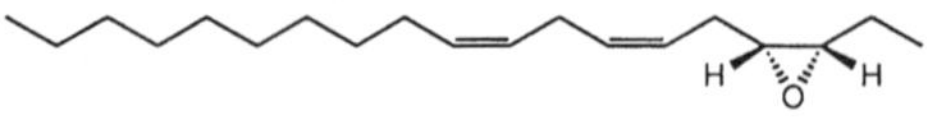 H O H | *Ascotis selenaria* |
| 5.1.90 | 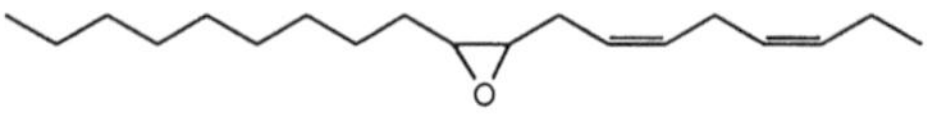 O | *Synaxis jubararia* |
| 5.1.91 | 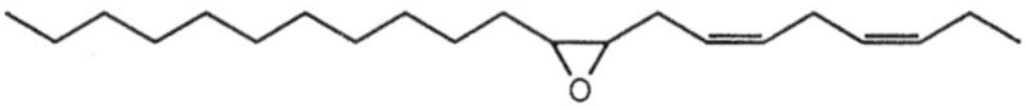 O | Saltmarsh caterpillar (*Estigmene acrea*) |
| 5.1.92 | 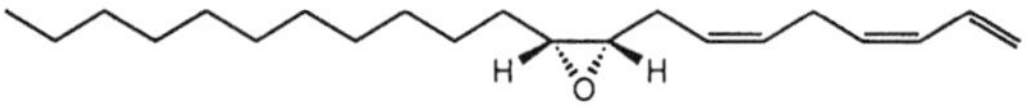 H O H | Fall webworm (*Hyphantria cunea*) |

**5.1.5 Hydrocarbons**

| Entry | Compound | Insect |
|---|---|---|
| 5.1.93 |  | Whitemarked tussock moth (*Orgyia leucostigma*) |
| 5.1.94 | 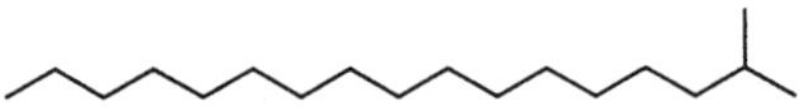 | Tiger moth (*Holomelina lamae*) |
| 5.1.95 | 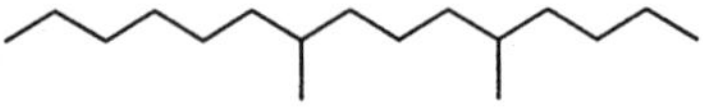 | *Perileucoptera coffeala* |
| 5.1.96 | 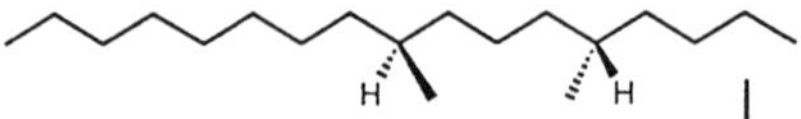 H H | *Leucoptera scitella* |
| 5.1.97 | 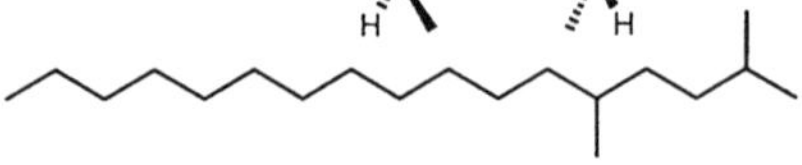 | *Lambdina fiscellaria fiscellaria* |
| 5.1.98 | 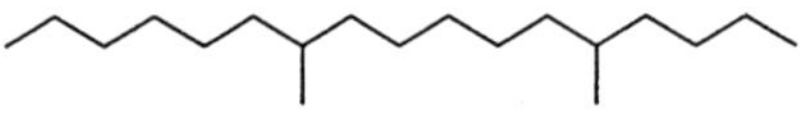 | *Lambdina fiscellaria fiscellaria* |
| 5.1.99 | 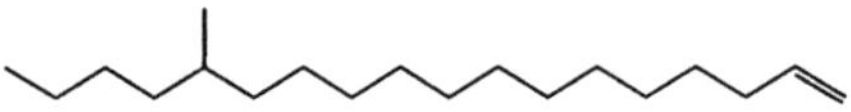 | Peach leafminer (*Lyonetia clerkella*) |
| 5.1.100 |  | Omnivorous looper (*Sabulodes caberata*) |
| 5.1.101 |  | *Mocsis latipes* |
| 5.1.102 |  | *Lomographa semiclarata* |
| 5.1.103 |  | Fall cankerworm (*Alsophila pometaria*) |
| 5.1.104 |  | Velvetbean caterpillar (*Anticarsia gemmatalis*) |
| 5.1.105 |  | Velvetbean caterpillar (*Anticarsia gemmatalis*) |
| 5.1.106 |  | Winter moth (*Operophtera brumata*) |
| 5.1.107 |  | Fall cankerworm (*Alsophila pometaria*) |
| 5.1.108 |  | Fall cankerworm (*Alsophila pometaria*) |
| 5.1.109 |  | Bella moth (*Ytetheisa ornatrix*) |

common are dodecan-1-ol, tetradecan-1-ol and hexadecan-1-ol, the acetates corresponding to these three alcohols, and the aldehydes, tetradecanal and hexadecanal). Class 5.1.1a covers the mono-unsaturated compounds. The di-unsaturated compounds (class 5.1.1b) have been further subdivided into those in which the two double-bonds are conjugated, i.e. separated by just one single bond, and those in which they are further apart. Class 5.1.1c contains tri-unsaturated compounds and the last compound in this class (entry 5.1.71) contains the unusual **alkyne** group, in which two of the carbons are joined by a triple-bond. This group is very uncommon among pheromones, although it occurs widely in many plant species.

Figure 5.1.2 illustrates the other kinds of esters – those in which the acid group is not acetic acid. Here we observe three types:

- those compounds in which both the alcohol and the acid components are straight chain, exemplified by entry 5.1.73: (Z)-7-dodecen-1-ol butanoate, the pheromone of the soybean looper, *Pseudoplusia includens*;
- those compounds in which the acid is a straight chain fatty acid and the alcohol component is non-terminal, e.g. pentan-2-ol decanoate (entry 5.1.74), used by the bagworm moth, *Thyridopteryx ephemeraeformis*;
- those compounds in which the alcohol is not branched and the acid component is more complex, e.g. (Z)-5-decen-1-ol 3-methylbutanoate (entry 5.1.76), used by the pine emperior moth, *Nudaurelia cytherea cytherea*.

Esters of acids with five carbons or fewer are often still referred to by the trivial names of the older literature. Hence esters of $CH_3CH_2CH_2COOH$ are called butyrates, those of $CH_3CH_2CH_2CH_2COOH$ are called valerates, those of $(CH_3)_2CHCOOH$ are isobutyrates and those of $(CH_3)_2CHCH_2COOH$ are isovalerates.

Figure 5.1.3 consists of compounds where the principal functionality is the **ketone** group, >C=O, which resembles the aldehyde group, has similar chemical properties, but occurs always at a non-terminal position. It is given the ending **-one**, and the compound should be numbered from the end closest to the ketone group as shown for structure (**5.1.27**), the pheromone of the Douglas-fir tussock moth, *E*-6-heneicosen-11-one (entry 5.1.83).

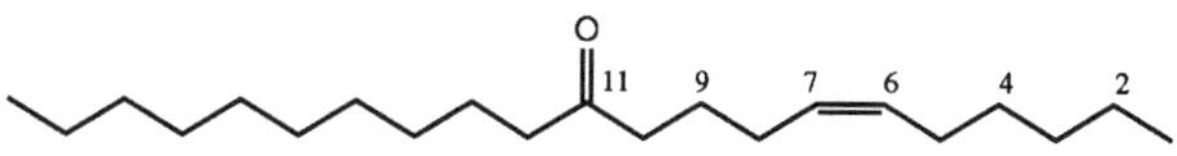

O 11 9 7 6 4 2 (**5.1.27**)

However, the double-bond is often given preference and, for

example, entry 5.1.81 is commonly named (Z)-7-nonadecen-11-one instead of (Z)-12-nonadecen-9-one.

The last of the oxygen-containing functionalities is represented by the pheromone of the gypsy moth, *Lymantria dispar* (entry 5.1.85), in Fig. 5.1.4 and contains the **epoxide** group, in which the oxygen is bound to two adjacent carbon atoms to form a three-membered ring. Here, as with double-bonds, there is the possibility of geometric isomerism, with the hydrocarbon chains being disposed either on the same or on opposite faces of the ring. As for double-bonds, the prefixes Z- and E- are used for denoting the geometry and the prefix **epoxy** is used to show the presence of the epoxide. (The older geometric designators of *cis-* for the same face and *trans-* for opposite faces are still in common use. Provided that there are only two chains attached to the ring, these give unambiguous assignments: see section 5.3.1.) This compound is also the first that we have encountered in which the hydrocarbon chain is branched and requires that the size and position of the branch be shown in its name. Again the oxygen function takes precedence for the numbering and the compound is called (Z)-7,8-epoxy-2-methyloctadecane.

The final group of compounds (Fig. 5.1.5) includes those which do not contain any hetero atoms, i.e. are hydrocarbons. Here we may note the common occurrence of 'skipped' polyenes, where the double-bonds are separated one from another by a single methylene group. This type of structure is characteristic of that found in the polyunsaturated fatty acids. In addition this group contains a number of branched chain compounds (entries 5.1.94 to 5.1.99), which form the basis of the next section.

The most comprehensive listing of lepidopteran sex pheromones and related attractants is to be found in the compilation by Arn, Tóth and Priesner (1992) made under the aegis of the International Organization for Biological Control by the Working Group on the use of pheromones and other semiochemicals in integrated control.

## 5.2 BRANCHED CHAIN STRUCTURES

Once we start to consider the pheromones of orders other than Lepidoptera, or even the compounds that are produced by males of lepidopteran species, it soon becomes apparent that the majority of structures are not simple straight chain ones. The smallest modification is one in which the chain is branched as in the pheromones of the tiger moth, *Holomelina lamae*, and the gypsy moth, *Lymantria dispar* (Fig. 5.1, entries 5.1.94 and 5.1.85), where a single carbon lies off the main chain. Both the position and the number of carbons in the branch need to be indicated, the former by a numeral and the latter by taking the hydrocarbon root and adding the ending **-yl**: methyl from meth-ane, pentyl from pent-ane and so on. Examples are the pheromones of the dermestid beetles (*Trogoderma* spp.) and the tsetse fly, *Glossina morsitans morsitans* – structures (**5.2.1**) and (**5.2.2**).

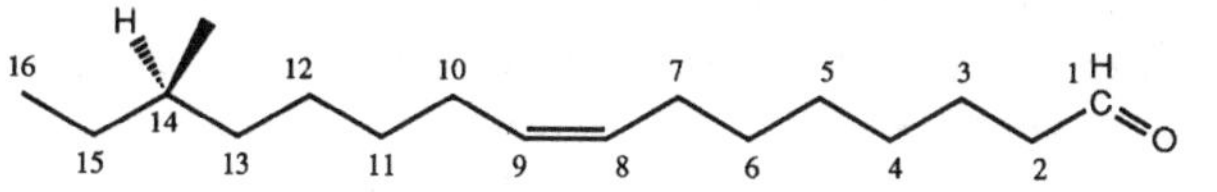

(5.2.1)

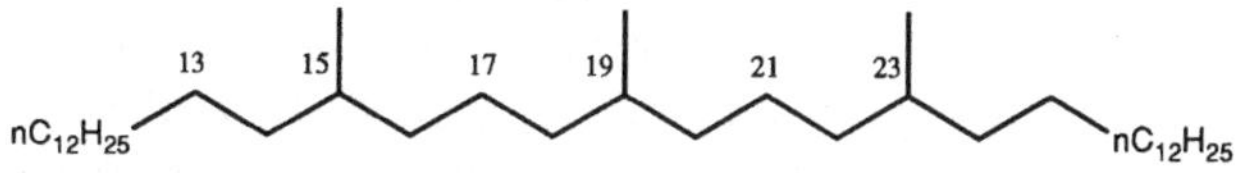

(5.2.2)

The former is 14-methyl-(Z)-8-hexadecenal while the latter is 15,19,23-trimethyl-heptatriacontane. Here the fact that three methyl groups are present requires the use of the multiplier **tri-** immediately before the branch size name. In structure (**5.2.2**) as shown, the straight chain portion at each end has been indicated with the abbreviation $C_{12}H_{25}$ with the prefix *n*- to show that there are no branches in this part.

An additional and new factor in structure is introduced by the position of the branch in compound (**5.2.1**) and this is the one of **chirality**, or handedness; just as two hands are not identical, but are mirror images of each other, so compound (**5.2.1**) is not the same as its mirror image (**5.2.3**), below.

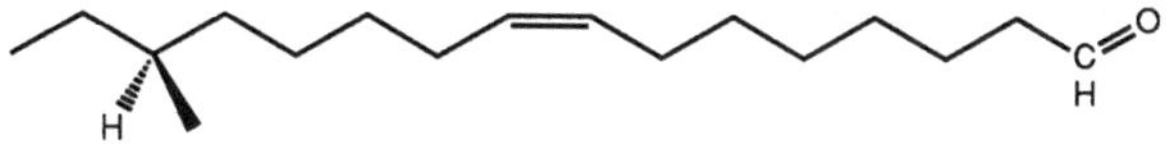

(5.2.3)

As a result of this difference, each of the two forms of the compound has the ability to rotate the plane of polarization of a beam of polarized light and it is this fact that originally led to the discovery of this form of structural isomerism. Hence we speak of this type of isomerism as **optical isomerism**.

### 5.2.1 Optical activity : enantiomers, absolute configuration, diastereoisomerism

Optical isomerism arises when the structure of a molecule is asymmetric and may be recognized by the application of one of two simple tests:

- either the structure cannot be superimposed on its mirror image;
- or the molecule does not possess a plane or centre of symmetry.

In most compounds such asymmetry will arise if four different groups are bonded to the same carbon atom and we may recognize this feature in structure (**5.2.1**) at C-14, which is bonded to the four different groups: hydrogen, a methyl, an ethyl (C-15—C-16) and the long chain. Such a carbon is usually referred to as a **chiral carbon** or a **chiral centre**.

The two mirror image isomers are called **enantiomers** and each will rotate the plane of plane polarized light to the same extent but with opposite handedness. With the exception of this effect on polarized light, the physical properties of the two enantiomers are identical, as are the vast majority of their reactions towards chemical reagents. In fact it is only in their reactions with a single enantiomer of another compound that enantiomers differ. This fact is significant for their interactions with enzymes, which are themselves chiral and single enantiomers, and almost certainly accounts for the selectivity in pheromone perception. Provided that we know the exact shape of each enantiomer in space, we can designate them by an absolute configurational assignment either (*R*) or (*S*). Unfortunately absolute configuration is not related to the direction in which the enantiomers rotate plane polarized light in a manner that is readily predictable.

An equal mixture of the two enantiomers is called a **racemic mixture** or **racemate** and is optically inactive. The majority of chemical reactions which produce a chiral centre will do so to give a racemate, and the synthesis of compounds in one of the single enantiomeric forms presents a formidable challenge. When a compound exists as a single enantiomer, it is referred to as being **homochiral**. The majority of pheromones that contain chiral centres are used in homochiral form by the insects that produce them.

### (a) *Absolute configuration*

The general system for designating absolute configuration in space is based on the Cahn–Ingold–Prelog priority system. Each group attached to the chiral carbon is assigned a priority number from 1 (the highest priority) to 4 (the lowest). The molecule is then viewed from the side remote from the group of lowest priority. The order of decreasing priority (1→2→3) of the remaining groups represents either a clockwise or a counterclockwise sequence. If the sequence is clockwise, the molecule is assigned the configuration (*R*), from *rectus* = 'right'; if the sequence is counterclockwise, the molecule is assigned the configuration (*S*), from *sinister* = 'left'.

The priority of the groups is assigned on the basis of the **priority rules**:

*Sequence rule 1*

If the four atoms attached to the chiral centre are all different, then the priority depends on the **atomic number**, with the atom of higher atomic number being given the higher priority. (If two isotopes of the same element are present, the atom of higher mass number has the higher priority.) Thus for the molecule of 2-bromoethane sulphonic acid in structure (**5.2.4**), the atomic numbers are Br (35), C (6), H (1) and S (16), leading to the priority order Br > S > C > H.

2-bromoethane sulphonic acid 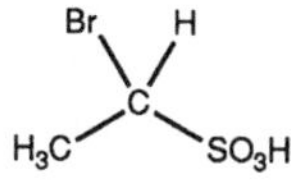 (**5.2.4**)

*Sequence rule 2*

If the relative priority of two groups cannot be decided by Rule 1, then it is determined by a similar comparison of the next atoms in the groups (and so on, if necessary, working outwards from the chiral centre). In other words, if two atoms attached to the chiral centre are the same, we compare the atoms attached to each of these first atoms.

This is the situation that most frequently occurs among natural products and needs to be applied both to (**5.2.1**) and the simpler case of (**5.2.5**), one of the constituents of the pheromone of the males of the grape borer, *Xylotrechus pyrrhoderus*. The compound is numbered as shown.

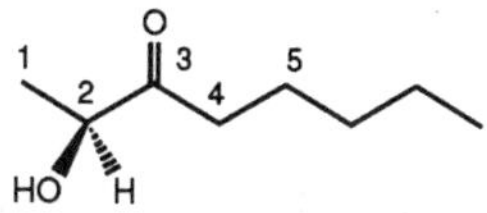

(**5.2.5**)

Here it is easy to assign O(H) as priority 1 (atomic no. 8) and H as 4. Of the two carbons, C-3 has an oxygen attached and therefore takes precedence over C-1, which has only hydrogens. Hence the sequence is HO > C(=O) > C($H_3$) > H and, viewing the molecule as in (**5.2.6**), we assign the absolute configuration as (*S*).

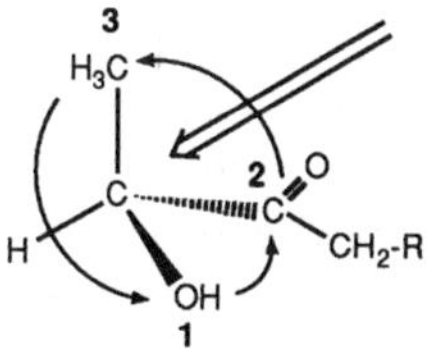

(**5.2.6**)

We name the compound as (2*S*)-2-hydroxyoctan-3-one, with the doubly bonded ketone group appearing as the ending and the alcohol group as a prefix: **hydroxy**. (IUPAC rules require that only the principal functional group appears as a suffix. Appendix 5B lists the prefixes and suffixes for the most commonly encountered groups.)

In compound (**5.2.1**), the three carbons attached to the chiral centre need to be considered in turn: the methyl group has only hydrogens attached and therefore ranks lower than either C-13 or C-15, which are each bonded to another carbon (C-12 and C-16). Considering each of these two, we need to compare C-12 and C-16, where we can make the distinction that C-12 (bonded to C-11) ranks higher than C-16 (bonded only to hydrogens). Hence the priority is C-13>C-15>methyl>H and the absolute configuration is (*R*). Of these two enantiomers, (14*R*)-14-methyl-*Z*-8-hexadecenal (**5.2.1**) is attractive to *Trogoderma* species, while (**5.2.3**), the (14*S*)-isomer, is biologically inactive.

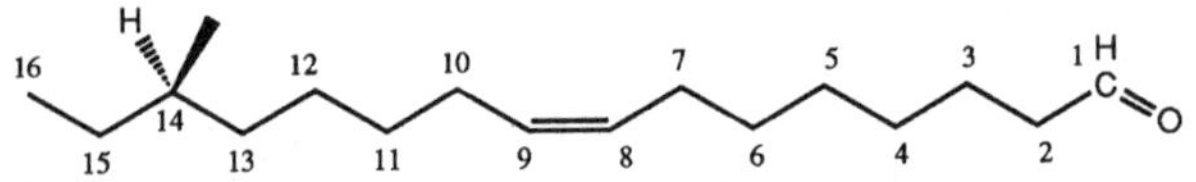

(5.2.1)

The third sequence rule is rather infrequently required for pheromone structures (but ipsdienol (see below) is an example) and is given here for completeness.

*Sequence rule 3*

Where there is a double or a triple bond, both atoms are considered to be duplicated or triplicated. Thus:

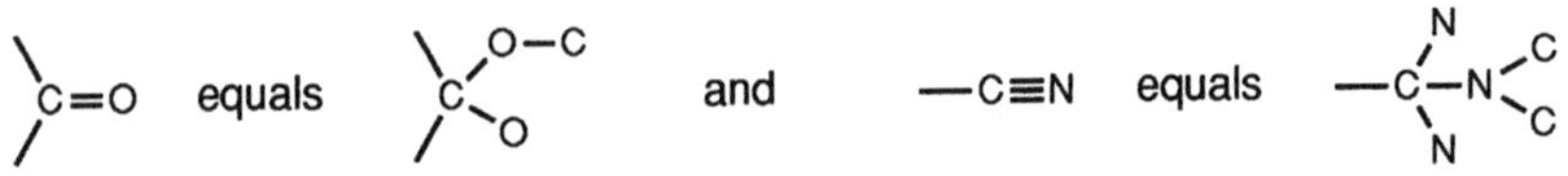

C=O equals ... and —C≡N equals ...

This is significant for compounds such as glyceraldehyde, structure (**5.2.7**), in order to distinguish between the CH=O and the $CH_2OH$ groups.

(*R*)-glyceraldehyde (5.2.7)

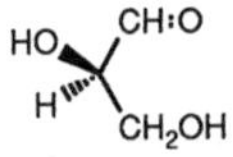

Among branched chain compounds we also find many examples where compounds possess double-bonds which are polysubstituted. In the consideration of these compounds with tri- and tetra-substituted double-bonds, we employ exactly the same set of sequence rules as for absolute configuration in order to determine the precedence for the geometric designators *E*- and *Z*-. At each end of the double-bond, we assign priority to the two atoms (or groups) present. When the two groups of higher priority lie on opposite sides, the stereochemistry is *E*-; and it is *Z*- when they are on the same side.

## (*b*) *Diastereoisomerism*

When a molecule contains more than one chiral centre, as in 2,3-octandiol (**5.2.8**), (a compound also produced by the grape borer, *Xylotrechus* *pyrrhoderus*), then the possibility exists for a new form of stereoisomerism. The stereo-centres in (**5.2.8**) may exist in the following four forms: (2*R*, 3*R*), (2*S*, 3*S*), (2*R*, 3*S*) and (2*S*, 3*R*) as shown.

The four optical isomers of 2,3-octandiol (**5.2.8**)

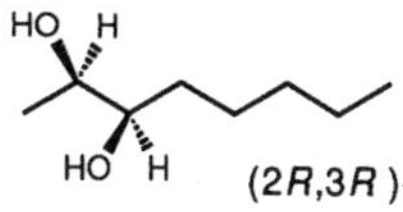

(2*R*,3*R*)-(**5.2.8**) 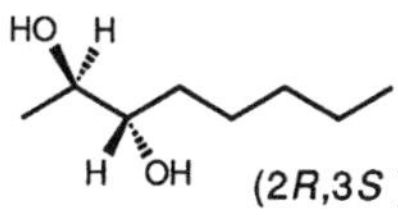 (2*R*,3*S*)-(**5.2.8**)

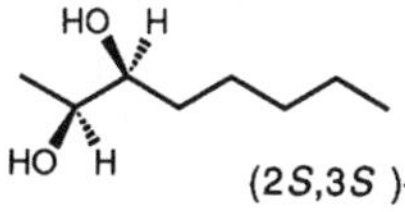

(2*S*,3*S*)-(**5.2.8**) 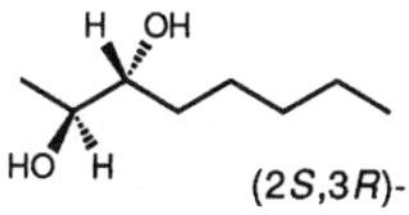 (2*S*,3*R*)-(**5.2.8**)

While it is clear that the (2*R*, 3*R*) and the (2*S*, 3*S*) forms are mirror images of each other and are therefore enantiomers, the (2*R*, 3*R*) form is **not** the mirror image of either the (2*R*, 3*S*) form or the (2*S*, 3*R*) one. Their similarity is evident and the difference is still one of shape. We call this form of isomerism **diastereoisomerism** and speak of the two forms (2*R*, 3*R*) and (2*R*, 3*S*) as **diastereomers**. A study of the four structures shows that the (2*R*, 3*S*) and (2*S*, 3*R*) ones are enantiomers of each other and both are diastereomers of the other pair of enantiomers – (2*R*, 3*R*) and (2*S*, 3*S*). Hence we may define diastereomers as stereoisomers that are not enantiomers.

Diastereoisomerism is extremely common among branched compounds and normally only one of the possible diastereomers is biologically active. Thus *X. pyrrhoderus* females are attracted to (2*S*, 3*S*)-2,3-octandiol, but are unaffected by the other three isomers. Note that where two (out of the three) of the groups attached at each chiral carbon are the same, as here, an additional method for designating the particular diastereomer is often used. The two terms are **threo**- and **erythro**- and they are applied thus: the (*R*,*R*)/(*S*,*S*) pair are called the threo-isomers and the (*R*,*S*)/(*S*,*R*) pair the erythro-isomers.

The number of possible stereoisomers that may exist when a compound has $N$ chiral centres is $2^N$. Of these isomers, one half will be the enantiomers of the other half, i.e. there will be $2^{(N-1)}$ diastereomeric compounds, each of which will consist of an enantiomeric pair. As an example we may take 4,6-dimethyl-7-hydroxynonan-3-one (**5.2.9**), which contains three chiral centres and has eight possible isomers, which may be divided into four pairs of enantiomers. Out of these eight compounds, it is only the (4*S*,6*S*,7*S*) isomer that is produced by females of the cigarette beetle, *Lasioderma serricorne*, and that is attractive to males of the species.

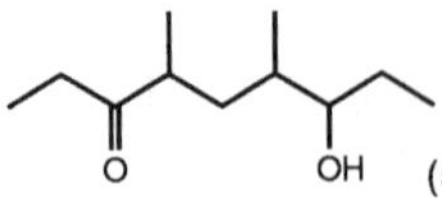

O OH (**5.2.9**) O HO H (4*S*,6*S*,7*S*)-(**5.2.9**)

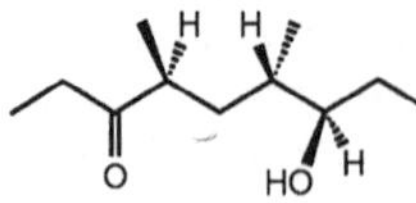

Unlike enantiomers, which differ only in their effect on plane polarized light, diastereomers have different physical properties from one another and may be separated by crystallization, distillation and chromatographic methods.

Figure 5.2 is illustrative of the pheromones of the males of lepidopteran species and of both males and females of other Orders. Where the absolute configuration of a compound is known, it is as given in the figure.

Figure 5.2.1 contains compounds with straight and simple branched chains. Amongst the straight chain compounds, the male sex pheromone of the bean weevil, *Acanthoscelides obtectus* (entry 5.2.10) is noteworthy for the uncommon allene group, >C=C=C<, which also provides a centre of chirality because the groups at either end lie in planes that are perpendicular to each other. As a result, methyl (*S*)-tetradeca-4,5-dienoate is **not** identical with its mirror image. The natural, active pheromone is the (*S*) enantiomer shown.

Among the pheromones of the females of Lepidoptera, it should be pointed out that the epoxides of Fig. 5.1.4 are also chiral compounds, the two carbons of the ring being the centres of chirality. Again the absolute configuration, where this has been determined, is as given in Fig. 5.1.4. In addition, those hydrocarbon pheromones in Fig. 5.1.5 which bear methyl groups at positions along the chain, i.e. not C-2 (entries 5.1.95 to 5.1.99) also possess chiral centres at these branching points. In two instances the absolute configuration of the bioactive compound has been determined (entries 5.1.96 and 5.1.99). Interestingly in both cases the stereochemistry is (*S*).

One final point of note arises when considering the pheromone (**5.2.2**) of the tsetse fly, *Glossina morsitans morsitans*. The active diastereomer is not yet known, but the compound is symmetric about the central carbon C-19 which therefore exhibits **pseudo**-chirality. Four isomers are possible: one pair of enantiomers – (15*R*, 23*R*) and (15*S*, 23*S*) – and two isomers that have mirror planes at the central atom – (15*R*, 19ψ*R*, 23*S*) and (15*R*, 19ψ*S*, 23*S*). Compounds such as this which are symmetric overall, although they possess chiral centres, are known as **meso** isomers and are sometimes said to be internally compensated, because the two chiral centres are of opposite handedness.

### 5.2.2 Terpenoids and the isoprene rule

Terpenoids, originally discovered in turpentine oil (hence the name), constitute a very well known and widespread class of natural products.

**Fig. 5.2 Acyclic pheromones of other Orders and of males of Lepidoptera (Key: M. = male; F. = Female; aggr. = aggregation)**

***5.2.1 Straight chains and simple branched-chain compounds***

*(a) Straight chains*

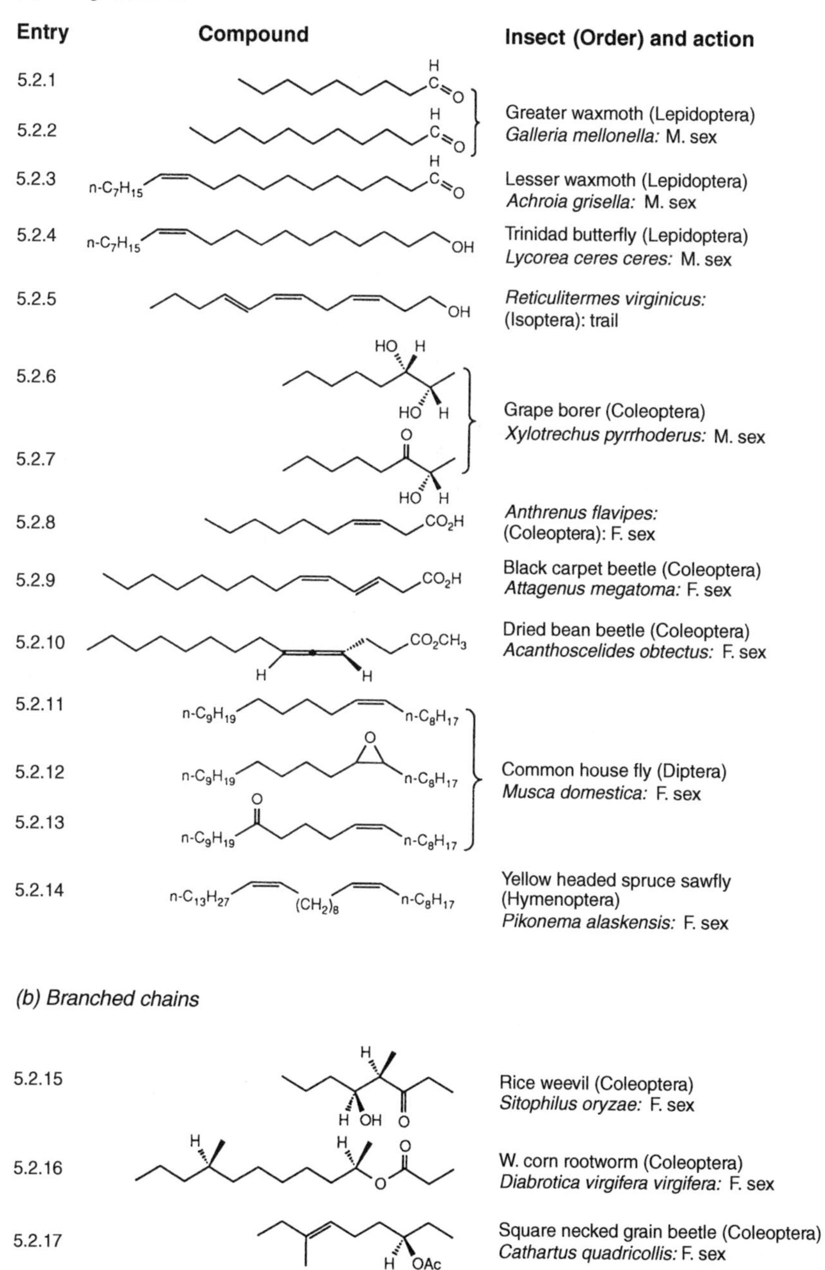

| Entry | Compound | Insect (Order) and action |
| --- | --- | --- |
| 5.2.18 | | Cigarette beetle (Coleoptera) *Lasioderma serricorne:* F. sex |
| 5.2.19 | | Lesser grain borer (Coleoptera) *Rhyzopertha dominica:* M. aggr. |
| 5.2.20 | | Greater grain borer (Coleoptera) *Prostephanus truncatus:* M. aggr. |
| 5.2.21 | | *Lardoglyphus konoi* (Acari): aggr. |
| 5.2.22 | | S. corn rootworm (Coleoptera) *Diabrotica undecimpunctata howardi:* F. sex |
| 5.2.23 | OAc | Red headed pine saw fly (Hymenoptera) *Neodiprion lecontei:* F. sex |
| 5.2.24 | OH | *Trogoderma inclusum* (Coleoptera): F. sex |
| 5.2.25 | O | *Trogoderma inclusum* (Coleoptera): F. sex |
| 5.2.26 | R = H; R--$(CH_2)_{18}$ | German cockroach (Orthoptera) *Blatella germanica:* F. sex |
| 5.2.27 | R = HO | German cockroach (Orthoptera) *Blatella germanica:* F. sex |

### *5.2.2 Acyclic terpenoids*

*(a) Monoterpenoids*

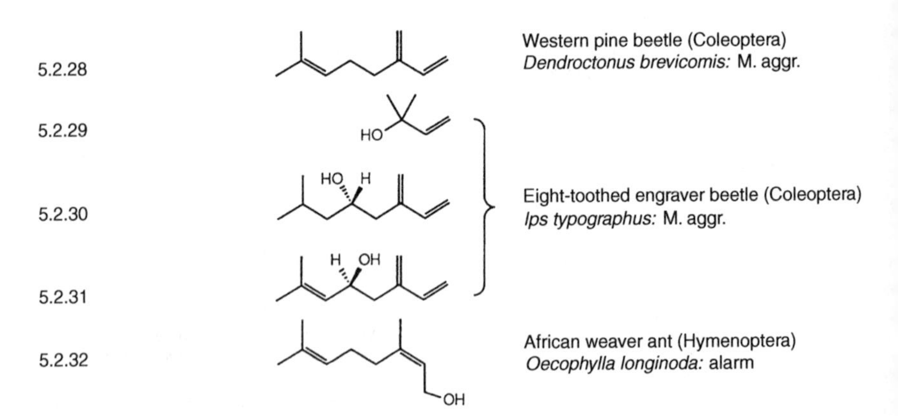

| Entry | Insect (Order) and action |
| --- | --- |
| 5.2.28 | Western pine beetle (Coleoptera) *Dendroctonus brevicomis:* M. aggr. |
| 5.2.29 | Eight-toothed engraver beetle (Coleoptera) *Ips typographus:* M. aggr. |
| 5.2.30 | Eight-toothed engraver beetle (Coleoptera) *Ips typographus:* M. aggr. |
| 5.2.31 | Eight-toothed engraver beetle (Coleoptera) *Ips typographus:* M. aggr. |
| 5.2.32 | African weaver ant (Hymenoptera) *Oecophylla longinoda:* alarm |

| Entry | Compound | Insect (Order) and action |
|---|---|---|
| 5.2.34 | 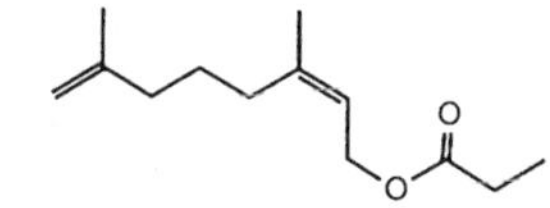 | San Jose scale (Homoptera) *Quadraspidiotus perniciosus:* F. sex |
| 5.2.35 | 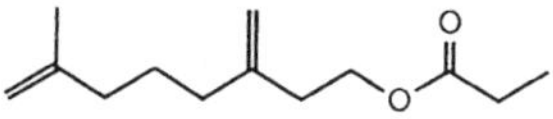 | San Jose scale (Homoptera) *Quadraspidiotus perniciosus:* F. sex |
| 5.2.36 | 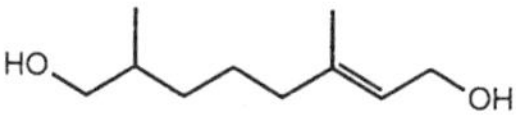 | African Monarch (Lepidoptera) *Danaus chrysippus:* M. sex |
| 5.2.37 | 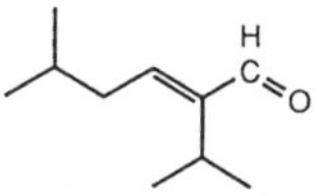 | *Calomyrmex* species (Hymenoptera): trail |
| 5.2.38 | 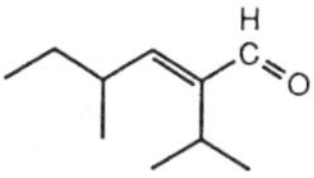 | *Calomyrmex* species (Hymenoptera): trail |
| 5.2.39 | 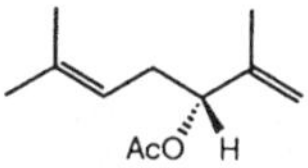 | Comstock mealybug (Homoptera) *Pseudococcus comstockii*: F. sex |
| 5.2.40 | 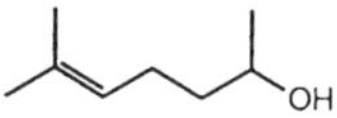 | *Gnathotricus sulcatus* (Coleoptera): M. sex and aggr. |

*(b) Sesqui- and diterpenoids*

| Entry | Compound | Insect (Order) and action |
|---|---|---|
| 5.2.41 | 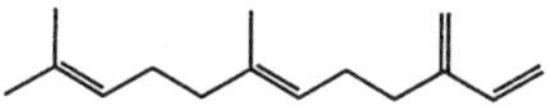 | Many aphid species (Hemiptera): alarm *Solenopsis* spp. (Hymenoptera): alarm |
| 5.2.42 | 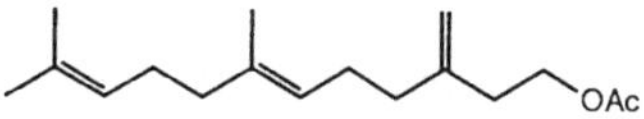 | Western click beetle (Coleoptera) *Agriotes ustulatus:* F. sex |
| 5.2.43 | 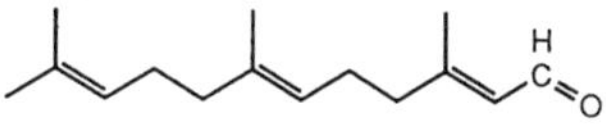 | Rice moth (Lepidoptera) *Corcyra cephalonica:* M. sex |
| 5.2.44 | 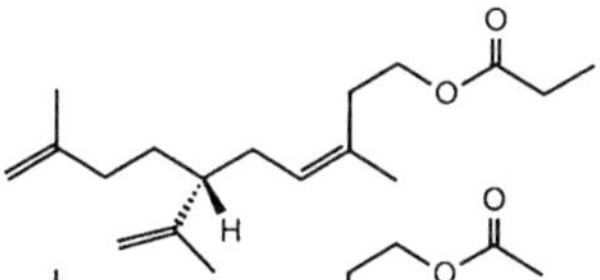 | White peach scale (Homoptera) *Pseudaulascaspis pentagona:* F. sex |
| 5.2.45 | 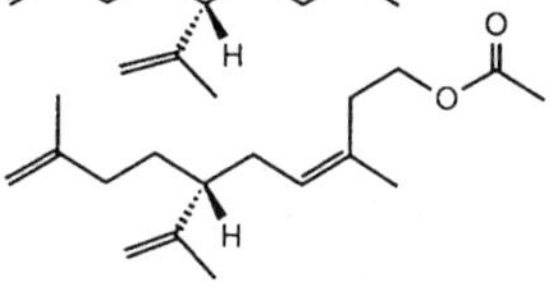 | California yellow scale (Homoptera) *Aonidiella citrina:* F. sex |
| 5.2.46 | 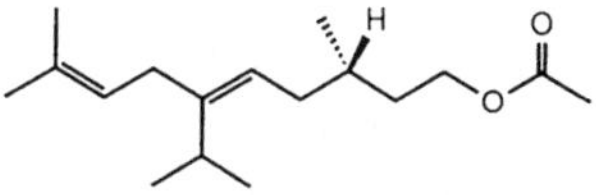 | California yellow scale (Homoptera) *Aonidiella citrina:* F. sex |

| Entry | Compound | Insect (Order) and action |
|---|---|---|
| 5.2.47 | 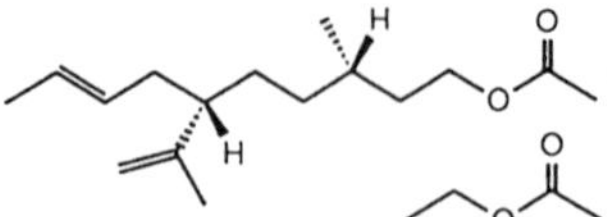 | California red scale (Homoptera) *Aonidiella aurantii:* F. sex |
| 5.2.48 | 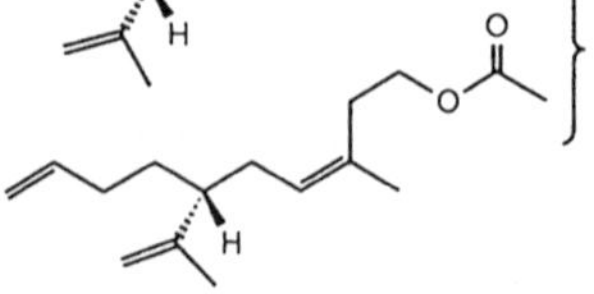 | |
| 5.2.49 | 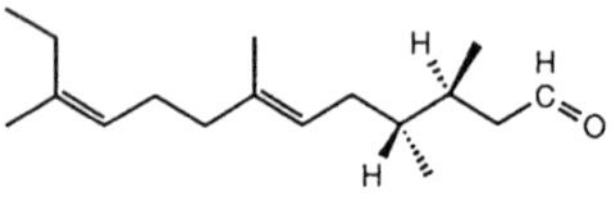 | Pharaoh's ant (Hymenoptera) *Monomorium pharaonis:* trail |
| 5.2.50 | 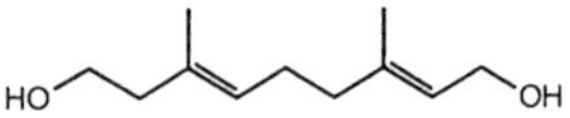 | Queen butterfly (Lepidoptera) *Danaus gilippus:* M. sex |
| 5.2.51 | 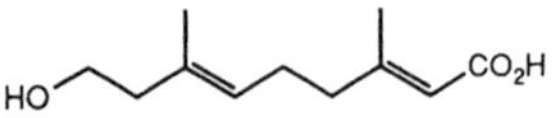 | Monarch butterfly (Lepidoptera) *Danaus plexippus:* M. sex |
| 5.2.52 | 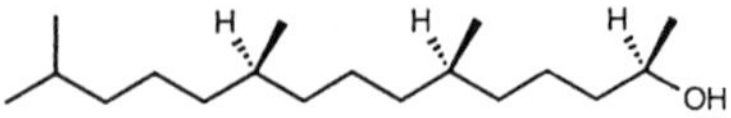 | Rice moth (Lepidoptera) *Corcyra cephalonica:* F. sex |

Therefore it is not surprising to find that they are represented among the pheromonal structural types. The simplest terpenes have 10 carbons and two branches as in geraniol, (**5.2.10**).

Although their biosynthesis occurs through the intermediacy of mevalonic acid, a six-carbon compound, (**5.2.11**), this involves the loss of one carbon and they may be considered as having been built up by the linking together of a number of the five-carbon units of isoprene, (**5.2.12**), in a head-to-tail fashion.

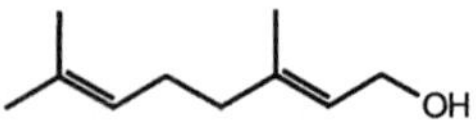

Geraniol (**5.2.10**) 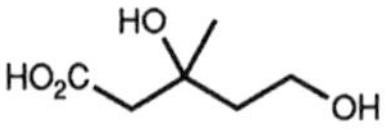 Mevalonic acid (**5.2.11**) 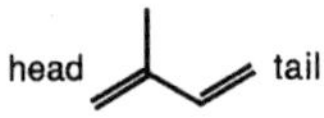 Isoprene (**5.2.12**)

Compounds in this class are found which contain 2, 3, 4, 6 and 8 isoprene units. The $C_{10}$ group of compounds are called monoterpenes, the $C_{15}$ group are sesquiterpenes, the $C_{20}$ ones are diterpenes and the $C_{30}$ and $C_{40}$ groups are ti- and tetraterpenes, respectively. Thus, as shown in Scheme 5.1, geraniol is a monoterpene, consisting of two isoprene units (solid lines) linked in the middle, with the addition of water (dashed lines), and farnesol (**5.2.13**) is a sesquiterpene (Scheme 5.1).

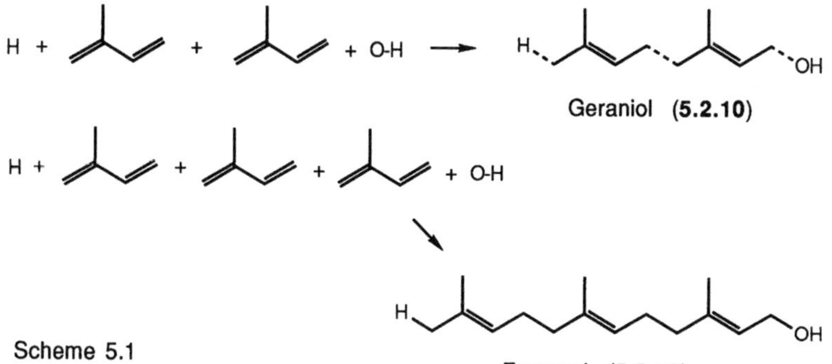

H + + + O-H → H OH

Geraniol (**5.2.10**)

H + + + + O-H

H OH

Scheme 5.1

Farnesol (**5.2.13**)

This recognition of the pattern of biogenesis of these compounds is what constitutes the **isoprene rule** and it is now customary to include, in the category of terpenoids, compounds that have this common biogenesis even though they may not correspond to the strict head-to-tail pattern or may contain one or two too many or too few carbon atoms in their skeletons to fall into the original $C_{10}$, $C_{15}$, etc., category.

In addition to open chain compounds, the terpenoids include many cyclic structural types and these will be discussed in section 5.3

### 5.2.3 Acyclic terpenoid pheromones

A representative selection of acyclic terpenoids is shown in Fig. 5.2.2. The majority of the compounds fall obviously into the normal biogenetic pattern discussed. Note that the eight-toothed engraver beetle, *Ips typographicus*, uses the simple isoprenoid, 2-methyl-3-buten-2-ol, (**5.2.14**), as well as the monoterpene alcohols ipsenol, (**5.2.15**), and ipsdienol, (**5.2.16**).

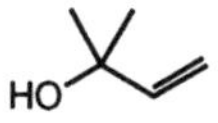

HO HO H H OH

(**5.2.14**) 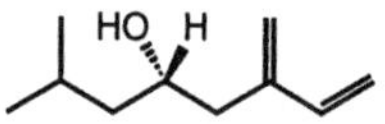 (**5.2.15**) 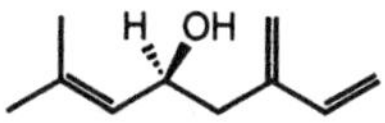 (**5.2.16**)

Abnormal terpenoids include the pheromone from the California yellow scale insect, *Aonidiella citrina*, which show a tail-to-middle connection, (**5.2.17**). The same pattern is evident in the two acetates (**5.2.18**) and (**5.2.19**) produced by the red scale insect, *A. aurantii*, but this time the methyl group on the left-hand isoprene unit is missing.

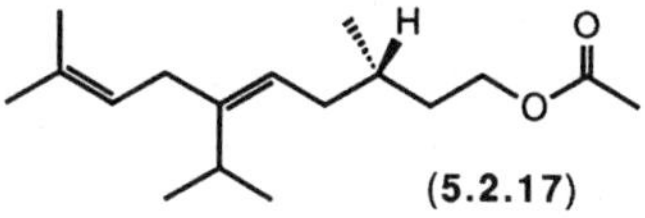

H O O

(**5.2.17**)

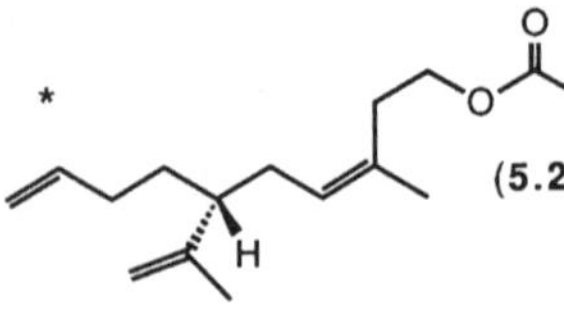

(5.2.18)

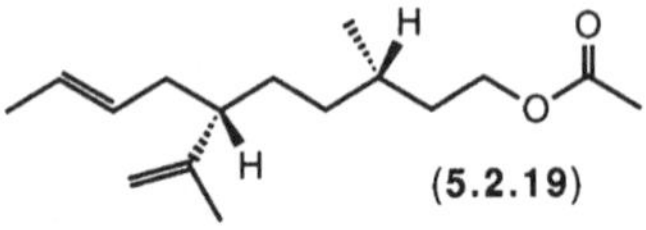

(5.2.19)

The pheromone of the comstock mealybug, *Pseudococcus comstockii* (Fig. 5.2.2 entry 5.2.39) also has one carbon too few, at the right-hand end, and that from the scolytid wood borer, *Gnathotricus sulcatus* (entry 5.2.40) is missing two carbons from the same position. In a similar vein, compound (**5.2.20**), the pheromone from the red flour beetle, *Tribolium castaneum*, is more probably a sesquiterpenoid with three carbons cut off than a monoterpene with two extra carbons.

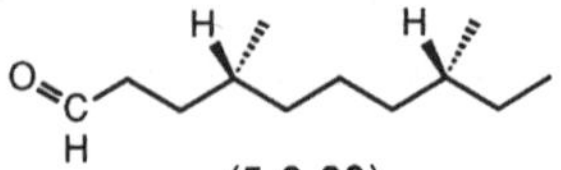

(5.2.20) cf.

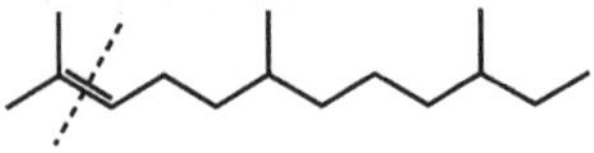

This is in contrast to the trail pheromone, (**5.2.21**), from Pharaoh's ant, *Monomorium pharaonis*, which contains 17 carbons: the two additional carbons are indicated by asterisks.

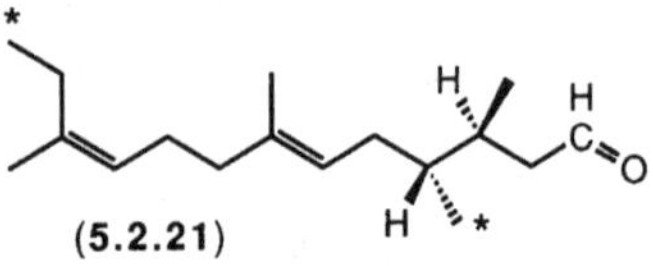

(5.2.21)

Particularly interesting is a (so far) unique example of a **female** lepidopteran sex pheromone which is found in this group: entry 5.2.52. This would appear to be a modified fully reduced diterpenoid that is missing two carbons from the tail (right-hand end as drawn). The male moth also employs a terpene, the sesquiterpene (*E,E*)-farnesal, (entry 5.2.43).

Because many terpenoid compounds were isolated and purified when the task of structure determination was considerably more difficult than now, there is a tradition of giving them trivial names derived from the source material and this has been continued with many of these terpenoid pheromones. Normally these trivial names are used for brevity, but they do not always make much difference. Thus *'faranal'* for (**5.2.21**) is obviously significantly shorter than (3*R*,4*R*)-3,4,7,11-tetramethyl-(*E,Z*)-

6,10-tridecadienal, but 'sulcatol' (entry 5.2.40, used by *G. sulcatus*) is not much different from 6-methyl-5-hepten-2-ol. At times the practice of using a trivial name can be positively misleading, as with 'ipsenol' (**5.2.15**), which is a diene, and 'ipsdienol' (**5.2.16**), which is a triene.

## 5.3 CYCLIC STRUCTURES

Compounds in which the atoms are bonded to form a ring are commonly divided into carbocyclic compounds, in which the ring consists entirely of carbon atoms, and heterocyclic compounds, in which one (or more) atom(s) of a different element (O, N and S are the most common) is incorporated into the ring.

### 5.3.1 Carbocyclic compounds

The ring may be of any size, with five- and six-membered ones being the most common. The compounds are named from the ring size, with the prefix **cyclo-** before the root name: cyclopentane, cyclodecane and so on. Simple groups attached to the ring are treated as with branches on acyclic structures, but if this group contains a priority substitutent (e.g. OH, CH=O,$CO_2H$), then the ring is treated as branch on this group. Thus structure (**5.3.1**) is a propylcyclopentanol, whereas (**5.3.2**) is a cyclopentylpropanol.

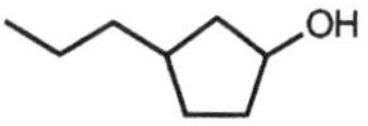

(5.3.1) 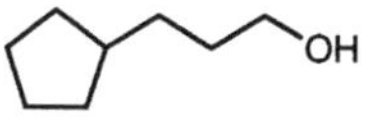 (5.3.2)

The atoms on the ring are numbered from the principal substituent (if it is directly on the ring) or from the point of attachment of the principal chain. Thus (**5.3.1**) is 3-propylcyclopentanol and (**5.3.2**) is 3-(cyclopentyl)propanol. The position of the principal substituent is always understood as 1 if it is not specified.

A benzene ring is always treated as a branch on the aliphatic chain. thus (**5.3.3**) is 2-phenylethanol and is found in various *Mamestra* species. The only exceptions to this are the simple compounds benzyl alcohol, (**5.3.4**), and benzaldehyde, (**5.3.5**), which are used by the males of a number of lepidopteran species.

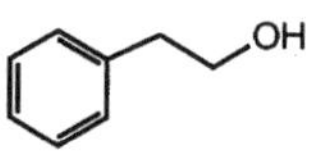

(5.3.3) 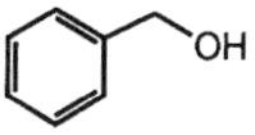 (5.3.4) 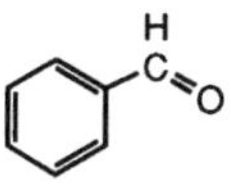 (5.3.5)

In the same manner as for double-bonds, rings introduce the factor of geometrical constraint, preventing free rotation of substituents relative to one another, and this gives rise to geometric isomerism. The particular geometry is indicated in the same way as with double-bonds, using *Z*- to indicate that groups are on the same face of the ring and *E*- to show that they are on opposite faces. Because a cyclic compound may have several substituent groups, these designators are determined by the relationship of the attached group to the principal substituent. Which of those present is to be regarded as the principal substituent is decided using the same priority rules given above for absolute configuration (section 5.2.1a). When only two groups are attached to the ring, then one finds that the old nomenclature of *cis* (on the same side) and *trans* (opposite sides) is frequently still employed.

Males of the oriental fruit moth, *Grapholita molesta*, produce (**5.3.6**) and (**5.3.7**), which are methyl *Z*-(3-oxo-2-(*Z*-2-pentenyl)cyclopentyl) ethanoate and ethyl *E*-3-phenyl-2-propenoate. Again the use of trivial names is quite common: (**5.3.6**) is jasmone, first isolated from jasmine oil, and (**5.3.7**) is ethyl *E*-cinnamate, found in cinnamon.

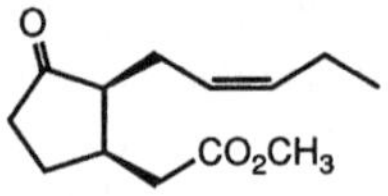

O $CO_2CH_3$ $CO_2$-$CH_2CH_3$

(5.3.6) 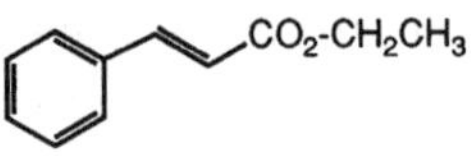 (5.3.7)

Figure 5.3.1 contains examples of the known types of carbocyclic structures.

## 5.3.2 Heterocyclic structures: ethers and lactones

The introduction of a single oxygen into the ring gives rise to a cyclic ether. The naming of monocyclic compounds is based on the parent ring systems of furan, (**5.3.8**), and 4H-pyran, (**5.3.9**), for the five- and six-membered rings, with the numbering starting from the hetero-atom. Simple examples are: (**5.3.10**), 3-(4′-methylpent-3′-enyl)-furan, a trail pheromone from ants of the *Tetramoriini* species; (**5.3.11**), (2*R*)2,3-dihydro-2-ethyl-6-methyl-4H-pyran-4-one, isolated from males of the ghost moth, *Hepialus californicus*; and (**5.3.12**), its double-bond isomer, 5,6-dihydro-2-ethyl-6-methyl-4H-pyran-4-one, which comes from the scale brushes of males of the swift moth, *H. hecta*).

O O O O O H O O H

(5.3.8) 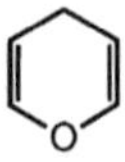 (5.3.9) 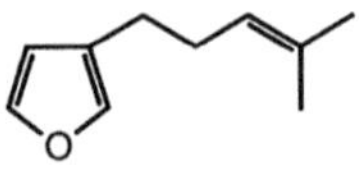 (5.3.10) 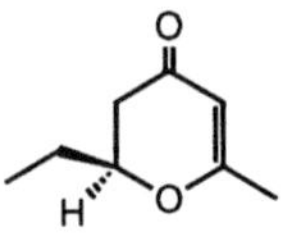 (5.3.11) 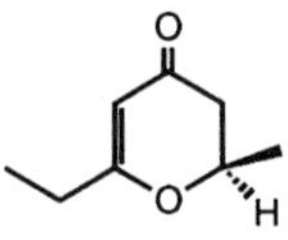 (5.3.12)

**Fig. 5.3 Cyclic pheromones of other Orders and of males of Lepidoptera (Key: M. = Male; F. = female; aggr. = aggregation)**

***5.3.1 Carbocyclic structures***

| Entry | Compound | Insect (Order) and action |
| --- | --- | --- |
| 5.3.1 | 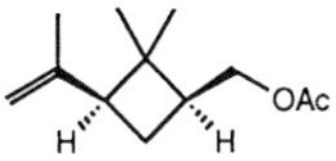 | Citrus mealybug (Homoptera) *Planococcus citri:* F. sex |
| 5.3.2 | | |
| | | Oriental fruit moth (Lepidoptera) *Grapholita molesta:* M. sex |
| 5.3.3 | | |
| 5.3.4 | | |
| 5.3.5 | | Boll weevil (Coleoptera) *Anthonomus grandis:* M. aggr. |
| 5.3.6 | | |
| 5.3.7 | 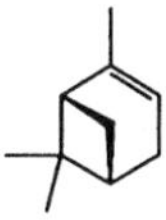 | *Myrmicaria eumenoides* (Hymenoptera): alarm |
| 5.3.8 | 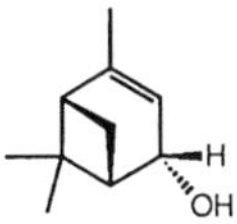 | Pine bark beetle (Coleoptera) *Ips paraconfusus:* M. aggr. |
| 5.3.9 | 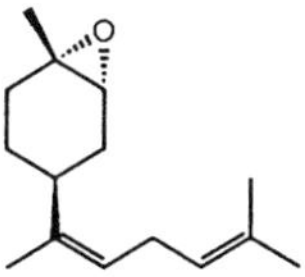 | Green stinkbug (Coleoptera) *Nezara viridula:* M. sex |
| 5.3.10 R = H | | *Ephestia kuehniella* (Lepidoptera): larvae alarm |
| 5.3.11 R = OH | | |
| 5.3.12 | 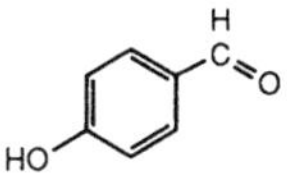 | African sugar-cane borer (Lepidoptera) *Eldana saccharina:* M. sex |

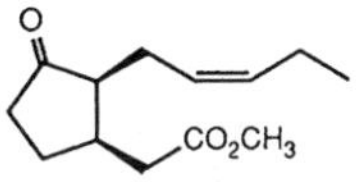

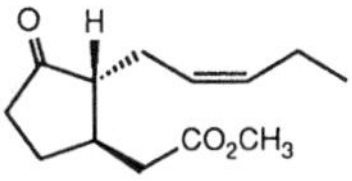

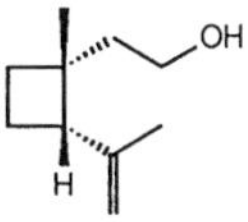

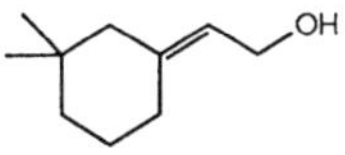

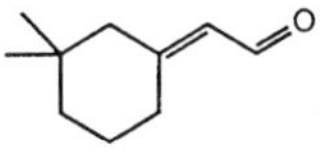

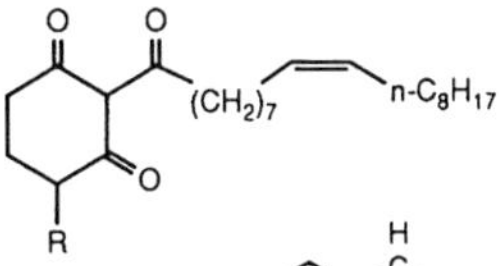

| Entry | Compound | Insect (Order) and action |
|---|---|---|
| 5.3.13 | 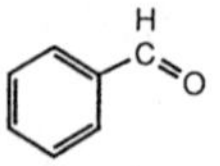 | Mamestra species (Lepidoptera): M. sex |
| 5.3.14 | 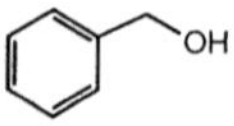 | |
| 5.3.15 | 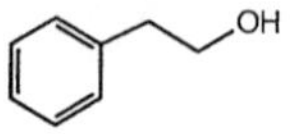 | |
| 5.3.16 | 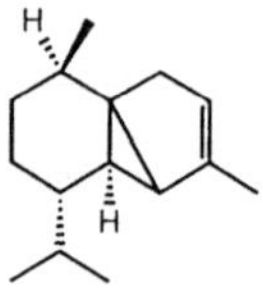 | Large elm bark beetle (Coleoptera) *Scolytus scolytus:* host derived aggr. |
| 5.3.17 | 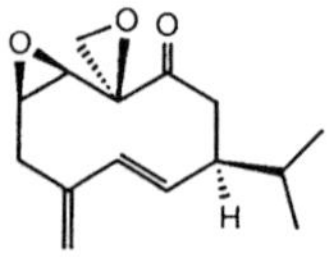 | American cockroach (Coleoptera) *Periplaneta americana:* F. sex |
| 5.3.18 | 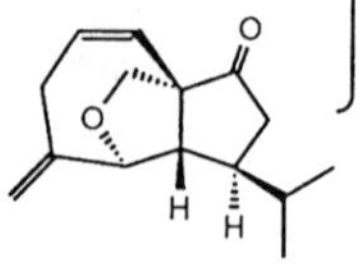 | |
| 5.3.19 | 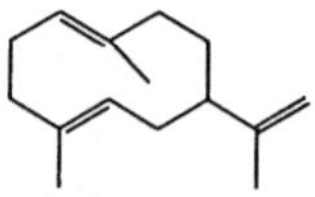 | Aphid species (Hemiptera): alarm |
| 5.3.20 | 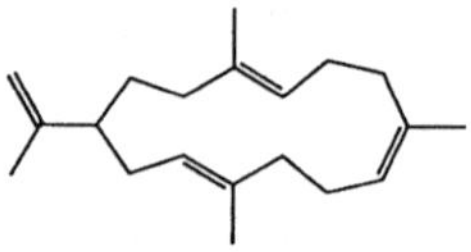 | Australian termite (Isoptera) *Nasutitermes exitiosus:* trail |

### *5.3.2 Heterocyclic structures: monocyclic ethers and lactones*

| Entry | Compound | Insect (Order) and action |
|---|---|---|
| 5.3.21 | 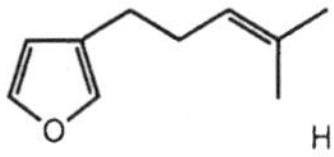 | *Tertramoriini anguilinode* (Hymenoptera): trail |
| 5.3.22 | 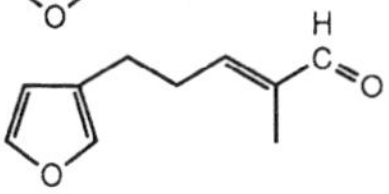 | European pine sawfly (Hymenoptera) *Neodiprion sertifer:* M. aggr. |
| 5.3.23 | 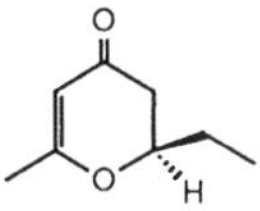 | Ghost moth (Lepidoptera) *Hepialus californicus:* M. sex |
| 5.3.24 | 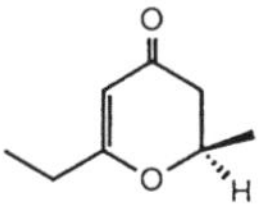 | Swift moth (Lepidoptera) *Hepialus hecta:* M. sex |
| 5.3.25 | R = $CH_3$ 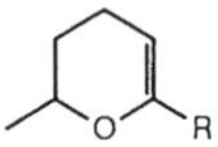 | *Apis mellifica* (Hymenoptera) |
| 5.3.26 | R = n-$C_3H_7$ | *Scaptotrigona bipunctata* (Hymenoptera) |
| 5.3.27 | R = n-$C_5H_9$ | *Plebia droryana* (Hymenoptera) |
| 5.3.28 | R = n-$C_7H_{15}$ | *Partamona cupira* (Hymenoptera) |
| 5.3.29 | R = n-$C_9H_{19}$ | *Partamona cupira* (Hymenoptera) |
| 5.3.30 | 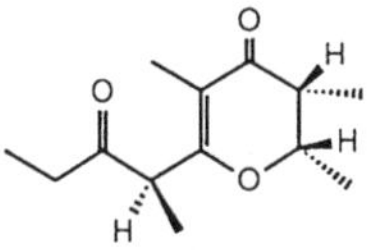 | Drugstore beetle (Coleoptera) *Stegobium paniceum:* F. sex |
| 5.3.31 | 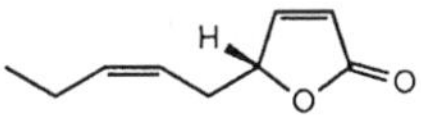 | Bumble-bee waxmoth (Lepidoptera) *Aphomia gularis:* M. sex |
| 5.3.32 | 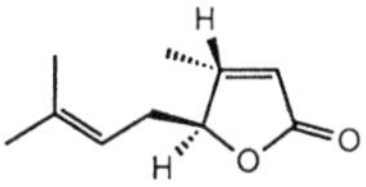 | African sugar-cane borer (Lepidoptera) *Eldana saccharina:* M. sex |
| 5.3.33 | 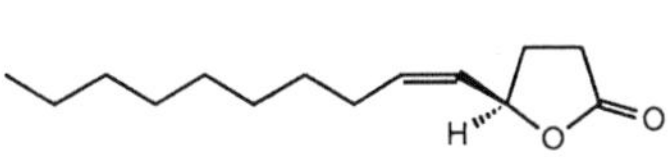 | Japanese beetle (Coleoptera) *Popillia japonica:* F. sex |
| 5.3.34 | 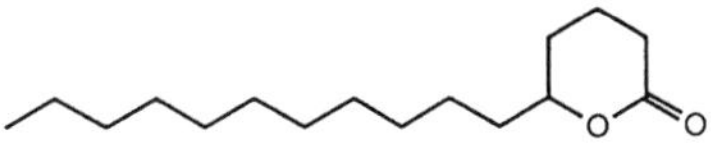 | Oriental hornet (Hymenoptera) *Vespa orientalis:* F. sex |
| 5.3.35 | 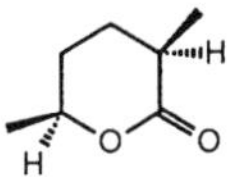 | Carpenter beetle (Hymenoptera) *Xylocopa hirutissima:* M. sex |

| Entry | Compound | Insect (Order) and action |
|---|---|---|
| 5.3.36 | 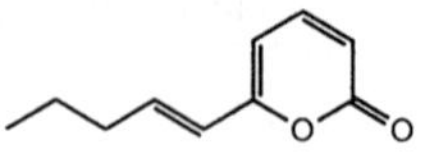 | Red fire-ant (Hymenoptera)<br>*Solenopsis invicta:* queen recognition |
| 5.3.37 | 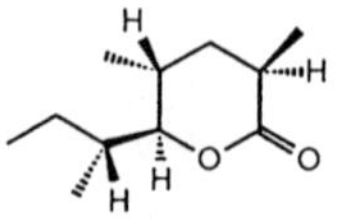 | |
| 5.3.38 | 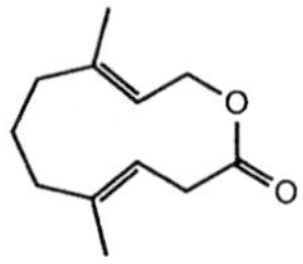 | Caribbean fruit fly (Diptera)<br>*Anastrepha suspensa:* F. sex |
| 5.3.39 | 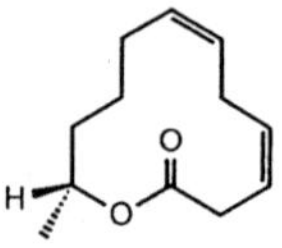 | Saw-toothed grain beetle (Coleoptera)<br>*Oryzaephilus surinamensis:* both aggr. |
| 5.3.40 | 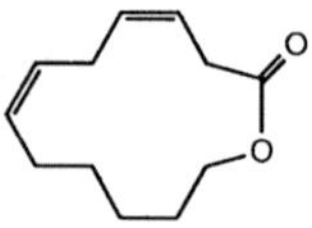 | |
| 5.3.41 | 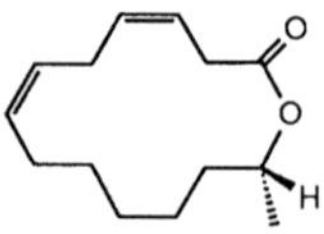 | |
| 5.3.42 | 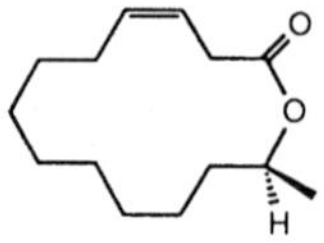 | Flat grain beetle (Coleoptera)<br>*Cryptolestes pusillus:* both aggr. |

### 5.3.3 Heterocyclic structures: polycyclic compounds and ketals

*(a) Fused rings*

| Entry | Compound | Insect (Order) and action |
|---|---|---|
| 5.3.43 | 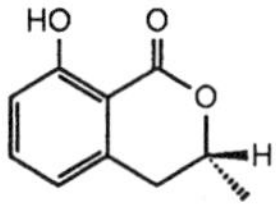 | Oriental fruit moth (Lepidoptera) *Grapholita molesta:* M. sex |
| 5.3.44 | 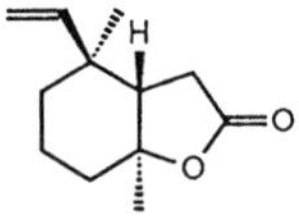 | Mexican fruit fly (Diptera) *Anastrepha ludens:* F. sex |
| 5.3.45 | 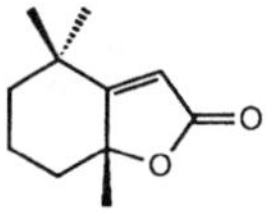 | Red fire-ant (Hymenoptera) *Solenopsis invicta:* queen recognition |
| 5.3.46 | 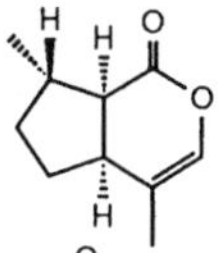 | Vetch aphid (Hemiptera) *Megoura viciae:* F. sex |
| 5.3.47 | 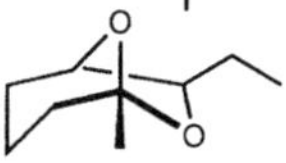 | Western pine beetle (Coleoptera) *Dendroctonus brevicomis:* F. aggr. |
| 5.3.48 | 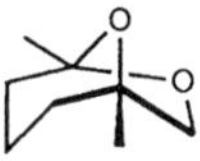 | Southern pine beetle (Coleoptera) *Dendroctonus frontalis:* F. aggr. |
| 5.3.49 | 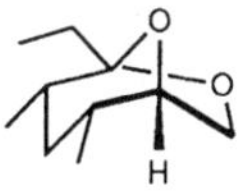 | Smaller elm bark beetle (Coleoptera) *Scolytus multistriatus:* F. aggr. |
| 5.3.50 |  | Striped ambrosia beetle (Coleoptera) *Tyrpodendron lineatum:* F. aggr. |
| 5.3.51 | 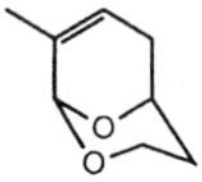 | Swift moth (Lepidoptera) *Hepialus hecta:* M. sex |
| 5.3.52 | 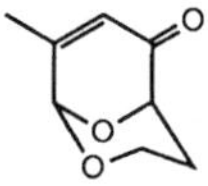 | Swift moth (Lepidoptera) *Hepialus hecta:* M. sex |

*(b) Spirocyclic rings*

| Entry | | Compound | Insect (Order) and action |
|---|---|---|---|
| 5.3.53 | | 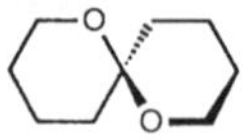 | |
| 5.3.54 | | 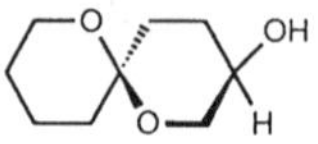 | Olive fruit fly (Diptera) *Dacus oleae:* F. sex |
| 5.3.55 | | 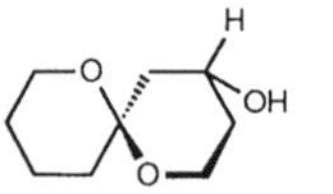 | |
| 5.3.56 | | 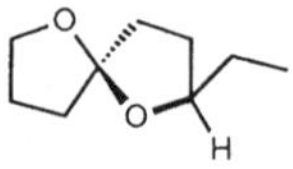 | *Pityogenes chalcographus* (Coleoptera): M. sex |
| 5.3.57 | R = $CH_3$ | 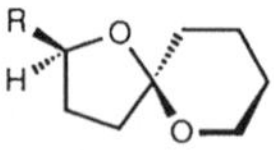 | Common wasp (Hymenoptera) |
| 5.3.58 | R = $C_2H_5$ | | *Paravespula vulgaris:* F. sex |
| 5.3.59 | | 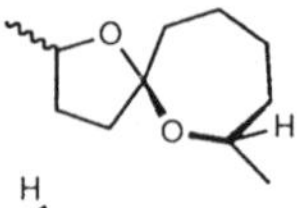 | |
| 5.3.60 | | 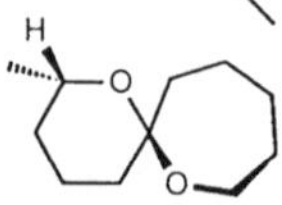 | *Andrena haemorrhoa* (Hymenoptera): F. sex |
| 5.3.61 | | 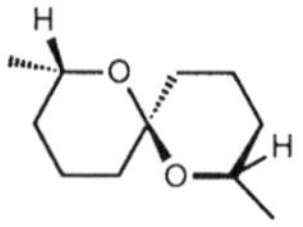 | *Andrena wilkella* (Hymenoptera): F. sex |

When the ring oxygen has a carbonyl group (C=O) adjacent, then the compound is a cyclic (or internal) ester and is called a **lactone**. Systematic nomenclature gives the ending **-olide** to these structures and the numbering system starts from the carbonyl group. The male attractant produced by females of the oriental hornet, *Vespa orientalis*, is (**5.3.13**), hexadecan-5-olide; the numeral -5- indicating the position of attachment of the oxygen to the otherwise straight chain.

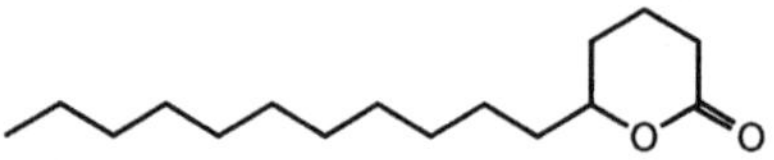

(**5.3.13**)

Where there are other functionalities as in (**5.3.14**), the male-produced sex pheromone of the bumble-bee waxmoth, *Aphomia sociella*, then these are indicated in the normal way: (Z,Z)-2,6-nonadien-4-olide for (**5.3.14**) and 2,4,6-trimethyloctan-5-olide for (**5.3.15**), one of the components of the queen recognition complex from the fire ant, *Solenopsis invicta*.

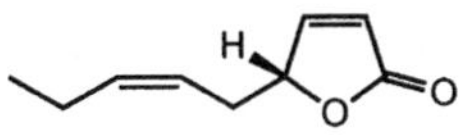

(5.3.14) 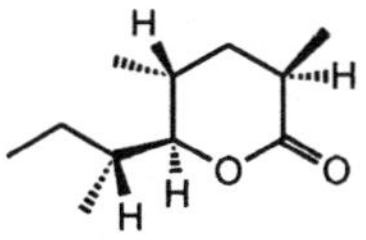 (5.3.15)

Figure 5.3.2 contains a selection of the simple heterocyclic structural types that have been found. It is of interest to note that a number of large-ring lactones are represented amongst them; macrolides of similar ring sizes are found also among a significant number of lactonic antibiotics. Among the pheromones we find compounds such as (**5.3.16**), (13*R*)-(Z,Z)-5,8-tetradecadien-13-olide, produced by the saw-toothed grain beetle, *Oryzaephilus surinamensis*, where the basis is a straight chain hydroxy-acid, and (**5.3.17**), 4,8-dimethyl-(*E,E*)-3,8-decadien-10-olide, isolated from the Caribbean fruit fly, *Anastrepha suspensa*. In this compound the branching pattern reveals a terpenoid origin.

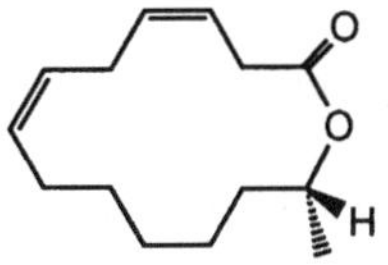

(5.3.16) 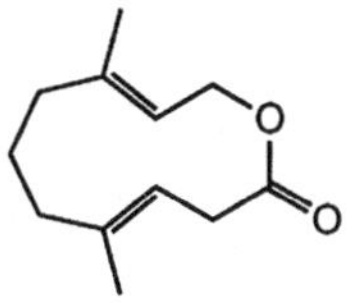 (5.3.17)

### 5.3.3 Polycyclic compounds and ketals

Unlike carbocyclic pheromones, which are mainly monocyclic, there is a significant number of oxygen-containing compounds which have bi- and tricyclic skeletons. For convenience they may be divided into two groups: those with **fused rings**, in which the rings share two or more atoms between them, and those which have only one atom in common – **spirocyclic** compounds. A representative selection is shown in Fig. 5.3.3.

#### *(a) Fused rings*

Examples of the first group are (**5.3.18**), also part of the queen recognition group from *Solenopsis invicta*, and (**5.3.19**), the female aggregation pheromone of the smaller European elm bark beetle, *Scolytus multistriatus*. Each contains two rings and for naming purposes is treated initially as though the oxygen(s) in the ring(s) were carbons. The presence of the oxygen atom(s) is then shown by a position numeral and the prefix **oxa-**. Starting from one of the two bridgehead atoms, the ring atoms are numbered sequentially around the periphery, beginning with the largest ring (longest chain between bridgehead atoms) and continuing with the next largest and so on. The root name for this **total** number of atoms is used; the length of each chain, not counting the bridgehead atoms, joining the bridgeheads is shown and the number of rings is indicated by **bicyclo-, tricyclo-** and so on.

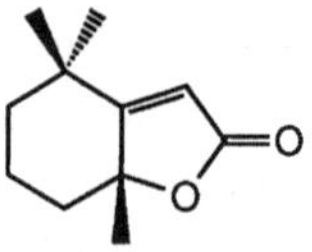

(**5.3.18**) 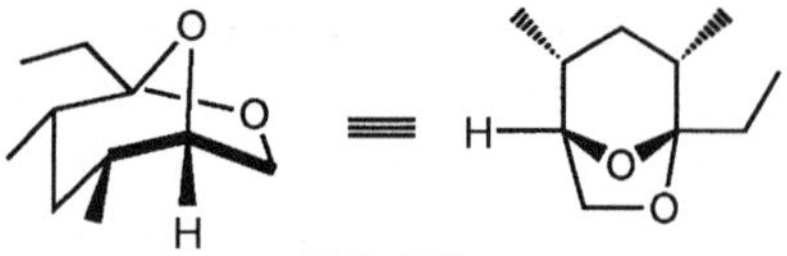 (**5.3.19**)

Thus (**5.3.18**) has the [4:3:0]-bicyclononane skeleton and (**5.3.19**) is a representative of the [3:2:1]-bicyclo-octane family. The position of the oxygen atoms and the other functionalities are then added to this basic name to give, for (**5.3.18**), 2,2,6-trimethyl-7-oxa-[4:3:0]-bicyclonon-1(9)-en-8-one; and for (**5.3.19**), 1-ethyl-2,4-dimethyl-7,8-dioxa-[3:2:1]-bicyclo-octane.

Although this nomenclature system is unambiguous, it often disguises the nature of the functionality present; for example, the lactone ring in (**5.3.18**) is only clear after one has translated the written name into a structural formula. This represents a situation where a suitable trivial name may be more helpful than the systematic one, and (**5.3.18**) is also known as invictolide.

### *(b) Ketals*

Although at first sight compound (**5.3.19**) would appear to be a simple ether (normally these are chemically unreactive), it is actually a representative of the functional group: the ketals. Ketals are compounds in which two ether type oxygens are bonded to the same carbon atom as C—O—C—O—C. This is C-1 in (**5.3.19**). A simpler example is 2,2-dimethoxypropane, (**5.3.20**). Ketals are derived by the elimination of water from two molecules of an alcohol and one of a ketone: Scheme 5.2.

Scheme 5.2 $(CH_3)_2C{=}O$ + $HOCH_3$ + $HOCH_3$ $\rightleftharpoons$ $H_2O$ + $(CH_3)_2C(OCH_3)_2$

(**5.3.20**)

This reaction is catalysed by acid and is completely reversible. The position of the equilibrium depends markedly on the structure of the ketal and the amount of water present; for example, to form (**5.3.20**), water must be removed to displace the equilibrium. In the presence of water and acid, 2,2-dimethoxypropane is rapidly transformed into propan-2-one (acetone) and methanol. On the other hand, (**5.3.19**) is apparently unaffected by aqueous acid, being recovered unchanged. However, under acidic conditions it will react with reagents which affect ketones, showing that the equilibrium of Scheme 5.3 is readily established.

Scheme 5.3 + $H_2O$

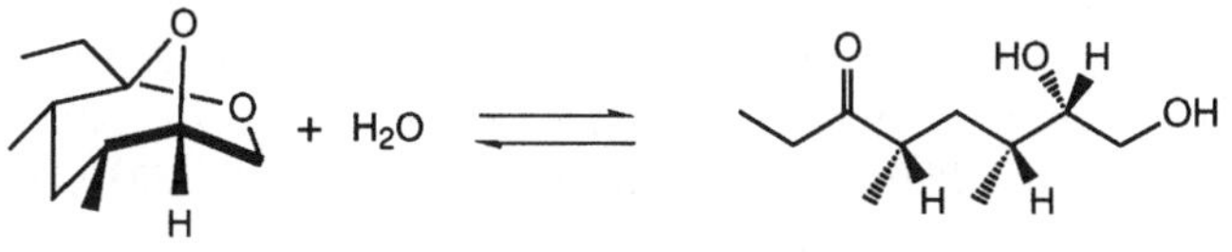

(**5.3.19**)

The ketal group is the significant functionality in the second group of bicyclic compounds found amongst pheromone structures: the spirocyclic group.

### (C) *Spirocyclic compounds*

Representatives of the group are the pheromones of the *Dacus* species–the olive fly, *D. oleae*, structure (**5.3.21**), and the Oriental fruit fly, *D. dorsalis*, (**5.3.22**)–and the common wasp, *Paravespula vulgaris*, compound (**5.3.23**).

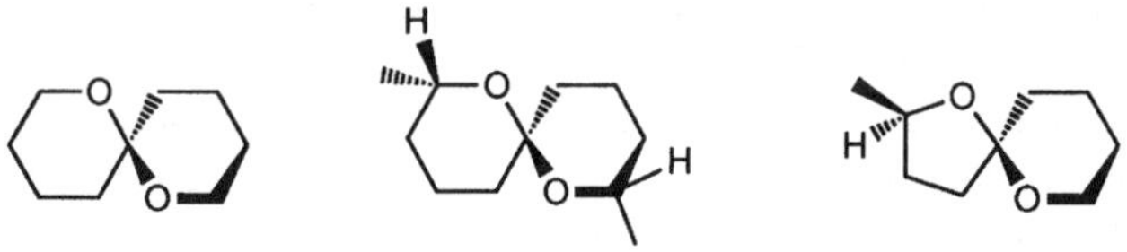

(**5.3.21**) (**5.3.22**) (**5.3.23**)

Exactly as with the fused rings, all the atoms are initially considered as carbons and are numbered sequentially from the atom adjacent to the common one (the **spiro** atom) starting with the larger of the two rings. The skeleton is then named by the prefix spiro- and the two numbers that show the length of the chain for each ring needed to return to the spiro atom. The olive fly pheromone is hence a spiro-[5,5]-undecane and the oxygen positions are shown by the same method as for fused rings, resulting in the name 1,7-dioxa-spiro-[5,5]-undecane for (**5.3.21**), 2,8-dimethyl-1,7-dioxa-spiro-[5,5]-undecane for (**5.3.22**) and 8-methyl-1,7-dioxa-spiro-[5,4]-decane for (**5.3.23**).

Although at first sight symmetric, the spirocyclic structure creates a centre of chirality at the spiro atom because the two rings are held in two planes that are perpendicular to each other. As a result, (**5.3.21**) can show optical activity. The two enantiomers are shown below.

(*R*)-(**5.3.21**) (*S*)-(**5.3.21**)

This chirality is an additional feature on top of the geometric isomerism that is inherent in the ring itself. The geometry is designated as indicated for simple rings, but with relation to the oxygen atom of the other ring, which is regarded as the principal substituent. For example, the *Paravespula vulgaris* pheromone (**5.3.23**) is (Z)-8-methyl-1,7-dioxa-spiro-[5,4]-decane.

## 5.4 NITROGEN-CONTAINING PHEROMONES

Although nitrogen-containing compounds form the basis of all proteins and nucleic acids, there are relatively few pheromones in which the element is found. The known classes of acyclic and carbocyclic compounds are shown in Fig. 5.4.1 and the heterocyclic ones in Fig. 5.4.2.

### 5.4.1 Acyclic compounds: amides

Almost without exception, the acyclic pheromones containing nitrogen do so in the form of the **amide** group, >N—C=O—, which is the nitrogen analogue of the ester group, —O—C=O—, and is the functionality that constitutes the backbone of proteins. Like esters, amides are formed from an acid and an amine (Scheme 5.4) and are named as derivatives of the acid part of the molecule.

Scheme 5.4 $NH_2$ + HO → N H

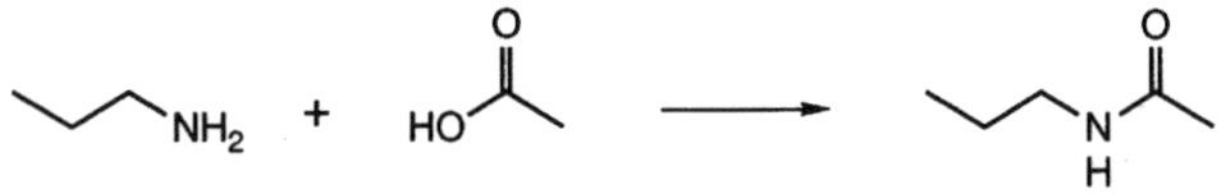

(**5.4.1**)

The group on the nitrogen is named in the usual way for radicals and its attachment is shown by using the indicator N-; thus (**5.4.1**) is N-propyl acetamide. Three amides have been isolated from males of the melon fly, *D. cucurbitae*; they are N-(2-methylbutyl) acetamide (**5.4.2**), N-(3-methylbutyl) acetamide (**5.4.3**) and N-(3-methylbutyl) 2-methoxyacetamide (**5.4.4**).

**Fig. 5.4 Cyclic pheromones of other Orders and of males of Lepidoptera (Key: M. = male; F. = female; aggr. = aggregation)**

### *5.4.1 Nitrogen-containing compounds*

*5.4.1.1 Acyclic and carbocyclic*

| Entry | Compound | Insect (Order) and action |
|---|---|---|
| 5.4.1 | 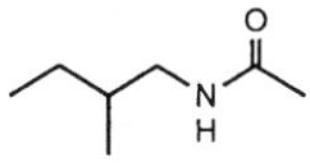 N-H, O | Melon fly (Diptera) *Dacus cucurbitae:* M. sex |
| 5.4.2 | 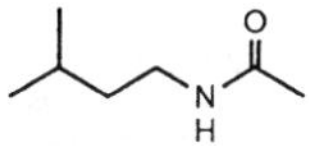 N-H, O | |
| 5.4.3 | 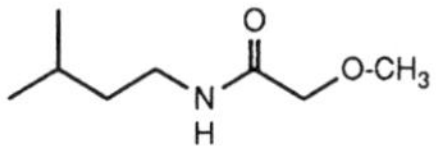 N-H, O, O-$CH_3$ | |
| 5.4.4 | 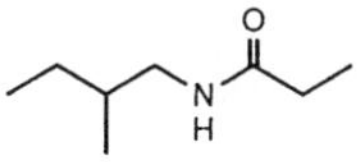 N-H, O | Queensland fruit fly (Diptera) *Dacus tryoni:* M. sex |
| 5.4.5 | 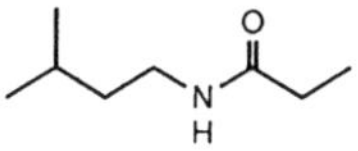 N-H, O | |
| 5.4.6 | 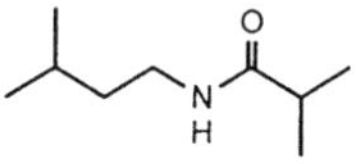 N-H, O | |
| 5.4.7 | 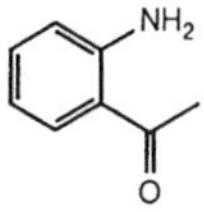 $NH_2$, O | Web-spinning larch saw fly (Hymenoptera) *Cephalcia lariciphila:* F. sex |

*5.4.1.2 Heterocyclic compounds*

| Entry | Compound | Insect (Order) and action |
|---|---|---|
| 5.4.8 | 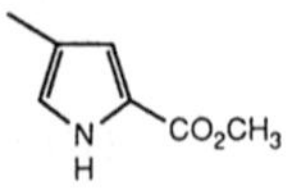 | Leaf-cutting ants (Hymenoptera)<br>*Atta texana:* trail<br>*Atta cephalotes:* trail |
| 5.4.9 | 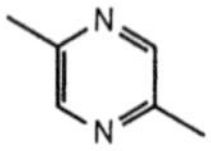 | *Tetramorium caespitum* (Hymenoptera): trail |
| 5.4.10 | 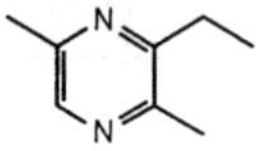 | Leaf-cutting ants (Hymenoptera)<br>*Atta rubropilosa:* trail<br>*Atta sexdens:* trail |
| 5.4.11 | 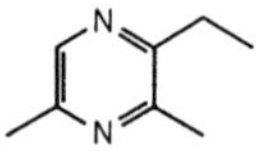 | *Odontomachus brunneus* (Hymenoptera): alarm |
| 5.4.12 | 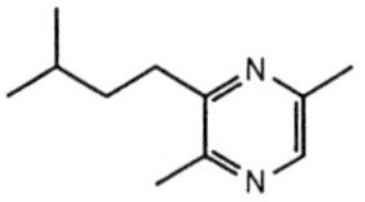 | |
| 5.4.13 | 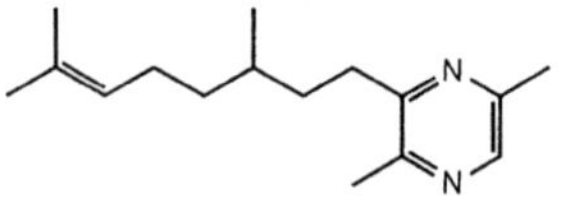 | *Rhytidoponera metallica* (Hymenoptera): trail |
| 5.4.14 | 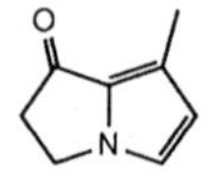 | Queen butterfly (Lepidoptera)<br>*Danaus gilippus:* M. sex |
| 5.4.15 | 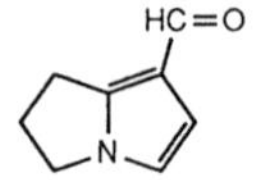 | Trinidad butterfly (Lepidoptera)<br>*Lycorea ceres ceres:* M. sex |
| 5.4.16 | 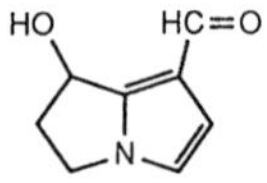 | |
| 5.4.17 | 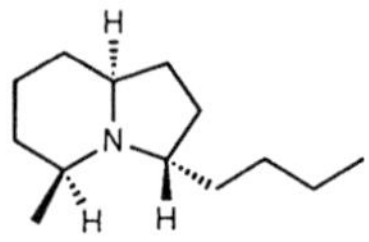 | Pharaoh's ant (Hymenoptera)<br>*Monomorium pharaonis:* trail |

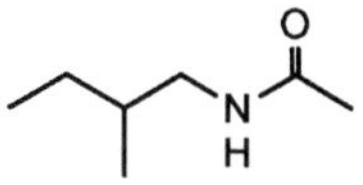

(5.4.2) 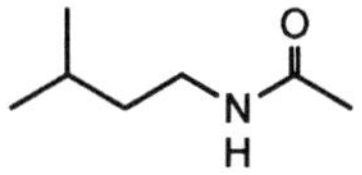 (5.4.3) 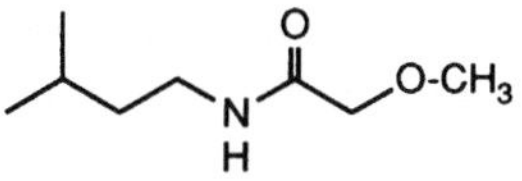 (5.4.4)

## 5.4.2 Nitrogen heterocycles

The widespread occurrence of heterocyclic nitrogen compounds in plant alkaloids and other natural products means that all the ring systems found in the pheromonal compounds have well used trivial names. The two monocyclic types are those based on pyrrole, (**5.4.5**), and pyrazine, (**5.4.6**), which are numbered as shown.

The leaf-cutting ant *Atta texana* uses methyl 4-methylpyrrole-2-carboxylate, (**5.4.7**), as a trail pheromone component and the ant *Rhytodoponera metallica* has been found to use 2,5-dimethyl-3-(3′,7′-dimethyl-6′-octenyl)-pyrazine, (**5.4.8**), in which we may recognize the terpenoid side chain at C-3.

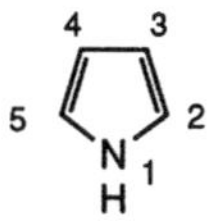

(5.4.5) 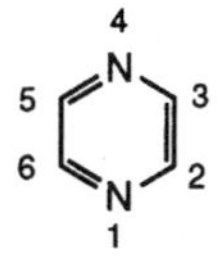 (5.4.6) 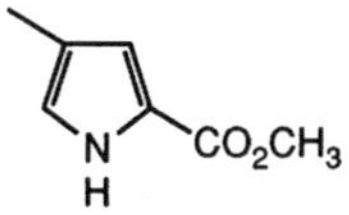 (5.4.7)

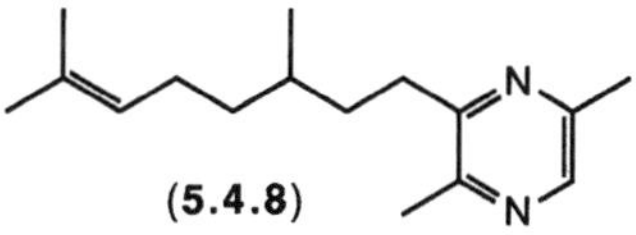

(5.4.8)

The two bicyclic skeletons are those derived from pyrrolizine, (**5.4.9**), and indolizine, (**5.4.10**), each of which has the nitrogen at the junction between the two rings.

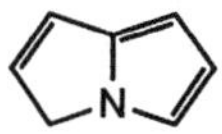

(5.4.9) 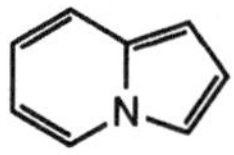 (5.4.10) 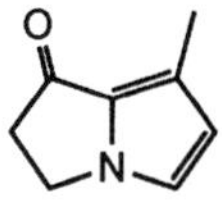 (5.4.11)

The first of these is the basis of a number of compounds employed by different members of the Danaid family as well as a number of other species. Particularly common is danaidone, (**5.4.11**): systematically 2,3-dihydro-7-methyl-1H-pyrrolizin-1-one.

Finally, the compound in Table 5.4 (entry 5.4.17) which was first thought to be the trail pheromone of the Pharaoh's ant, *Monomorium pharaonis*, is an example of the indolizine group. It now seems that this compound acts as a synergist for the true pheromone, which has been discussed in section 5.2.3, structure (**5.2.21**).

In conclusion, one can see that among the chemical compounds used as pheromones there is a very wide range of structural types. In general, it is found that within a family very similar compounds are used by the different species and this can be of considerable value in any structural investigation. However, as with all natural systems, there is still a wide variation and one needs to keep a wary eye open for the not infrequent exception.

## REFERENCES

Arn, H., Tóth, M. and Priesner, E. (1992) *List of Sex Pheromones of Lepidoptera and Related Attractants*, Organisation Internationale de Lutte Biologique, Section Régionale Ouest Paléarctique, Montfavet.

Baker, T. C. (1989) Sex pheromone communication in the Lepidoptera: new research progress. *Experientia*, **54**, 248–262.

Cahn, R.S. (1974) *Introduction to Chemical Nomenclature*, 4th edn, Butterworth, London.

Mayer, M. S. and McLaughlin, J.R. (1991) *Handbook of Insect Pheromones and Sex Attractants*, CRC Press, Boca Raton.

Tumlinson, J. H. (1988) Contemporary frontiers in insect semiochemical research. *J. Chem Ecol.*, **11**, 2109–2129.

## APPENDIX 5A NAMES OF HYDROCARBONS AND MULTIPLYING PREFIXES

### (a) Names of hydrocarbons

| | | |
|---|---|---|
| 1 methane | 11 undecane | 21 heneicosane |
| 2 ethane | 12 dodecane | 22 docosane |
| 3 propane | 13 tridecane | 23 tricosane |
| 4 butane | 14 tetradecane | 24 tetracosane |
| 5 pentane | 15 pentadecane | 25 pentacosane |
| 6 hexane | 16 hexadecane | 26 hexacosane |
| 7 heptane | 17 heptadecane | 27 heptacosane |
| 8 octane | 18 octadecane | 28 octacosane |
| 9 nonane | 19 nonadecane | 29 nonacosane |
| 10 decane | 20 eicosane | 30 triacontane |
| 40 tetracontane | 50 pentacontane | |

### (b) Multiplying prefixes

The multiplying prefixes are derived from the hydrocarbon names with the exceptions of:
1 mono
2 di
3 tri
4 tetra
From 5 onwards they are regular: 5 = penta, etc.

## APPENDIX 5B PREFIXES AND SUFFIXES FOR COMMON FUNCTIONAL GROUPS

| Functional group | Formula | Prefix | Suffix |
|---|---|---|---|
| Alcohol | —OH | hydroxy | -ol |
| Aldehyde | —CH=O | formyl | -al |
| Amine | $—NH_2$ | amino | -amine |
| Carboxylic acid | —COOH | carboxy | -oic acid |
| Ester | —COOR | R-oxycarbonyl | R -oate |
| Ketone | >C=O | oxo | -one |

# 6

# Isolation and structure determination

Before starting on any attempt to isolate an insect pheromone, it is important to demonstrate that such a chemical cue is actually involved and also to define the behaviour that is associated with its production and release. Examples of behaviour might be trail following, aggregation, alarm, or sexual arousal. Such studies must lead to the development of an appropriate bioassay based on the behavioural response of the insect to the pheromone. Bioassays are discussed in more detail in Chapter 4 but we may note here that among the problems associated with bioassays are:

- the innate sensitivity of the insect to the test;
- the variability of behaviour and response within a population;
- the temporal (diurnal) variation of the insects' response to the test;
- the discriminatory ability of the bioassay.

As well as, or alternatively to, a behavioural assay, electroantennography (EAG) may be used to detect those fractions of an extract which are biologically active. While EAG is both straightforward and readily used, one must bear in mind that it will identify **all** the components of a mixture to which the insect will respond and not only those which mediate the behaviour elicited by the pheromone being sought.

With a suitable bioassay in hand, one may then proceed to the isolation and analysis of the insect pheromone, a procedure which may be outlined by the flow chart shown in Fig. 6.1.

It is **essential** that each step in the process be monitored by the bioassay (or EAG). Any extract contains large numbers of compounds and to isolate, purify and identify all of them would involve an excessive amount of time and effort, mostly to no avail. The bioassay ensures that the maximum effort is concentrated on those fractions and those

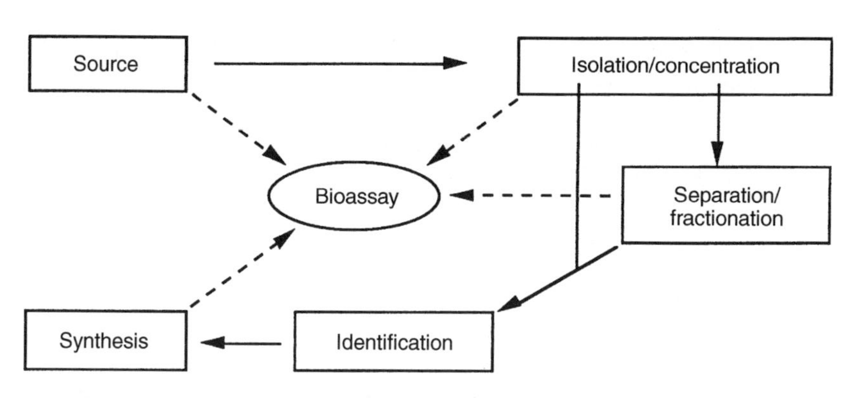


**Fig. 6.1** Flow chart of procedure for isolation of pheromones.

particular compounds that are involved in the behaviour one wishes to investigate.

## 6.1 ISOLATION

At the outset it is desirable to find out the source of the pheromone within the insect. The initial studies should have shown whether it is produced by both sexes or only by one of them, and also which one. Dissection of the insect and bio-testing of the various parts should be employed to determine:

- whether the pheromone is produced and/or stored in a specialized gland (e.g. ovipositor, mandibular), or is a cuticular component.
- whether it is present in the gland as the pheromone, or as a precursor.

Biological tests are also needed to decide if the pheromone is produced continuously or only at certain times of the day ('calling behaviour') and also to gain some idea of the quantities produced.

The exact choice of isolation method will depend on the insect being studied, the numbers available and the type of pheromone system (e.g. trail, alarm, oviposition). The major problems associated with all the methods are those of:

- **quantity**–the amounts of pheromonal material are always very small, commonly of the order of nanograms per insect, and hence the risk of loss of active material is high.
- **chemical change**, which may occur as a result of aerial oxidation, acid, base or enzymatic action.
- **contamination** of active material by minor impurities in the solvents used, the glassware employed and, most particularly, from plastic

materials, such as vials, caps and tubing, from which significant amounts of plasticizers are all too readily leached. Probably the most common compounds found in an extract are dibutyl- and dioctylphthalate. Such contamination may inhibit the activity of the pheromone and invariably makes isolation and identification more difficult.

### 6.1.1 Isolation methods

*(a) Solvent extraction*

This method involves the maceration of the insect (or part thereof) with a suitable organic solvent, followed by filtration and concentration (usually by distillation). Generally large numbers of insects are required, typically from a few hundred to 100 000. To avoid contamination, the use of pure re-distilled solvents and clean glassware is essential. Ideally, the choice of solvent will depend on the type of material to be extracted and the solvent's polarity should match that of the pheromone. In the initial phase of an investigation, this is not possible and therefore successive extraction with a range of solvents of increasing polarity needs to be employed. The most useful solvents have low boiling points and contain relatively few contaminants when received from the suppliers; of those that are readily available, pentane and dichloromethane are probably the ones that best fit these criteria and represent solvents of low and moderate polarity. However, any solvent should be carefully re-distilled using a fractionating column for the most exact work. An involatile contaminant that is present at only 1 ppm in the solvent means that evaporation of 1 litre of solvent will leave 1 mg of material – about 1000 times the amount that is likely to be obtained from the extraction of 1000 insects.

The advantages of solvent extraction are that it gives information on the amounts of the pheromonal components (and of any other products) that are present in the insect at the time of extraction and that it is simple to apply. Subsequent treatment of the residue may also give information about material held as a precursor (section 6.3.1a: codling moth).

There are three major ways in which solvent extraction has been employed:

**Whole body extraction**

In this the whole of the insect is 'blended' with solvent, typically in a liquidizer; the solution is filtered off and distilled to remove the solvent. The method is simple and straightforward, does not require any preliminary work to discover the source of the pheromone within the insect and can be applied to any system. It has the major disadvantage that large amounts of cuticular material are of necessity also present in the extract, entailing a corresponding increase in complexity of the separation process (section 6.3.1a).

**Excised gland extraction**

This requires prior identification of the gland producing the pheromone, dissection of the appropriate part of the insect and then maceration of the sample with solvent as above. It avoids most of the problems of the unwanted components involved in whole body extraction, but does require the time and skill needed for micro-dissection and is only feasible where the insect is of a reasonable size (section 6.3.1b).

**Ovipositor washing**

This method has been used extensively in the study of lepidopteran pheromones. It consists of squeezing the body of the female insect in such a way as to force the ovipositor and gland to extrude and then excising them both and rinsing them with solvent, most commonly hexane or heptane. The result is a mixture containing relatively few components, facilitating analysis, but it can only be applied to moths and similar large insects (section 6.3.1c).

*(b) Aeration*

This method involves trapping of the volatile components produced by the insect over a period of time either in a **cold trap** or on a **solid adsorbent**. Live insects are held in a suitable closed container and air is drawn over the insects and passed into either a cold trap or a small tube containing the adsorbent. The nature of the containing vessel, the exact conditions within it and the type of trap will depend very much on the particular insect under study and the numbers available. Figures 6.2 to 6.4 show typical experimental set-ups that have been used. Regardless of the method of trapping, this technique requires a completely uncontaminated air supply. This is more readily achieved if the air is drawn through the apparatus, rather than being supplied from a compressor. Typically, a large filter of activated charcoal is employed to ensure that any volatiles in the air supply are removed prior to its passage into the insect chamber.

The major advantages of aeration are that it provides a sample of the volatile compounds actually emitted by the insect and, because the insects are kept alive, collection of volatiles may continue over extended periods of time. This is particularly useful when the numbers available are limited or the amount of pheromone present per insect is very low. Where the insects are kept in the container for extended periods, then

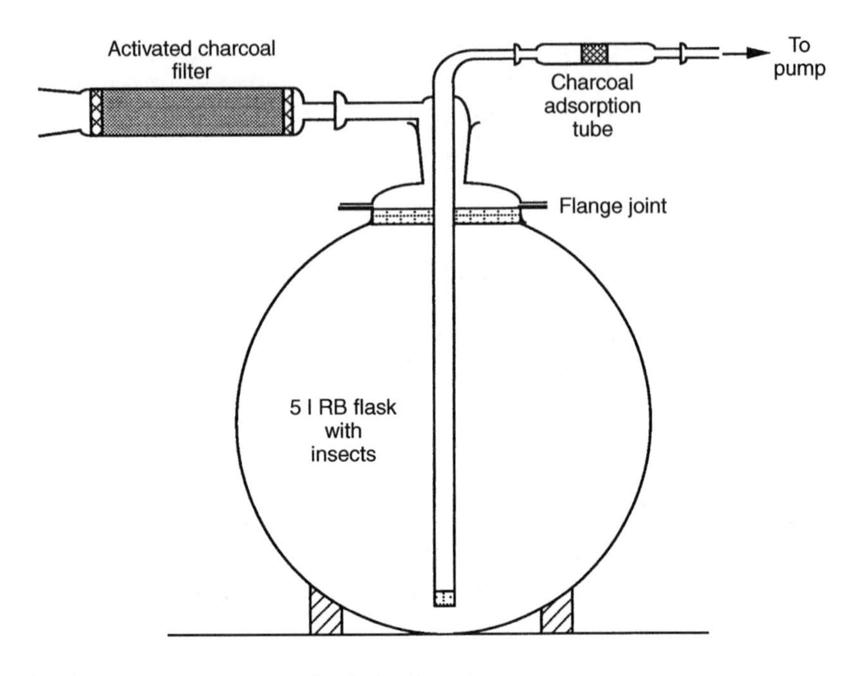


**Fig. 6.2** Aeration apparatus for flying insects.

both food and water need to be supplied. This requires that a suitable 'blank' run (in the absence of insects) be carried out in order to factor out the food volatiles that will be entrained with the pheromone.

Features and problems which depend on the method of trapping are:

**Cold trapping**

Depending on the volatility of the materials sought, the trap may be held at temperatures down to –195°C (liquid nitrogen). At such temperatures, rapid cooling of the air stream may result in aerosol formation, with consequent loss of material. This may be overcome by appropriate trap design. The other major problem is that large amounts of water are invariably condensed along with the pheromone and this must be removed in a subsequent step. Rinsing the trap with solvent should be employed to minimize adsorption on to the surface of the glass.

**Solid adsorbent**

A wide variety of solid adsorbents have been used, the most common being activated charcoal, glass and the synthetic porous polymers–Porapak-Q and Tenax. Desorption may be achieved either by washing the filter with solvent (either single pass or soxhlet extraction), or by

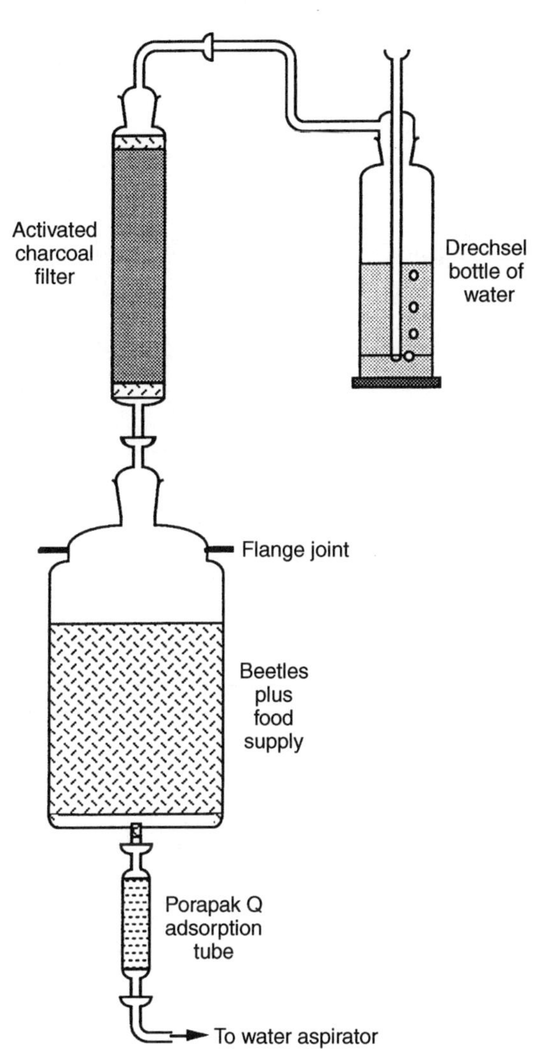


**Fig. 6.3** Aeration apparatus for grain beetles. (Drawn after Pierce *et al.*, 1984.)

thermal desorption directly into a gas chromatograph or solvent trap. The problems that arise are that adsorbed oxygen sensitive compounds may undergo oxidation; that isomerization of alkenes has been known to occur; that artefacts may be formed from the adsorbent, during both aeration and desorption (this is more a problem of the polymeric supports, which required extensive conditioning before use); that solvent desorption can introduce impurities (as with extraction); and

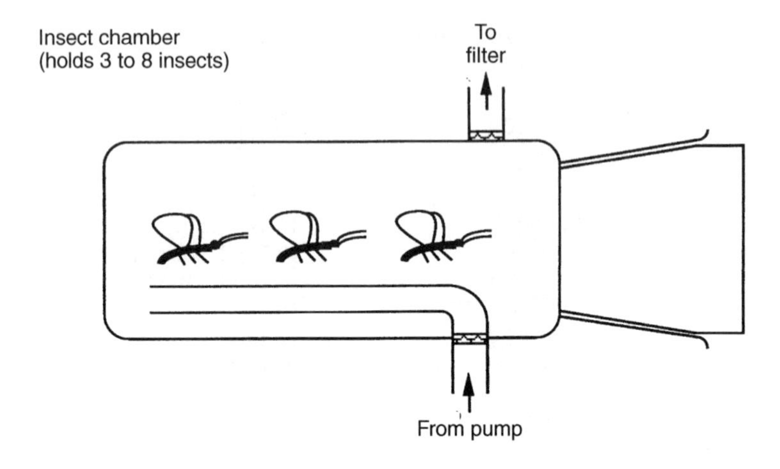


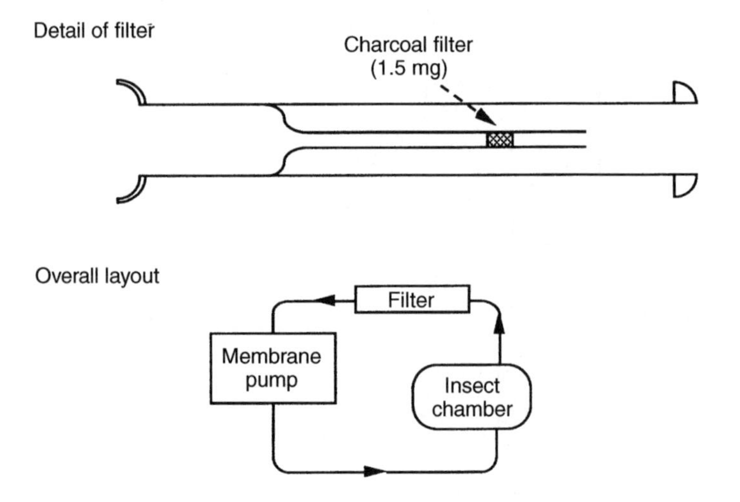


**Fig. 6.4** Closed loop aeration apparatus: (a) insect chamber (holds 3–8 insects); (b) detail of filter; (c) overall layout. (Redrawn from H.J. Bestmann, J. Erler and O. Vostrosky, 1988, *Experientia*, **44**, 797.)

that thermal desorption can induce structural changes in sensitive compounds.

### *(c) General and specific problems: solvents*

Most of the problems associated with isolation have been dealt with in the preceding sections and may be summarized in the advice: 'Employ careful cleanliness at all times and avoid contact of samples with plastic materials'.

Many different solvents have been employed for extraction purposes but the most generally useful are:

- **non-polar**: pentane, hexane (note that the latter requires more careful fractionation);
- **semi-polar**: dichloromethane;
- **polar**: methanol, carbon disulphide (although more toxic and more flammable than methanol, $CS_2$ is very effective for the recovery of compounds from solid adsorbents and also has the advantage that it gives very low signals on the flame ionization detector of gas chromatographs).

As previously mentioned, careful re-distillation of **all** solvents is an essential precaution which must not be neglected.

Finally, the use of scrupulously clean glassware and the minimization of the amount of grease used on the taps of separating funnels, or preferably Teflon coated taps, and its avoidance elsewhere cannot be overemphasized.

### 6.1.2 Separation methods

With the exception of distillation, all the separation methods are based on one form or another of chromatography, in which compounds are separated by differential partition between a mobile and a stationary phase. The nature of the two phases involved is used to define the different types of chromatography. Operational details for these methods are to be found in most texts on practical organic chemistry (e.g. Harwood and Moody, 1989; Casey *et al.*, 1990) and in more specialized books on chromatography (e.g. Braithwaite and Smith 1985) and will not be discussed here.

#### *(a) Distillation*

This is most frequently used for the removal of excess solvent from extracts in order to concentrate them for further study. If the pheromone is particularly volatile, then a fractioning column will be required when the volume of solvent gets down to about 1 ml. When the chemical nature of a pheromone has been established, then short path distillation in the presence of a suitable carrier may prove useful in order to separate the required compound from both more and less volatile materials before further purification. Steam distillation is useful in the early stages of a separation to remove the more volatile pheromones from fats and waxes. Small volumes of extracts in volatile solvents may conveniently be concentrated by placing them in a small tapered vial and gently blowing a stream of dry purified nitrogen over the surface, while maintaining the sample at room temperature. Some care is required to prevent water condensing into the sample.

### *(b) Thin layer and column chromatography*

Thin layer chromatography (TLC), so called because the stationary phase is held as a thin layer on the surface of a glass, aluminium or plastic plate, provides a rapid and easy method for evaluating a number of possible solvent systems for the separation of a mixture and also provides a useful way of following the progress of a column chromatogram. A solution of the sample (*c* 1 mg) is applied as a spot near one end of a TLC plate and solvent is then allowed to rise up the plate by capillary attraction. This elutes the components, whose positions may subsequently be visualized by one of a number of methods (e.g. fluorescence quenching, iodine, vanillin–sulphuric acid). TLC may also be used for rapid separations of quantities up to 100 mg (preparative TLC) by using wide plates and applying the material as a line. Components are recovered by scraping off the adsorbent in zones and subsequent solvent extraction.

Column chromatography is applicable on any scale from 100 mg up to 20 g with little modification. It has commonly been employed for preliminary separations of multi-gram amounts of material from large-scale whole body extractions. The mixture is applied at the top of a column of solid adsorbent (usually alumina or silica gel) which is packed, normally as slurry in a non-polar solvent, into a glass tube. The column is eluted by gravity (or moderate air pressure applied at the top–flash chromatography) with solvents of gradually increasing polarity to effect the separation. The solvent issuing from the column is collected in a number of fractions, which are monitored by thin layer chromatography and also by the bioassay. Biologically active fractions are then treated further.

### *(c) Gas–liquid chromatography*

In gas–liquid chromatography (GLC) the mobile phase is a gas stream and the stationary phase is an involatile liquid held on an inert porous support. The block diagram (Fig. 6.5) shows the essential features of a GLC apparatus.

The material to be separated is injected on to the column in a low boiling solvent and flash vapourized in the injector assembly (or on the top of the column). The mixture is carried into the column by the carrier gas stream (usually nitrogen or helium) at flow rates of the order

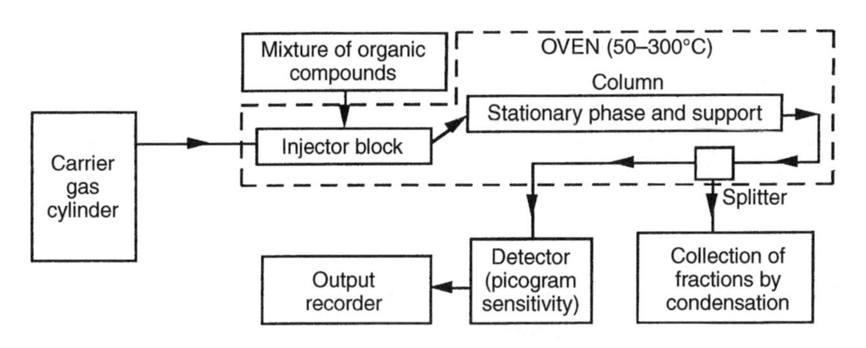


**Fig. 6.5** Block diagram of gas liquid chromatography apparatus.

of 5 to 50 ml/minute and it is there separated into its various components. Because the separation depends both on the vapour pressures (boiling points) of the components and on their relative solubilities in the stationary phase, different column packings (stationary phases) will give different orders of elution for the same mixture and will have different separating abilities for any pair of compounds. Hence it must be remembered that each separate detector signal does not necessarily represent a single compound. The most widely used detector for GLC work is the Flame Ionization Detector (FID), which is almost ideal for this purpose. It shows the following significant characteristics:

- a high sensitivity to virtually all organic compounds (down to picogram level);
- good quantitative linearity over a very wide sample range (*c.* $10^7$);
- little or no response to water, carbon dioxide and the common carrier gas impurities;
- it is not much affected by fluctuations in temperature, carrier gas flow rate or pressure and hence gives a stable base line.

Unfortunately the signal given by the FID is caused by the combustion of the eluted component and hence the material is destroyed in the detector.

The time taken for a particular component to elute is called the **retention time**. When measured relative to an internal standard under controlled conditions, this is a reproducible parameter which is of use in compound identification. Because compounds with long retention times give peaks that are broader in proportion to their retention time, it is normal for the oven temperature to be gradually raised during an analysis (temperature programmed run). This serves two purposes: it decreases total analysis time and it increases the height-to-width ratio of the peaks due to the slower running components, hence increasing the ease of their detection.

Because of its speed, resolution, sensitivity, quantitative precision and simplicity, GLC is ideally suited for the separation and analysis of pheromone mixtures, which are normally sufficiently volatile for this method. Figure 6.6 shows a typical GLC trace from a pheromone analysis.

The use of capillary columns in GLC provides a far higher degree of resolution than conventional packed columns. However, owing to their lower loading of liquid phase and small internal volume, the column capacity is very low and it is necessary to ensure that only the correct amount of sample is introduced. This is done by using a suitable injector splitter. The low capacity is compensated by the greatly increased resolution and the shorter retention times. The overall sensitivity is about the same as that for packed columns. Two types of capillary column are available:

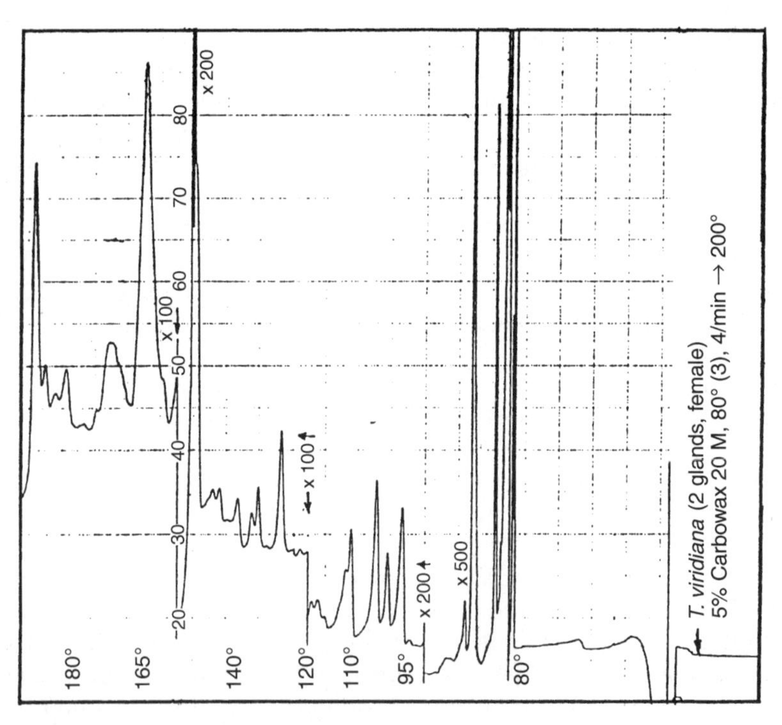


**Fig. 6.6** GLC trace of solid sample injection. Two glands from female European oak leaf-roller (*Tortrix viridiana*) on 5% Carbowax 30M, programme: 80°C (3 minutes) / 4°C/min to 200°C.

- **WCOT** (wall coated open tubular) columns, which have the liquid phase as a thin coating on the inner wall of the capillary. These generally have greater separating power than SCOT columns, but lower sample capacity.
- **SCOT** (support coated open tubular) columns, which have a thin layer of fine particulate support on the inner wall of the capillary and this support is coated with the liquid phase.

The recent introduction of wide-bore fused silica capillary columns has been a significant advance. These approach the ease of use and sample capacity of packed columns, and give better resolution and

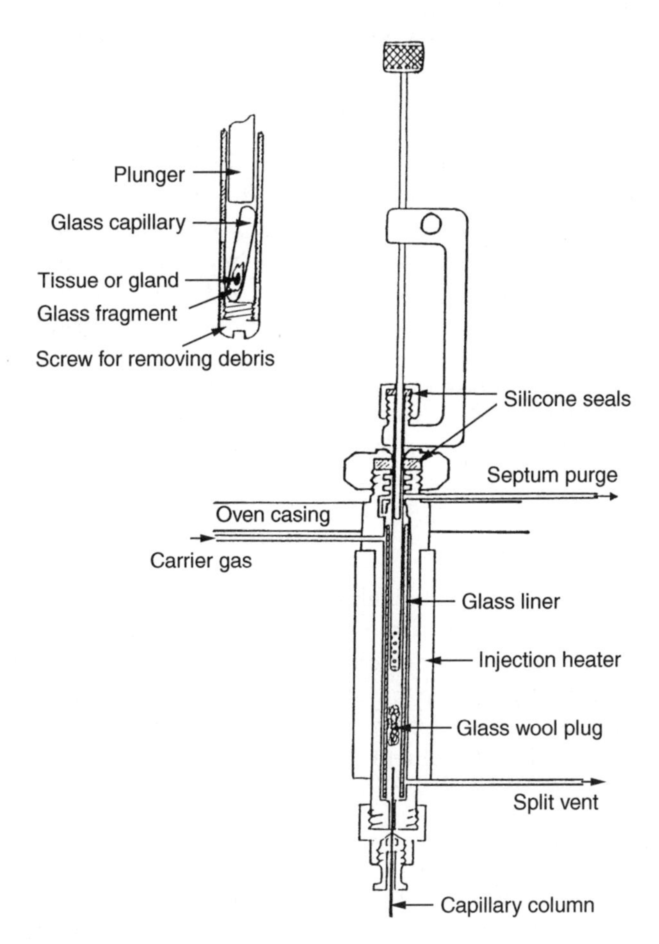


**Fig. 6.7** Solid sample injector assembly. Inset shows detail of the glass capillary held in tip of sampler before crushing.

reduced analysis times without the need for inlet splitting. They do not, however, perform as well as true capillary columns.

By means of a modified injector assembly (Fig. 6.7), solid samples, such as extirpated glands, may be introduced directly on to the GLC column (Morgan and Wadhams, 1972; Morgan, 1990). This can be extremely useful in preliminary work and when re-examining for minor components. The technique also reduces the problems of inadvertent contamination of samples.

If the column effluent is split before passing to the detector, then GLC may be used as a preparative method for the separation of quantities typically from one microgram to tens of milligrams, although with care as little as 1 ng may be handled (Attygalle and Morgan, 1988). The components may be trapped from the effluent gas by bubbling through a cold solvent, condensed in a cold trap, or a tube of adsorbent.

### *(d) High performance liquid chromatography*

Although similar to column chromatography in principle, involving partition between a liquid mobile phase and a solid stationary one, high performance liquid chromatography (HPLC) uses much smaller particle sizes for the solid (5–10 mm), which improves dramatically the efficiency and resolution but requires extremely high pressures (typically 5000 psi) to force the solvent through the resulting much smaller interstitial spaces. It is particularly useful for the separation and purification of compounds that cannot be gas chromatographed due to their involatility, polarity or thermal instability. The efficiency and resolution of HPLC is as great as that of GLC and, in some instances, it is possible to separate mixtures that cannot be resolved by GLC. However, its overall sensitivity is not as great. Two detector systems are generally available:

- Ultraviolet – in which the absorption of a beam of UV light is monitored. This requires that the compounds to be analysed possess a UV chromophore. The detection limit is about 10 ng and the sensitivity depends strongly on the structure of the compound.
- Refractive index – in which the difference in refractive index between the solvent and the effluent is monitored. All components of a mixture will be detected, with a detection limit of about 10 mg.

Four column types are commonly found:

- Normal – similar to column chromatography and separating on the basis of polarity; the column is polar and the solvent of varying polarity. Non-polar compounds elute first.
- Reverse phase – here a hydrocarbon chain (typically C18) is bonded

to the stationary phase, making it non-polar, and solvent of decreasing polarity is used for elution. Polar compounds elute first.
- Gel filtration – compounds are separated on the basis of molecular size.
- Ion exchange – separating on the basis of ion association; this type is not of significance in pheromone work.

HPLC is also useful on a preparative scale; samples up to 5 mg can be separated on an analytical column and from 50 mg to 50 g on wide-bore preparative columns. Because the detector is non-destructive and of very low volume, fractions of effluent are readily collected as each peak is eluted.

### 6.1.3 Bioassays and the separation problem

Each step in a separation need to be monitored by the bioassay. With liquid chromatographic methods this requires concentration and separate testing of each fraction – a time-consuming and often slow process. With GLC, however, it is possible to couple the separation and biological testing if EAG is used for the latter.

By splitting the effluent stream from the GLC column and arranging for part to be diverted to the EAG (Fig. 6.8) each separated component may be tested at the same time as it is detected by the FID (Moorhouse *et al.*, 1969). Rather than being continuous, the gas stream to the EAG is pulsed. This allows relaxation of the antennal receptor between pulses and provides a blank signal against which the output due to a 'bioactive' component may be measured.

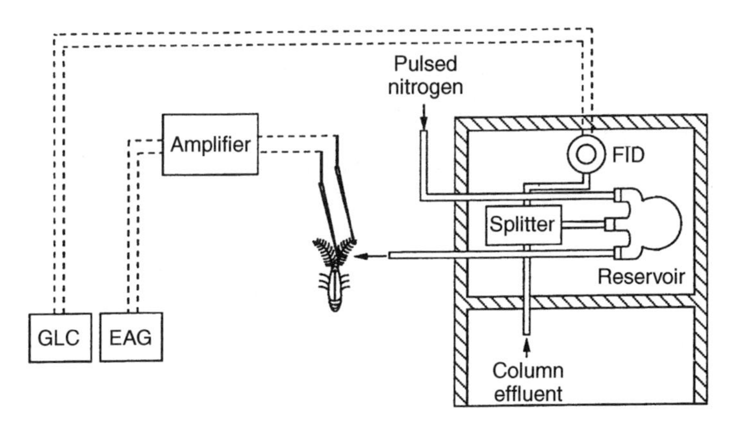


**Fig. 6.8** Splitter assembly for coupled GLC–EAG. (Redrawn after Moorhouse *et al.*, 1969.)

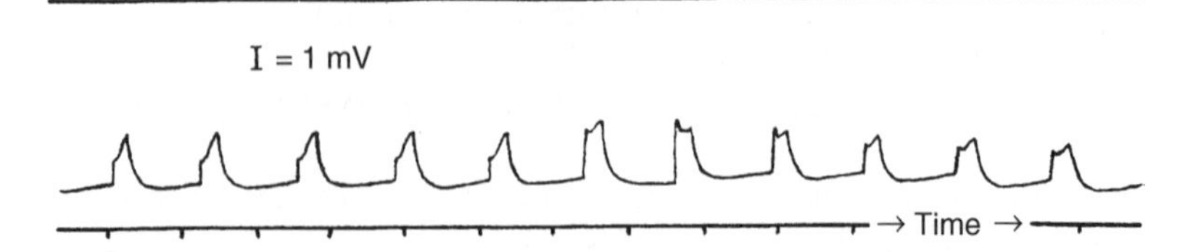


**Fig. 6.9** GLC-EAG output from male red bollworm (*Diparopsis castanea*). Stimulus source 1 ng 9,11-dodecenyl acetate.

Figure 6.9 shows the output from a male red bollworm (*Diparopsis castanea* Hampson) caused by a 1 ng sample of 9,11-dodecadienyl acetate as it emerges from the coupled GLC. The use of EAG as part of the detection system in GLC has led to the method being called the electroantennographic detector (EAD). For pheromonal constituents, EAG detection is frequently more sensitive than the conventional FID as a result of the 'tuned receptor' of the insect's antenna.

The advantage in speed resulting from the coupling of EAG and GLC more than outweighs the disadvantages of the EAG method in showing up even those components which may have an inhibitory effect on the behaviour under investigation.

Finally, the system may be employed to look for and identify minor components in a gland extract (say) by the expedient of employing, as the detecting antenna, males of a different species that use the compound that one wishes to identify as the major component of their own pheromonal system (Arn *et al.*, 1980).

## 6.2 STRUCTURE DETERMINATION

The majority of analytical methods require purified compounds (typically >95% pure) for reliable results, but in certain cases the coupling of a separation step in the analysis allows information to be obtained directly on each component of a mixture. Both physical (spectroscopic) and chemical methods are available and will be discussed in that order, because the former are, with the exception of mass spectrometry, non-destructive, thus permitting further examination of the sample.

### 6.2.1 Physical methods

#### *(a) Mass spectrometry–coupled GC–MS*

In terms of micro-scale methods, mass spectrometry (MS) is undoubtedly the most sensitive and it provides information at three levels:

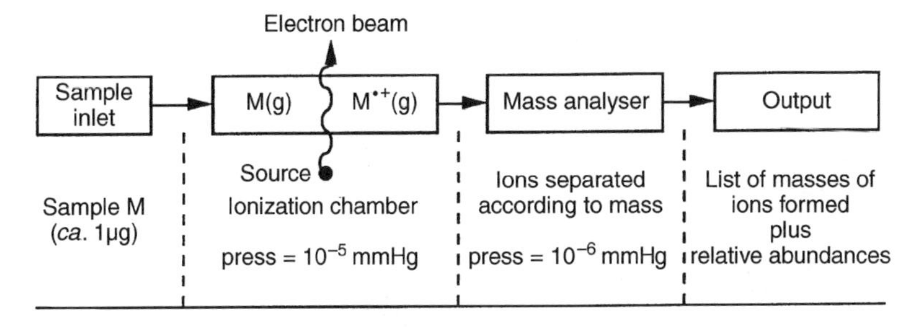


**Fig. 6.10** Block diagram of mass spectrometer system.

molecular mass, elemental composition and structural. The block diagram of Fig. 6.10 indicates the four significant parts of the MS system and the processes taking place in each of them.

In the ionization chamber the sample molecules, now in the gas phase, are bombarded with a beam of electrons (usually with an energy of 70 eV). This causes an electron to be knocked out of the molecule (M) giving a radical–cation ($M^{\cdot+}$). For example, with acetone:

$$H_3C{-}C({=}\ddot{O}{:}){-}CH_3 + e^- \longrightarrow \left[H_3C{-}C({=}\dot{O}{:}){-}CH_3\right]^+ + 2e^-$$

Equation 6.1 Radical cation formation from acetone

This method of forming the radical–ion is called **electron impact** (EI). Once formed, the radical–cation passes through the mass analyser to give rise to a peak with mass/charge ratio (m/e) of 58 atomic mass units (amu). Hence we may infer that the molecular mass of acetone is 58.

The **molecular mass** of the compound being analysed is given by the mass of the molecular ion $M^{\cdot+}$. This is normally the ion of highest molecular mass observed in the spectrum, but for certain types of chemical structures the molecular ion may be very weak or even absent. In such cases its mass may often by inferred from the other peaks observed.

Provided that we measure them accurately enough, all the isotopes of all the chemical elements known have non-integral atomic masses. The scale used is based on the carbon isotope of mass 12 ($^{12}C$) being defined as having atomic mass 12.000 000. On this scale, hydrogen ($^{1}H$) has mass 1.007 825 and oxygen ($^{16}O$) has mass 15.994 915. Hence the exact molecular mass of acetone ($C_3H_6O$) is calculated as:

$$\begin{aligned} 3\times C &= 36.000\,00 \\ 6\times H &= 6.046\,95 \\ 1\times O &= \underline{15.994\,91} \\ &\qquad\qquad = 58.041\,86 \end{aligned}$$

and that of butane ($C_4H_{10}$):

$$\begin{aligned} 4\times C &= 48.000\,00 \\ 10\times H &= \underline{10.078\,25} \\ &\qquad\qquad = 58.078\,25 \end{aligned}$$

Although both these compounds have whole number molecular masses that are the same, if they are measured to an accuracy of 1 ppm then they are clearly different and are readily distinguishable. Furthermore, given the exact mass of a molecular ion, it is possible to calculate the number of atoms of each element that must be present in the compound which gave rise to the particular ion. For large molecules, this is only feasible because of the high speed computers that are now available.

**High resolution mass spectrometry** (HRMS) achieves the level of precision and resolution required for such analysis by making use of two mass analysers (often of different types) in tandem in the same instrument. It thus affords precise masses of the ions and hence enables computation of the **elemental composition** of the compound being analysed.

The third level of information–**structural**–is obtained as a result of the energy imparted to the molecular ion in consequence of the impact of the high energy electrons in the ionization chamber. This energy is more than enough to cause the rupture of bonds within the molecular ion, resulting in the formation of molecular fragments by the loss of groups of atoms. Thus a molecular ion A-B-C-D$\cdot^{+}$ will fall apart into a mixture of cations and radicals (sometimes molecules and radical–cations) by one or more paths and the resultant cations will themselves fragment further:

$$\begin{aligned} &\text{A-C-B-D}\cdot^{+} \rightarrow \text{D}\cdot + \text{A-B-C}^{+} \rightarrow \text{A}\cdot + \text{B-C}^{+} \\ &\text{A-C-B-D}\cdot^{+} \rightarrow \text{C-D}\cdot + \text{A-B}^{+} \rightarrow \text{A}\cdot + \text{B}^{+} \\ &\text{A-C-B-D}\cdot^{+} \rightarrow \text{A-B} + \text{C-D}\cdot^{+} \rightarrow \text{D}\cdot + \text{C}^{+} \end{aligned}$$

The radicals, bearing no charge, are not detected by the system and each ion resulting from the fragmentation gives rise to a peak, whose abundance reflects its ease of formation.

Fragmentation occurs in chemically intelligible ways and hence we are able to gain **structural information** and **fragmentation patterns**. Bond cleavage occurs preferentially to give cationic species which are resonance stabilized as shown for 6-methoxyheptan-2-one (Fig. 6.11).

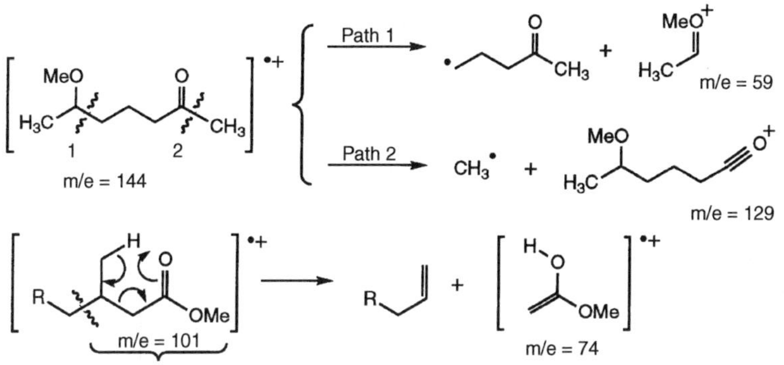

**Fig. 6.11** Mass spectral fragmentation of 6-methoxyheptan-2-one.

The structural information here points to the presence of carbonyl, methyl and methoxy-ether groups.

Three factors dominate the fragmentation processes:

- Weak bonds tend to be broken.
- Stable fragments tend to be formed.
- Some fragmentation processes depend on the ability of the molecule to assume a cyclic transition state; e.g. McLafferty rearrangements.

Common types of fragmentations are described in all texts on spectroscopy of organic compounds, which also given tables of common fragment ions and their masses (e.g. Kemp, 1991; Williams and Fleming, 1989). An example of the type of fragmentation pattern seen with pheromonal compounds is given in Fig. 6.12, which shows the mass spectrum of dehydrolinalool (4,8-dimethylnon-7-en-1-yn-3-ol), together with the associated HRMS computer report on the ions formed.

The comments for Fig. 6.12 have been added later and show the fragmentations involved. First of all we may note the very low abundance of the molecular ion, a feature that is typical of alcohols, which readily lose water. In this instance, the base peak arises from the loss of water and a methyl group. The most interesting feature is the peak at 69 amu, which on the bar chart looks like a single fragment. However, high resolution shows it to consist of the two peaks reported, corresponding to two very facile cleavages: the one at the higher mass occurring between C-4 and C-5, which gives the cation $(CH_3)_2C{=}C{-}CH_2^+$, where the charge is stabilized by resonance with the double-bond; and the other affording the cation $HC{\equiv}C{-}C^+(OH){-}CH_3$, in which the charge is stabilized by both the alkyne and the hydroxyl group, involving breakage of the C-3 to C-4 bond. It should be noted that not all of the fragments are readily interpreted.

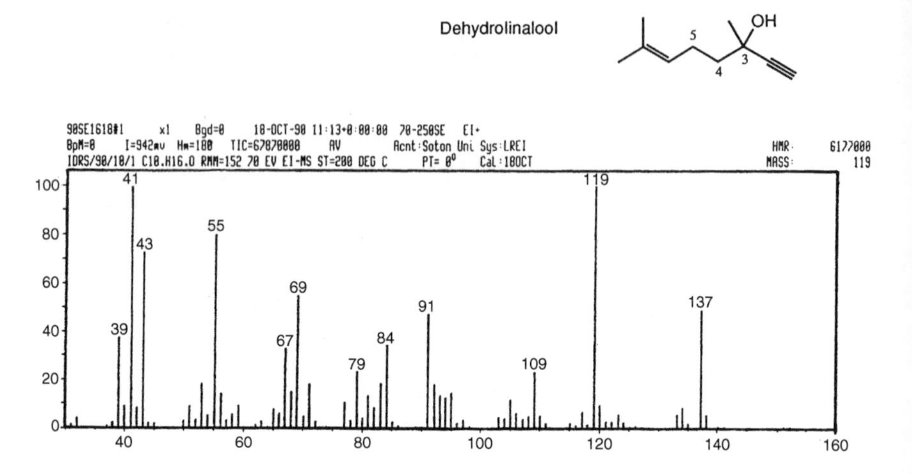


Dehydrolinalool: Mass Intensity Report

| Measured Mass | % Base | C | H | O | Dev'n | Comments |
|---|---|---|---|---|---|---|
| 152.1176 | 0.35 | 10 | 16 | 1 | -2.5 | molecular ion |
| 137.0989 | 48.3 | 9 | 13 | 1 | 2.3 | loss of $CH_3$ |
| 119.0887 | 100.0 | 9 | 11 | 0 | 2.6 | loss of $H_2O$ and $CH_3$ |
| 109.0610 | 23.4 | 7 | 9 | 1 | -4.3 | |
| 93.0736 | 13.0 | 7 | 9 | 0 | 3.2 | |
| 91.0583 | 47.1 | 7 | 7 | 0 | 3.5 | |
| 83.0515 | 17.9 | 5 | 7 | 1 | 1.8 | loss of $C_5H_9$, cleavage at C-4---C-5 |
| 71.0547 | 17.6 | 4 | 7 | 1 | 5.0 | |
| 69.0698 | 21.8 | 5 | 9 | 0 | -0.6 | cleavage at C-4---C-5 to give allylic cation $(CH_3)_2C{=}C{-}CH_2^+$ |
| 69.0334 | 32.8 | 4 | 5 | 1 | -0.6 | cleavage at C-3---C-4 to give stabilized cation $HC{\equiv}C{-}C^+(OH){-}CH_3$ |
| 55.0539 | 79.8 | 4 | 7 | 0 | -0.8 | |

**Fig. 6.12** Mass spectrum of dehydrolinalool.

Fragmentation patterns are not completely consistent from one MS instrument to another, particularly as regards the relative abundance of the ions formed; however, on any given instrument, the fragmentation pattern observed is highly characteristic of the particular molecular structure and acts as a fingerprint for the compound being investigated. Thus comparison of an 'unknown' with a synthetic sample of the putative structure can be used to confirm identity, which will be certain if the compounds show identical fragmentation patterns.

In some instances, the ease of fragmentation is such that the molecular ion is even weaker than for the example above, or may not be observable at all. In part this is due to the high energy of the electron beam involved in the EI method and in such cases alternative methods of forming the molecular ion may be used. The simplest is to lower the energy of the bombarding electron stream, but this lowers the overall abundance of ions formed and the electron energy used must be very low. The most commonly useful technique is chemical ionization (CI). Fast atom bombardment (FAB) and secondary ion mass spectrometry (SIMS) are alternative procedures which are of use particularly for very involatile materials. All of these techniques result in a greater abundance of the peak due to the molecular ion $M^{\cdot +}$, or the M+1 or M–1 peaks, and also give rise to simpler fragmentation patterns.

By combining the separating power of the gas chromatography with the analytical capability of the mass spectrometer, we arrive at the **coupled GC–MS**, probably the most useful tool in the armoury of the pheromone analyst. The major problem of interfacing the two instruments – GLC operating at above atmospheric pressure and MS at high vacuum – has been overcome by the development of suitable enrichment devices which remove the excess carrier gas used in the GLC without removing the compound being detected. With capillary column GLC, the use of very high speed pumping systems allows direct coupling of the GLC output into the MS inlet system. In operation, a sample is injected into the GLC and as each separated component is eluted its mass spectrum is measured. The data is normally acquired and stored on a computer, which allows detailed analysis and computer matching of unknowns against a mass spectral library also held on a computer file.

HRMS runs would then be carried out on those components identified as of interest from bioassay or EAG.

### *(b) Infra-red spectroscopy*

Infra-red (IR) spectra show the absorption of radiation by a molecule caused by excitation of bond stretching and bond bending within the molecule. The spectral region of interest lies between 4000 and 650 $cm^{-1}$ (2.5 and 15 $\mu m$). Skeletal vibrations, which involve a majority of the atoms within the molecule, give absorptions with the range 1400 to 650 $cm^{-1}$. This region is known as the 'fingerprint region' and the pattern is characteristic of the molecule, but the absorptions cannot, in the main, be assigned to particular functional groups. Characteristic group vibrations caused by recognizable functional groups are found between 4000 and 1500 $cm^{-1}$. These absorptions occur at positions which are not strongly affected by the overall molecular structure and allow the ready recognition of the type of functional group which causes them; for example, hydroxyl (OH) at 3600 $cm^{-1}$, nitrile (cyano, CN) at 2250 $cm^{-1}$ and carbonyl (C=O) from 1800 to 1600 $cm^{-1}$. The exact wave-number of absorption of a carbonyl group can indicate whether it is present as a lactone, ketone, ester or aldehyde.

IR spectra require that the sample be purified beforehand and the

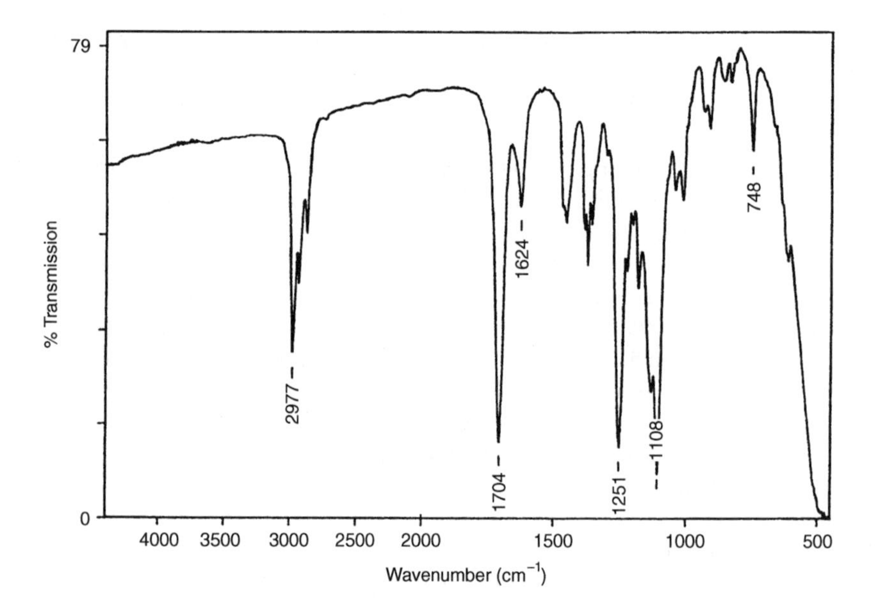


| Peak ($cm^{-1}$) | Inference |
|---|---|
| 2977 | C-H stretch of $CH_3$/$CH_2$ (two bands confirm) |
| 1704 | C=O stretch of conjugated ester, low wave-number shows likely diene |
| 1624 | C=C stretch of double-bond (conjugated) |
| 1251 } | C-O of ester, two bands confirm |
| 1106 } | |
| 748 | possible C-H bend in tri-substituted alkene |

**Fig. 6.13** Infra-red spectrum of 1-methylethyl *E*,*E*-2,4-dimethylhepta-2,4-dienoate, structure (6.2.1).

sample size needed is usually from about 60 $\mu$g upwards. Figure 6.13 shows a typical infra-red spectrum and the assignments made from it.

With the introduction of Fourier transform (FT) methods, recent advances in instrumentation have dramatically decreased the sample size required for IR spectrometry and have therefore enabled the coupling of GLC instruments directly to IR spectrometers. This combination, GC–FTIR, enables the IR spectrum of each component to be acquired as it emerges from the GLC column and, together with library spectral searching methods and the use of GC–MS, promises to speed enormously the task of compound identification.

### *(c) Ultraviolet–visible spectroscopy*

Ultraviolet–visible (UV–Vis) spectroscopy shows the absence or presence of a conjugated system within the molecule and gives some indication of the type of conjugation (e.g. diene, $\alpha,\beta$-unsaturated carbonyl, aromatic ring). It is very sensitive for conjugated systems with a detection limit of about 100 ng.

### *(d) Nuclear magnetic resonance spectroscopy*

Although an extremely powerful technique, nuclear magnetic resonance (NMR) spectroscopy has the drawback that it is inherently a very insensitive method requiring correspondingly large samples of material. Improvements in instrumentation and the methods used to obtain spectra have, over the last 20 years, reduced the sample size required from 1–2 mg down to 5–20 $\mu$g for proton spectra, with corresponding improvements for carbon-13 spectra, down from 100 mg to 200 $\mu$g.

As applied to pheromone analysis, two nuclei are studied by NMR spectrometry: protons ($^{1}$H nmr) and carbon-13 ($^{13}$C nmr), as these are always present in organic compounds. The sample, dissolved in a suitable solvent (usually deuteriochloroform, $CDCl_3$), is placed into a strong magnetic field and the absorption of radio frequency electromagnetic radiation by the nucleus under study is measured. The nucleus of an atom will be shielded from the applied magnetic field by the electrons surrounding it. Because the resonant (absorption) frequency depends on the magnetic field experienced by the nucleus, both will be influenced by the extent of this shielding. The shielding effect of the electrons around the nucleus depends, in turn, on the chemical environment of the nucleus. Hence, the resonant frequency of a given nucleus depends on its chemical environment and in consequence on the chemical structure of the molecule.

The change in absorption frequency of a nucleus with respect to some defined internal standard – e.g. tetramethylsilane, $(CH_3)_4Si$, (TMS) – is

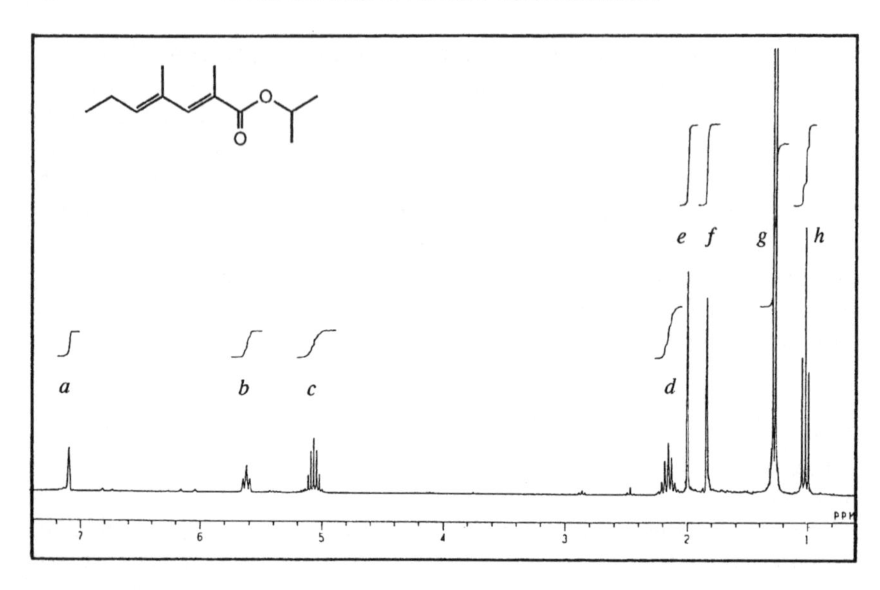


**Fig. 6.14** Compound (6.2.1), 1-methylethyl *E,E*-2,4-dimethyl-hepta-2,4-dienoate: 270 MHz $^1$H nmr spectrum.

called the **chemical shift** and can be used to assign a particular chemical environment to the given proton (say).

Figure 6.14 shows the $^1$H nmr spectrum of 1-methylethyl (*E,E*)-2,4-dimethylhepta-2,4-dienoate, the second constituent of the pheromone of the greater grain borer, *Prostephanus truncatus*, measured at 270 MHz, structure (**6.2.1**).

(**6.2.1**)

Structure (**6.2.1**) 1-methylethyl E,E-2,4-dimethylhepta-2,4-dienoate

The horizontal scale is the chemical shift ($\delta$) in parts per million with respect to TMS. Eight separate signals are seen, with different chemical shifts centred at 7.09, 5.62, 5.06, 2.15, 1.99, 1.83, 1.27 and 1.01 ppm, corresponding to the eight sets of protons (*a* to *h*, respectively), each with a different chemical environment. The chemical shifts of *a* and *b* tell us that these protons are attached to sp$^2$ hybrid (vinyl) carbons; the shifts of *d*, *e* and f indicate that, while they are on saturated (sp$^3$ hybrid) carbon, the adjacent carbon is part of a double-bond, and the shifts of *g* and *h* show that they are on sp$^3$ hybrid carbon attached to saturated carbons. The shift position of *c* is ambiguous, for it could be caused by a hydrogen on a vinyl carbon (as *a* and *b*), or one on a saturated carbon bearing a very electronegative group, such as an oxygen that is part of an ester.

The total intensity of each signal is directly proportional to the number of protons contributing to it and this is shown on the **integration** trace above each signal; *a*, *b* and *c* each corresponding to 1H; *d* to 2H; *e*, *f* and *h* to 3H each; and *g* to 6H.

It may be seen that, out of the eight, each of five signals (*b*, *c*, *d*, *g* and *h*) consists of a number of lines with a regular pattern of intensity (multiplicity). This multiplicity is connected with the number of protons on the adjacent carbon atoms which interact with the proton(s) giving rise to the signal; this gives structural information on the way the carbons are connected. In this instance, the multiplicity shows that *c* and *g* are situated on adjacent atoms and that *d* lies between *b* and *h*, which itself must be at the end of a chain.

Taken together, these three sources of information allow one to deduce the structure, but not the stereochemistry, of the compound. The signals are assigned as shown in Fig. 6.15. In this instance, because the double-bonds are trisubstituted, $^{1}$H nmr cannot readily distinguish between *E* and *Z* stereoisomers, but when the double-bonds are disubstituted, then the magnitude of the coupling between the hydrogens at each end of the bond enables this assignment to be made with relative certainty.

The interpretation of more complex spectra is rather more difficult than this brief outline, but $^{1}$H nmr is a very powerful structural analytical tool.

A glance at the structure of compound (**6.2.1**) shows that not all the carbon atoms bear hydrogens and hence $^{1}$H nmr gives us no information about them. This information is obtained from the $^{13}$C nmr spectrum, which is shown in Fig. 6.16, where it may be seen that each of the 11 different carbon atoms of the molecule gives a separate signal in the spectrum. The assignment of each signal to the structural element involved is made on the basis of chemical shift and spectral editing techniques which enable one to determine the number of hydrogens

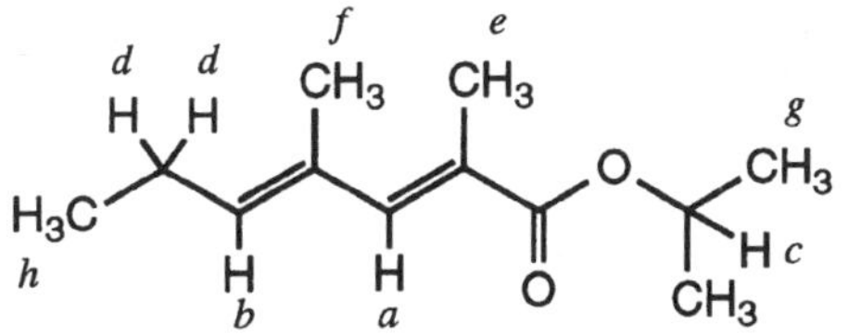

**Fig. 6.15** H nmr spectral assignment for compound (6.2.1).

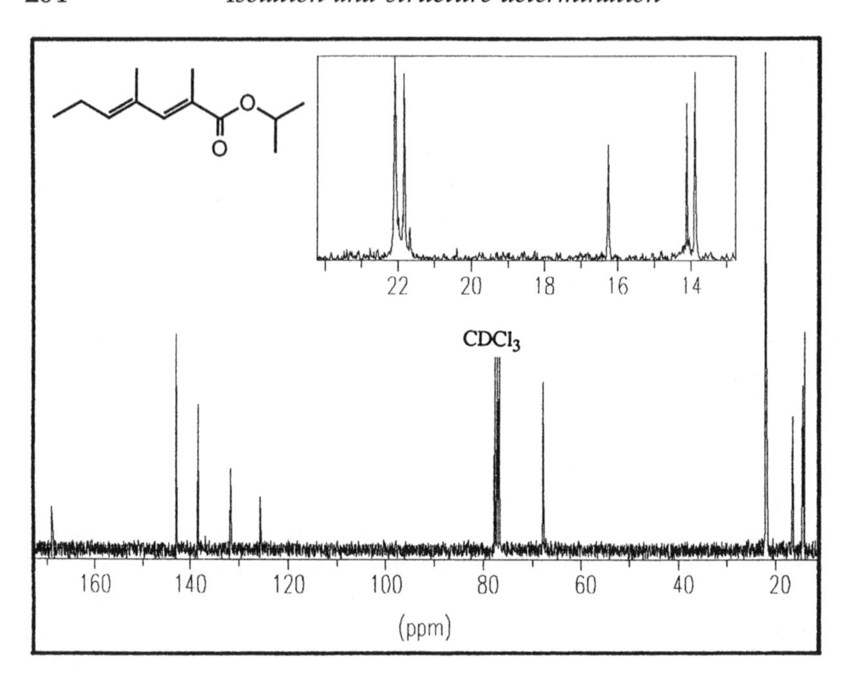


**Fig. 6.16** Compound (6.2.1), 1-methylethyl *E,E*-2,4-dimethyl-hepta-2,4-dienoate: 75 MHz $^{13}C$ nmr spectrum.

directly attached to the carbon giving rise to the signal in question. In this instance, the signals at $\delta$ 168.9 and 67.8 ppm confirm the presence of the ester group: 168.9 is the C═O carbon and 67.8 the C(H)—O one. The three closely spaced signals at 77 ppm are due to the $CDC1_3$ solvent.

Because of the greater spread of chemical shift (200 ppm as opposed to 10 ppm), $^{13}C$ nmr spectra will almost always show a separate signal for each chemically distinct carbon, whereas this is often not the case in $^{1}H$ nmr spectra of large molecules. The additional information, on the types of different chemical environments obtained, compensates for the substantially larger sample size required. Unlike proton spectra, $^{13}C$ spectra do not provide any information on connectivity between the carbons, but can be very helpful in determining stereochemistry for trisubstituted double-bonds.

For more information on the interpretation of NMR spectra, the reader is referred to the many excellent texts on spectroscopy, which contain tables of typical chemical shifts, methods for predicting them, coupling constants and other useful data (Kemp, 1991; Williams and Fleming, 1989).

*(e) Non-spectroscopic methods*

Use of the Kovats system of GLC retention indices can provide information on the functional groups, polarity and molecular size of an unknown. The retention index indicates where a compound will elute on a chromatogram relative to the *n*-alkanes, using adjusted retention times. GLC may also be used to compare compounds isolated from an insect with those of synthetic standards with the aim of showing identity. This is normally done by co-injection of the standard with the unknown. It is important, therefore, to note that if two compounds are resolved by GLC then they are definitely different. Identical retention times on a given column does not prove identity, but non-resolution on three columns of differing polarities points to a high probability of the two compounds being the same.

The use of liquid crystal stationary phases in GLC allows the resolution of *E*- and *Z*-isomers of long chain acetates and alcohols, often difficult on other columns, and enables one to deduce the double-bond geometry where this needs to be determined.

### 6.2.2 Chemical methods

The major difficulties associated with chemical methods of structure determination are those of purity and of the handling of the very small samples involved. In addition, the methods do 'destroy' the compound involved by changing its chemical structure. Nevertheless, many reactions can be carried out on a micro-scale by making use of GLC and MS methods for analysis of the products. With care, as little as one microgram may be used in a quantitative manner.

When used in conjunction with the bioassay, chemical tests can be used at an early stage to assess the nature of the functional groups that are present in the active component. Thus a solution of the biologically active component is subjected to chemical treatment followed by concentration. Repetition of the bioassay on the solution after reaction shows whether the activity has been lost or retained. Knowledge of the functionalities affected by the reagents used then allows one to deduce whether or not they are present. Quantitative tests of this nature can often be carried out on one 'female equivalent' of material, frequently as little as 1 ng. The majority of the chemical reactions described below are suitable for such qualitative tests.

*(a) Reductive processes: hydrogenation, metal hydrides*

**Catalytic hydrogenation**

Catalytic hydrogenation of a compound may be carried out by passing gaseous hydrogen through a solution of the compound in ethanol to which either palladium on charcoal or platinum oxide (Adams' catalyst) has been added. Subsequent GC–MS of the solution gives the change in molecular formula involved. Alternatively, by using a pre-column containing the catalyst and hydrogen as the carrier gas, the sample may be hydrogenated directly in the GLC apparatus (Beroza and Sarmient, 1966).

Catalytic methods result in the reduction of carbon–carbon double and triple bonds and, depending on the catalyst employed, may lead to the opening of epoxide rings. Comparison of the molecular formula of the reduction product with that of the starting material gives the number of double-bonds reduced. Comparison with the molecular formula of the alkane, containing the same number of carbon atoms, enables one to deduce the number of rings and carbon–oxygen double-bonds remaining and hence present in the original material.

For example, catalytic hydrogenation of structure (**6.2.2**) neocembrene-A, the trail pheromone from *Nasutitermes exitiosus*, formula $C_{20}H_{32}$, gave a product of molecular mass 280 and formula $C_{20}H_{40}$, indicating the addition of eight hydrogens and hence the presence of four carbon–carbon double-bonds. The corresponding alkane would have the formula $C_{20}H_{42}$ and the difference of two hydrogens shows that the hydrogenation product must contain **one** ring. Hence neocembrene-A is a monocyclic compound with four double-bonds.

(**6.2.2**) Neocembrene-A

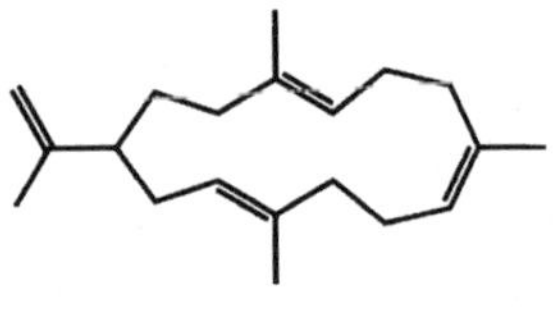

Structure (**6.2.2**) Neocembrene-A

**Metal hydride reduction**

Metal hydride reduction, on the other hand, reduces carbon–oxygen double-bonds without affecting C=C bonds. Lithium aluminium hydride reduces all carbonyl compounds (aldehydes, ketones, esters and acids) to the corresponding alcohols. In the case of esters, the acid part is reduced to a primary alcohol and cleaved off from the original alcoholic portion of the molecule. Thus Z-5-decenyl 3-methylbutanoate (the pine emperor moth pheromone) was treated with $LiAlH_4$ in ether followed by aqueous sodium hydroxide. Analysis by GLC showed the formation of a 1:1 mixture of Z-5-decen-1-ol and 3-methylbutan-1-ol. Sodium borohydride is more selective and reduces only aldehydes and ketones, enabling a distinction to be made.

### *(b) Oxidative process: ozone, Cr(VI), Mn(IV)*

*Ozonolysis*

Micro-scale ozonolysis causes the cleavage of carbon–carbon double-bonds with the formation of aldehydes and ketones at the positions where the double-bond(s) were found in the starting molecule (Beroza and Bierl, 1967; Moore and Brown, 1971). A stream of ozonized oxygen is passed into the sample (10–100 $\mu$g) dissolved in either carbon disulphide, ethyl acetate or pentyl acetate (50–200 $\mu$l) at –78°C for 30 seconds to 5 minutes. After quenching the cold solution with triphenyl phosphine, the resultant mixture may be analysed directly by GLC or GC–MS. Addition of an inert internal standard, such as tetradecane, before the ozonolysis, enables the method to be applied quantitatively. Thus neocembrene-A, structure (**6.2.2**), was shown to afford two equivalents of 4-oxopentanal (equation 6.2) by cleavage at the four points indicated.

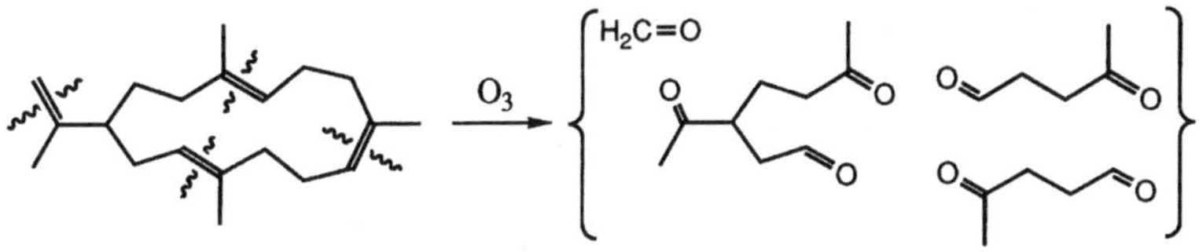

$H_2C{=}O$ $\xrightarrow{O_3}$

Equation 6.2 Ozonolysis of neocembrene-A

Ozonolysis is the most useful of the oxidative methods for it normally results in the formation of a number of smaller molecules, which are more easily recognized, and hence allows the formal 'reconstruction' of the compound under investigation.

*Chromium (VI)*

The use of chromium (VI) as an oxidant enables one to detect the presence of primary and secondary alcohols, aldehydes and other readily oxidizable groups. It does not oxidize ketones, C—C double-bonds, or aromatic rings.

*Manganese dioxide*

Activated manganese dioxide is a very specific oxidant, for it reacts only with primary and secondary hydroxyl groups which are allylic to (i.e. one carbon away from) a carbon–carbon double-bond as shown in equation 6.3.

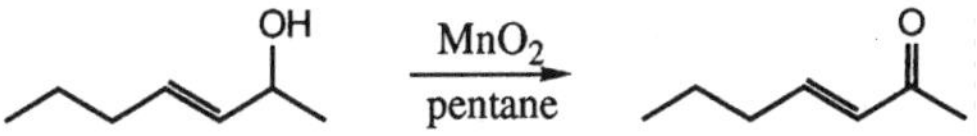

OH $\xrightarrow[\text{pentane}]{MnO_2}$ O

Equation 6.3 Manganese dioxide oxidation of allylic alcohol

When the double-bond is further away from the alcohol, then no reaction occurs.

### *(c) Addition reactions: methoxymercuration, dimethyldisulphide, Mn(VII), bromine*

*Methoxymercuration*

Methoxymercuration enables the position of a double-bond in a chain to be determined on a micro-scale (Abley *et al.*, 1970). The compound in methanol is treated with mercuric acetate, $Hg(OAc)_2$, and then with sodium borohydride, $NaBH_4$; subsequent GC–MS allows straightforward analysis. These two chemical steps result in the addition of the elements of methanol across the double-bond, as shown in equation 6.4, which illustrates the reaction of methyl *Z*-9-hexadecenoate with these reagents. The product is a mixture of the two ethers given, in which a methoxy group has been added to either one end or the other of the original double-bond, i.e. at either C-9 or C-10.

methyl (*Z*)-9-hexadecenoate

(i) $Hg(OAc)_2$, $CH_3OH$
(ii) $NaBH_4$

Equation 6.4 Methoxymercuration of methyl (*Z*)-9-hexadecenoate

These methoxy groups then direct the subsequent mass spectral fragmentation of the products with the formation of pairs of prominent fragments differing by 14 amu (Fig. 6.17). The peaks at 129 and 143 have exact masses corresponding to $C_8H_{17}O$ and $C_9H_{19}O$, respectively, while those at 201 and 215 have molecular formulae of $C_{11}H_{21}O_3$ and $C_{12}H_{23}O_3$. The first pair shows that either seven or eight carbons are cleaved with the methoxy group, corresponding to the methyl (left-hand) end of the molecule, cleavage *a*. The second pair correspond to the carboxyl end of the molecule and show that this part is cleaved with either nine or 10 carbons of the original chain, cleavage *b*. Together these two pieces of information show that the double-bond must be between C-9 and C-10 of the chain.

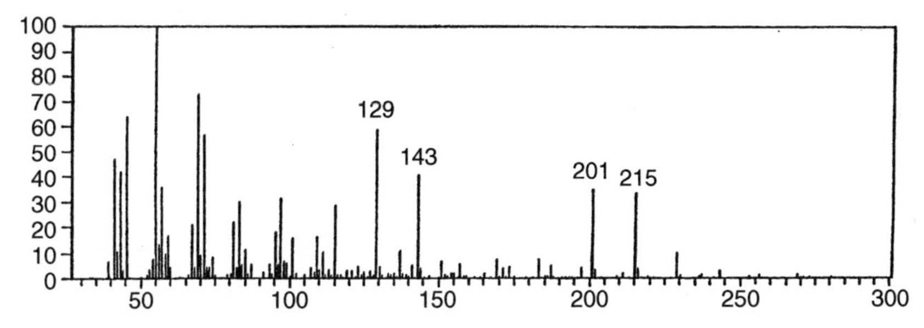


**Fig. 6.17** Mass spectrum of methoxymercuration product from methyl Z-9-hexadecenoate.

*Dimethyldisulphide*

Addition of dimethyldisulphide provides an alternative method for the determination of double-bond position, which has been applied particularly to straight chain acetates and alcohols (Buser *et al.*, 1983). The alkene in hexane is reacted with dimethyldisulphide (DMDS) and a catalytic amount of iodine. Addition occurs with a methylthio group becoming attached to each carbon of the original double-bond, as shown in equation 6.5.

O-Ac

$CH_3S\text{-}SCH_3$
15 h., 40°C, $I_2$

$CH_3S$ $SCH_3$ O-Ac

Equation 6.5 Reaction of dimethyldisulphide with (*Z*)-6-tetradecenyl acetate

Removal of the iodine and concentration is followed by GC–MS analysis. Even with terminal acetates, a recognizable molecular ion is formed and cleavage at the bond indicated gives the two major fragments (**A**) and (**B**), in equation 6.6.

$CH_3S$ $SCH_3$ $H_3C$ $(CH_2)_p$ $(CH_2)_q$ O-Ac → $CH_3\dot{S}^+$ $H_3C$ $(CH_2)_p$ (**A**) + $^+\dot{S}CH_3$ $(CH_2)_q$ O-Ac (**B**)

Equation 6.6 Mass spectral cleavage of dimethyldisulphide adduct

Fragment (**A**) gives a peak at m/e = (61 + 14*p*) and fragment (**B**) one at m/e = (119 + 14*q*), where *p* and *q* are the number of methylene groups in the two parts of the original chain. This method has been applied for the analysis of samples isolated directly from the aeration of just a few individual females.

*Mn(VII) and bromine*

Both permanganate [**Mn(VII)**] and **bromine** are used in a qualitative manner to confirm the presence of unsaturation in a biologically active compound. Addition of a dilute solution of bromine in dichloromethane (*c*. 1 ml of a 1 mM solution) to a solution of the active compound is followed by removal of the excess bromine with sodium bisulphite. The resultant solution may then be tested for biological activity, either directly or after concentration. Dilute aqueous potassium permanganate solution may be used in a similar fashion. Both reagents add on to carbon–carbon double and triple bonds, but do not react with aromatic (benzene) rings. Bromine will also react with aldehydes and, much more slowly, with ketones.

### (d) Other chemical reactions

The conversion of one functional group into another (derivatization) can also be a useful process in the detection of functionality at an early stage, making use of the bioassay in conjunction with the chemical reaction for monitoring purposes.

Alcohols and amines will react with acetic anhydride (or acetyl chloride) in pyridine (or dimethylformamide) to give the corresponding acetates (acetamides). Treatment with hexamethyldisilazane will convert alcohols to trimethylsilyl ethers and acids to trimethylsilyl esters, decreasing their polarity and increasing the ease with which they may be chromatographed. For compounds containing a number of alcoholic groups, this will often enable them to be gas chromatographed when they would otherwise be unsuitable due to their involatility.

The conversion of C—C double-bonds into epoxides with *m*-chloroperbenzoic acid has been used to help in assigning stereochemistry (equation 6.7). The reaction occurs with retention of stereochemistry and not only are the resultant epoxides more readily separated by GLC but also their relative retention indices are more characteristic of the *E*- or Z-stereochemistry than those of the alkene isomers from which they are derived. The epoxides also give MS fragmentation patterns that assist in determining double-bond position.

$$R^1\text{CH=CHR}^2 \xrightarrow[\text{CH}_2\text{Cl}_2\text{, 0°C}]{\text{mClC}_6\text{H}_4\text{CO}_3\text{H}} \text{epoxide (H, R}^1\text{; H, R}^2\text{; O)}$$

Equation 6.7 Epoxidation of double-bond with *m*-chloroperbenzoic acid

Reaction gas chromatography can be used to remove compounds containing a specific functional group from an injected sample by making use of a pre-column containing a suitable reagent. For example, phosphoric acid selectively removes epoxides. The same principle may be applied after the separation has taken place by interposing a post-column of reagent between the analytical column and the detector.

### 6.2.3 The human element

By far the largest factor in a successful structure determination is the analytical power of the human brain. The investigator must always be prepared to keep an open mind and not allow preconceptions as to the expected type of compound to colour the experimental evidence. The temptation to ignore an 'inconvenient' fact is ever present and must always be resisted.

## 6.3 EXAMPLES OF ISOLATION AND STRUCTURAL ELUCIDATION

### 6.3.1 Isolation

*(a) Whole body extraction*

**Codling moth** (*Cydia pomonella* L.) (McDonough *et al.*, 1969)
The bioassay that was used in this isolation consisted of 10 male moths held in a glass jar and conditioned to light. The test materials were introduced on the end of a glass rod. A positive response was recorded if the moths took flight and attempted to copulate with the sample. Preliminary tests showed that maximal pheromone release occurred when the virgin females were 3 days old. Newly hatched moths contained only 10% of the amount of pheromone.

About 200 000 female moths were homogenized with dichloromethane (enough to cover plus 50%) in a blender. The homogenate was filtered (Buchner funnel) and the filtrate evaporated (rotary evaporator, 30°C, water aspirator) to give a solid (84 g). The solid residue from the filtration was extracted with methanol (2 × 200 ml), centrifuged and evaporated to give a further solid (45 g). The combined solids were boiled with 10% methanolic sodium hydroxide (1.5 l), cooled and partitioned between ether and water. Further extraction with ether was followed by washing the ether layer successively with water and dilute hydrochloric acid. After drying over sodium sulphate, removal of the solvent gave a solid (9 g). This solid was dissolved in pentane and chromatographed on alumina (300 g). The column was eluted with pentane; 1:1 pentane–ether; ether; 1:1 ether–dichloromethane; dichloromethane; and finally 10% methanol in dichloromethane. The active fractions (1:1 ether–dichloromethane) were concentrated and re-chromatographed on alumina (100 g) using the same solvent series. The active fractions (10% methanol in dichloromethane) were concentrated, dissolved in 4:1 pentane–ether (2 ml) and subjected to preparative gas chromatography [5% Carbowax 20M on GasChrom Q, 2 m × 4 mm, isothermal at 200°C]. Trapping of each peak in a stainless-steel capillary cooled to –78°C and bioassay gave an active fraction (Kovats Index 2235) which was again gas chromatographed, [5% Apiezon L on GasChrom Q, 2 m × 6 mm, isothermal at 200°C], on which the active component had Kovats Index 1515. This procedure afforded the pheromone (7 $\mu$g).

Tests showed that the GLC trapping was 80% efficient and hence that the pheromone content per moth was 0.55 ng. Separate examination of the dichloromethane and methanol extracts showed that about one half of the pheromone was present in the latter as an ester and hence biologically inactive. This was the reason for the hydrolysis step immediately after the extraction.

Qualitative chemical tests were carried out on the pheromone in solution, using aliquots of 10 ml containing one 'female equivalent' (FE), and the product was bioassayed after reaction.

- 1 FE in dichloromethane was treated with bromine. Product was inactive.
- 25 FE in ethanol were hydrogenated with Pt and hydrogen. Product was inactive.
- 1 FE in ether was reacted with $LiAlH_4$. Product was still active.

The first two tests showed that the active component has at least one C—C double-bond in its structure and the third that it is not an ester, ketone or aldehyde, but probably an alcohol.

**Australian termite** (*Nasutitermes exitiosus* Hill) (Moore, 1966; Birch *et al.*, 1972)

The compound sought here was the trail pheromone of the termite. Preliminary tests on crude extracts showed that in trail-following bioassays the termites reacted to the pheromone only if its concentration lay between $10^{-8}$ and $10^{-5}$ g/ml. This meant that each extract had to be tested at several different dilutions.

A few thousand termites were extracted with pentane and subjected to chromatography on alumina (Activity 1). The active component was eluted before β-carotene. Preparative TLC of this fraction on silica gel showed 5 bands, of which the one of Rf 0.4 contained the active compound. GLC [10% silicone oil on Celite, 1.2 m × 4 mm, 200°C isothermal] showed this to contain two components with Kovats indices indicating they were C-15 and C-20 compounds. Collection and bioassay showed that only the latter was active in trail tests. The amount per termite was estimated at 0.25 ppm. Micro-scale testing showed that the compound was stable to alkaline hydrolysis.

The termites were stored in 95% ethanol in the freezer until *c.* 12 kg had been accumulated. The whole was then homogenized and centrifuged. The solvent was discarded and the solid extracted for 24 hours with petroleum ether (bp < 40), filtered and re-extracted. Removal of the pentane from the combined extracts gave a mainly lipid residue, which was saponified with potassium hydroxide in methanol. The mixture was extracted with petroleum ether (bp < 40) and washed with water to remove the soap. After concentration, the solution was cooled to 4°C and the precipitated soap centrifuged off. The petroleum ether (bp < 40) solution was reduced to small volume and chromatographed on alumina (100 g), using petroleum ether (bp < 40) as eluant, and all the material which emerged before the β-carotene was retained. Removal of the solvent yielded a hydrocarbon fraction (1.5 g). To this *n*-eicosane (200 mg) was added as carrier and the material subjected to short-path distillation (Kugelrohr) at 0.05 mm Hg. The C-20 fraction (250 mg) was then chromatographed on silica gel. The unsaturated hydrocarbons were eluted in the 10% ether in petroleum ether (bp < 40) fractions, which were concentrated and subjected to preparative GLC [5% Carbowax 20M, 2 m × 6 mm, 175°C isothermal]. The trail pheromone (2–3 mg) (neocembrene-A) was collected.

The structure elucidation is described in section 6.3.2a.

### (b) Excised gland

**Pine emperor moth** (*Nudaurelia cytherea cytherea* Fabricius) (Henderson *et al.*, 1973)

The bioassay used was a standardized EAG on the male moth antennae. The two terminal segments were excised from $1.7 \times 10^4$ newly emerged female moths. These were frozen in liquid nitrogen, finely ground and homogenized with dichloromethane. After filtering through cotton wool and concentration (rotary evaporator) to about 60 ml, acetone (800 ml) was added and the solution cooled to –78°C to precipitate the waxes. The solid was filtered off and re-extracted by again dissolving in acetone and cooling. The combined solutions were concentrated to 150 ml and the de-waxing process repeated. Removal of solvent gave an oil (43 g), which was steam distilled until no further active material came over (*c.* 30 h). The distillate was extracted with dichloromethane and evaporated to give an oil (600 mg), which was chromatographed on Florisil (50 g). Pentane was used as eluant with gradual addition of acetone to increase the polarity (0, 0.2, 0.5,

1.0, 3.0, 5.0%) and finally pure acetone. The active fractions (eluted with 5% acetone in pentane) were concentrated to give an oil (100 mg), which was separated by GLC [2.5% FFAP on Chromosorb W (AW DMCS treated), 5.5 m × 3.5 mm, 12 min isothermal at 60°C, then 1.5°C/min to 145°C] into four fractions by collection in cooled (–78°C) pentane solution (1.5 ml). The active fraction (#3) was concentrated to 100 μl and further purified by preparative GLC [5% OV-25 (phenyl silicone) on Chromosorb W (AW DMCS), 5.5 m × 3.5 mm, isothermal at 150°C] and collected in carbon tetrachloride. This gave pure pheromone (170 μg), which was active to EAG and, after synthesis, in field tests. The average pheromone content per female was ~10 ng.

Chemical tests showed that the activity was destroyed by catalytic hydrogenation, lithium aluminium hydride reduction and bromine and also by alkaline hydrolysis.

**Swift moth** (*Hepialus hecta* L.) (Sinnwell *et al.*, 1985)
The pheromone in question is emitted by the male moths during 'calling' flights after sunset. The compound proved to be electrophysiologically active to both male and female antennae and hence EAG was used as the bioassay.

Male moths were caught during calling flights and the hindleg scale-brushes, which are the source of the pheromone, were dissected from the tibia. Extraction of the scale-brushes from 50 males with pentane gave a solution which showed three major compounds on capillary GLC (SE-54); estimated amounts of the three were 40 mg, 5 mg and 5 mg per individual. After concentration, the compounds were isolated by preparative GLC [3% SE-30 on Chromosorb W (AW DMCS), 2 m × 3 mm], temperature programmed: 70–180°C at 2°C/min. This afforded the three active components (**A**) (1.6 mg), (**B**) (*c.* 200 μg) and (**C**) (*c.* 200 μg).

The structure determination is described in section 6.3.2d.

*(c) Ovipositor washing*

**General** (Sower *et al.*, 1973)
The moths were chilled at –10°C to immobilize them. Finger pressure on the abdomen was used to extrude the ovipositor and gland, which was then clamped with microforceps and cut off close to the body. The ovipositor was dabbed on a paper towel to remove haemolymph, rinsed in ether (200 μl) and discarded. The ether was dried with magnesium sulphate (20 mg), filtered and concentrated to 3–5 μl by a stream of dry nitrogen. The solution was then examined by GLC.

**Corn earworm** (*Heliothis zea* Boddie) (Klun *et al.*, 1979, 1980)
Field tests were used for the bioassay. These were carried out on synthetic samples after the identification.

The ovipositors from 24- to 96-hour-old female corn earworm moths were excised transversely through the centre of the tergum of the 8th abdominal segment. Each ovipositor was extracted with heptane (3 μl) and the extract analysed directly by capillary GLC [60 m × 0.25 mm, SP-1000, WCOT]. The major component, (Z-11-hexadecenal, (92.4%), had been previously identified and the minor components were identified by GC–MS. Epoxidation followed by GC–MS was used to establish the double-bond positions. Comparison with known samples confirmed them as (Z)-7-hexadecenal (1.1%), (Z)-9-hexadecenal (1.7%) and hexadecanal (4.4%).

Field tests using this four component mixture showed all the activity of captive females, whereas the major component alone was insufficient.

*(d) Aeration*

**Rusty grain beetle** (*Cryptolestes ferrugineus* Stephens) (Pierce *et al.*, 1984)
The aeration apparatus of Fig. 6.3 was employed; the filter contained Porapak-Q (about 100 mg). The bioassay used was a dual-choice pitfall test.

The grain beetles (40 to 50 g, $\sim 1.5 \times 10^5$) were mixed with rolled oats (1.5 kg) and placed in the container. Air was drawn through the apparatus for 7 days at a flow-rate of about 2 l/min. The beetles and food supply were then removed and sieved to separate the beetles from eggs and larvae and the beetles and fresh food were replaced in the container. After one month, the Porapak-Q filter was removed and extracted with pentane and then dichloromethane by soxhlet extraction. After concentration, the active constituents were separated by preparative GLC (5% Carbowax 20M) into a cooled trap. This gives a mixture of (Z)-3-dodecen-11-olide and 4,8-dimethyl-(*E,E*)-4,8-decadien-10-olide (total *c.* 150 mg).

**False codling moth** (*Cryptophlebia leucotreta* Meyer)* (Bestmann *et al.*, 1988)
The apparatus used is shown in Fig. 6.4. Bioassay was by EAG. The main purpose of the investigation was to show diel periodicity of pheromone release.

Between three and eight 2–3-day-old female moths were placed in

*This species is variously placed in the genera *Argyroploce* and *Pseudogalleria*.

the insect chamber and the pump was run for 45 minutes. The filter was then changed. Pheromone was extracted from the filter by passing carbon disulphide (10 μl) back and forth (10 times) through the charcoal by alternately warming and cooling one end after closing it off. The solvent was then removed with a GLC syringe and the operation repeated twice more. The combined extracts were concentrated to about 1 μl in a nitrogen stream and analysed by GLC and GC–MS. The amounts recovered were ~33 ng per insect per night, with maximal emission of 8 ng per insect per hour.

Tests using standard samples in the insect chamber showed that 50–70% was adsorbed in the filter and that the recovery from the filter was >95%.

**Codling moth** (*Cydia pomonella* L.) (Arn *et al.*, 1985)
The bioassay used was EAG coupled to the GLC. Forty 4-day-old female moths were held in a carefully washed and dried Pyrex bottle (500 ml) from 5 h before dark. An airflow of 20 ml/min was maintained through the bottle. After 'calling' was complete, the moths were removed and the bottle rinsed with hexane (3 × 250 μl). The solution was filtered and concentrated to 75 μl at room temperature with a nitrogen stream. GC–MS analysis showed that the total emission was ~4 ng per insect and the structure was confirmed as the previously identified compound *E,E*-8,10-dodecadienol. Direct extraction of ovipositor glands gave about 2 ng of pheromone per insect.

Control experiments showed that the pheromone was more effectively trapped in this way (*c.* 50%) than by conventional aeration into a filter (<10%).

**Grape borer** (*Xylotrechus pyrrhoderus* Bates) (Sakai *et al.*, 1984)
In this insect, the males attract the females by pheromone release. The bioassay used was standard EAG. Normal aeration with trapping on Porapak-Q was found to be very poor. Whole body extraction gave active extracts, but also a complex mixture of compounds. The following procedure gave material that contained two major components as well as some alkanes. Twenty male grape borers were placed into each of 90 glass containers of 3 l capacity and maintained there for 3 days. After removal of the insects, the containers were rinsed with hexane (2 × 100 ml). Removal of the solvent gave a waxy solid (170 mg), which was separated by preparative GLC [CP Wax 51, 150°C] into the hydrocarbons and the EAG active components **[D]** (4.8 mg) and [E] (0.8 mg). Neither component alone was active in a cage test, but a mixture of the two in ratios from 20 : 1 to 4 : 1 elicited attraction.

### 6.3.2 Structure elucidation

*(a)* **Pink bollworm moth** (*Pectinophora gossypiella* Saunders) (Bierl *et al.*, 1974)

The active material was isolated by a whole body extraction similar to that described in section 6.3.1a; about 70 $\mu$g of pheromone was obtained from $1.5 \times 10^5$ female moths.

Saponification with methanolic sodium hydroxide was found to destroy the bioactivity, indicating the presence of an ester functionality, later confirmed by a peak in the IR spectrum at 1740 $cm^{-1}$. The activity was restored by acetylation with acetyl chloride and pyridine.

Catalytic hydrogenation, using a pre-column hydrogenator of 1% palladium on GasChrom P (25 mg), and GLC [5% SE-30 on Anakrom ABS, 0.9 m × 6.3 mm, 200°C] gave a peak corresponding to *n*-hexadecane.

GLC analysis on 3% OV-1 [1.5 m × 2 mm, 170°C isothermal] gave a single peak with a retention index corresponding to a straight chain acetate with an equivalent chain length of 15.6 carbons. Using a 15 m capillary EGGS-X (silicone polymer) SCOT column, the pheromone was resolved into two components of equivalent chain lengths of 16.70 and 16.78 carbons, consistent with a C-16 dienic acetate.

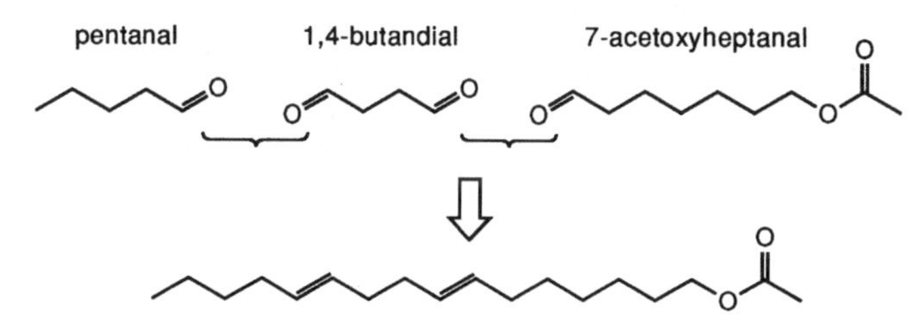


Equation 6.8 Retro-ozonolysis of pink bollworm moth pheromone

Micro-ozonolysis of pheromone (20 $\mu$g) in carbon disulphide (100 $\mu$l), followed by GLC analysis on 5% Carbowax 20M, gave three components identified as *n*-pentanal, 1,4-butandial and 7-acetoxyheptanal, by comparison with authentic samples. These three components can be 'linked' in only one way (equation 6.8) and establish the overall structure of the pheromone as hexadeca-7,11-dien-1-ol acetate. This can exist as four possible geometric isomers. In order to define the stereochemistry, all four isomers were synthesized and compared with the isolated material by GC–MS. Capillary GLC [EGGS-X SCOT] of the synthetic material showed that each of the four isomers gave a clearly resolved peak and that the isolated material was a mixture of the (Z,Z)- and (Z,*E*)-isomers in a 66 : 34 ratio. Subsequent field testing confirmed that the greatest attraction was observed with a 1:1 mixture of (Z,Z)- and (Z,*E*)-7,11-hexadecadien-1-ol acetate.

*(b)* **Australian termite** (*Nasutitermes exitiosus* Hill) (Birch *et al.*, 1972)

The extraction of the trail pheromone is described in section 6.3.1a. The mass spectrum of the isolate gave a molecular ion at 272 amu, corresponding to $C_{20}H_{32}$, and showed a significant fragment at [M-$CH_3$] and base peak at 68 amu ($C_5H_8$). This corresponds to the molecular formula of isoprene and strongly suggested a terpenoid structure for the pheromone, which would therefore be a diterpene. The molecular formula indicates **five units of unsaturation** (rings or double-bonds); $C_nH_{2n+2}$, would require 42 hydrogens.

In the IR spectrum, peaks at 3070, 1640 and 880/cm$^{1}$ confirmed the presence of double-bonds and suggested the structural element $C{=}CH_2$. In the UV spectrum, only end absorption was observed, showing that none of the double-bonds were conjugated.

Catalytic hydrogenation of the compound (250 $\mu$g) over pre-reduced platinum oxide (1 mg) in ethanol (1 ml) gave a compound with m/e at 280 corresponding to $C_{20}H_{40}$. The uptake of eight hydrogens demonstrated the presence of **four double-bonds** in the pheromone. Hence the overall structure must be **monocyclic**. Micro-scale ozonolysis (50 $\mu$g) in ethyl acetate (100 $\mu$l) and GLC [5% Carbowax 20M] showed that two molecules of 4-oxopentanal and one of formaldehyde were formed, accounting for 11 out of the 20 carbons, but leaving a 9-carbon fragment unknown. 4-Oxopentanal is a common ozonolysis product from terpenoid compounds and helped to confirm the putative diterpene structure.

H O
(6.3.1) Pyrophosphate
(6.3.2)

Equation 6.9 Postulated biogenesis of neocembrene-A (**6.3.2**) from farnesylfarnesyl pyrophosphate (**6.3.1**)

Based on the known biogenetic paths for the formation of diterpenes via the acyclic compound farnesylfarnesyl pyrophosphate, structure (**6.3.1**), the hypothesis of the monocyclic structure (**6.3.2**) was advanced (equation 6.9). This structure is closely related to that of the already known monocyclic diterpene, cembrene, structure (**6.3.3**), and hence was named 'neocembrene-A'.

The octahydropheromone formed by hydrogenation (above) was

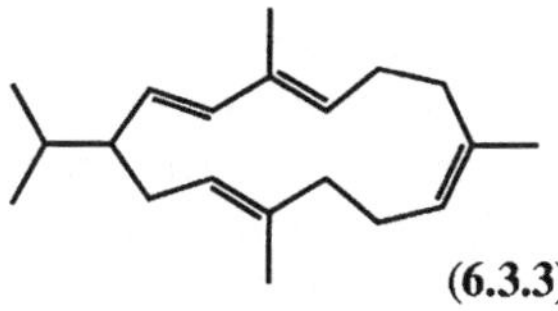

(6.3.3)

Structure (**6.3.3**) cembrene

compared with octahydrocembrene (obtained from cembrene) and shown to have identical GLC properties and to give an identical MS fragmentation pattern, which confirmed the overall skeleton.

The $^1$H nmr spectrum (100 MHz, $CDCl_3$) of neocembrene-A showed peaks at $\delta$ 5.2 to 4.8 (3H, broad) indicating three vinyl hydrogens, 4.68–4.51 (2H, multiplet) from the $C{=}CH_2$ group, and four 3H signals attributable to methyl groups on double-bonds at 1.63 (quartet, $J = 1.4$ Hz), confirming the partial structure $H_3C{-}C{=}CH_2$, and at 1.58, 1.56 and 1.55 from the three other methyl groups.

Confirmation of the skeleton and double-bond positions came from isomerization of neocembrene-A (2 mg) with the strong base sodium methylsulphonylmethide (0.11 g) in dimethyl sulphoxide (2 ml). This converted about 10% of the neocembrene-A into an isomer, isoneocembrene-A, which was separated by preparative GLC [5% Carbowax 20M] and submitted to micro-ozonolysis. The products were two equivalents of 4-oxopentanal and one each of glyoxal and 6-methylheptan-2,5-dione (equation 6.10). Together with the ozonolysis evidence from the pheromone itself, these lead to structure (**6.3.4**) for isoneocembrene-A and hence confirm the hypothesis of structure (**6.3.2**) for neocembrene-A itself.

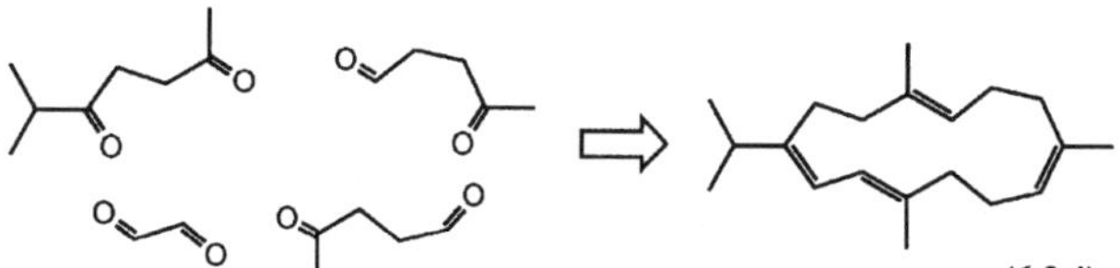

(6.3.4)

Equation 6.10 Retro-ozonolysis of isoneocembrene-A (**6.3.4**)

It should be noted that the stereochemistry of the double-bonds is not defined by this analysis nor is the chirality at the point of attachment of the isopropenyl group. These still await elucidation.

*(c)* **Pharaoh's ant** (*Monomorium pharaonis*) (Ritter *et al.*, 1977)

The trail pheromone (70 $\mu$g) was isolated from about $10^5$ worker ants by extraction with benzene, column chromatography and GLC, using trail-following tests for monitoring.

Peaks in the IR spectrum at 2820, 2710 and 1727 $cm^{-1}$ indicated the presence of a non-conjugated aldehyde group. The mass spectrum showed a parent ion at 250 amu and suggested a molecular formula of $C_{17}H_{30}O$. The $^1$Hnmr spectrum (300 MHz, $C_6D_6$) confirmed the presence of 30 hydrogens with peaks at ($\delta$) 9.42 (1H), 5.19 (1H), 5.16 (1H), 2.18 (2H), 2.09 (2H), 2.02 (2H), 1.98 (1H), 1.87 (1H), 1.86 (1H), 1.74 1H), 1.71 (1H), 1.70 (3H), 1.56 (3H), 1.28 (1H), 0.95 (3H), 0.74 (3H) and 0.72 (3H). Significant points about this spectrum are:

1. Signal at 9.42, a double-doublet (J = 2.5 and 1.5 Hz), confirming the saturated aldehyde and showing the partial structure $—CH_2—CH{=}O$;
2. Signals at 5.19 and 5.16 due to vinyl hydrogens; each is a triplet (J = 7 Hz) and shows the presence of two non-conjugated trisubstituted double-bonds demonstrating that the compound has a branched chain and contains, in two places, the structural element $—CH_2—CH{=}C{<}$;
3. Signals at 1.70 and 1.56 due to methyl groups on a double-bond; these extend the structural element shown by (2) to $—CH_2—CH{=}C(CH_3)—$.

Micro-scale ozonolysis yielded butan-2-one, 4-oxopentanal and 3,4-dimethylhexan-1,6-dial which were identified by GC–MS. These fragments may be reassembled in two ways and therefore the pheromone (10 $\mu$g) was reduced with sodium borohydride in ethanol and the resultant alcohol subjected to micro-ozonolysis. This gave 4-oxopentanal and 3,4-dimethyl-6-hydroxyhexanal, demonstrating that the 4-oxopentanal came from the middle of the chain (equation 6.11).

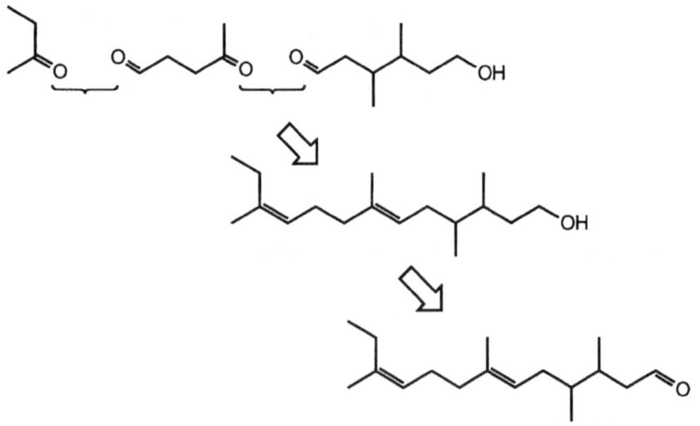

**(6.3.5)**

Equation 6.11 Retro-ozonolysis of faranol and structure of faranal (**6.3.5**)

These data establish the structure of the pheromone as 3,4,7,11-tetra-methyl-trideca-6,10-dienal, structure (**6.3.5**), and it was christened 'faranal'.

The stereochemistry of the double-bonds was assigned on the basis of $^1$H nmr chemical shift arguments in relation to the vinyl methyl groups – point 3 above – and double-resonance NMR experiments as 6*E* and 10*Z*. The final stereochemical point, the relative configuration of the methyl groups at C-3 and C-4, was deduced from the magnitude of the coupling constant between the hydrogens at C-3 and C-4 and it was concluded that faranal is the (3*S*, 4*S*) compound – or its mirror image, the (3*R*, 4*R*) isomer – as shown in structure (**6.3.6**).

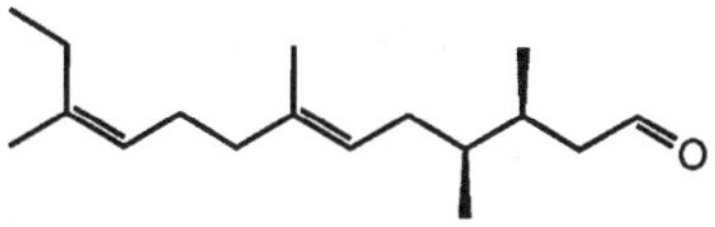

(**6.3.6**)

Structure and stereochemistry of Faranal (**6.3.6**)

*(d)* **Swift moth (***Hepialus hecta* L). (Francke *et al.*, 1985; Sinnwell *et al.*, 1985)

The isolation of the three major components (**A**), (**B**) and (**C**) are described in section 6.3.1b (swift moth). The total amounts of each of the components was (**A**): 1.6 mg, (**B**): 200 $\mu$g and (**C**): 200 $\mu$g.

The mass spectrum of compound (**A**) gave a parent ion at m/e = 140, for which HRMS gave the formula $C_8H_{12}O_2$. The $^1$H nmr spectrum had signals at $\delta$ 5.25 (1H), 4.44 (1H), 2.34 (2H), 2.22 (2H, q), 1.41 (3H, d) and 1.08 (3H, t). The MS cracking pattern was characteristic of a 4H-pyran-4-one; and the single vinyl hydrogen ($\delta$ 5.25), the methyl doublet at $\delta$ 1.41 ($H_3C$—C—O) and the triplet at $\delta$ 1.08 (indicating an ethyl group) were enough to confirm the structure as 2,3-dihydro-6-ethyl-2-methyl-4H-pyran-4-one.

Compound (**B**) showed a molecular ion at 182.1307 on HRMS, giving the molecular formula as $C_{11}H_{18}O_2$. It was not affected by $LiAlH_4$ (does not have a C=O), but was reduced by catalytic hydrogenation (5% Pd on C) to a pair of diastereoisomers, which were shown by HRMS to have the formulae $C_{11}H_{20}O_2$. This shows that (**B**) must be a compound with one C—C double-bond and two rings. The dihydro-(**B**) isomers showed prominent MS fragments at m/e 101 ($C_5H_9O_2$ by HRMS), corresponding to a known fragmentation mode of 1,3-dimethyl-2,9-dioxabicyclo-[3,3,1]-nonane. The rest of the mass spectrum was consistent with this carbon skeleton. The occurrence of a peak at [M-29], loss of an ethyl group, in the MS of both (**B**) and its dihydro-compounds accounted for the last two carbons and, in order to accommodate this fragmentation, it was placed at C-3.

The $^1$H nmr spectrum (400 MHz, CDC1$_3$) of (**B**) showed peaks at $\delta$ 5.75 (1H), 4.13 (1H), 3.77 (1H), 2.64 (1H), 1.90–1.30 (5H), 1.64 (3H), 1.41 (3H) and 0.88 (3H). The immediate conclusions that may be drawn are:

- the double-bond is trisubstituted; only one vinyl H ($\delta$ 5.75 signal);
- a methyl group is also on this double-bond ($\delta$ 1.64 peak);
- the ethyl group is confirmed by the $\delta$ 0.88 signal, which is a triplet;
- the methyl single at $\delta$ 1.41 is consistent with a C-1 methyl of the dioxabicyclononane skeleton.

The $^{13}$C nmr spectrum showed the 11 signals expected, of which the most significant were the two vinyl carbons at $\delta$ 133.0 and 123.4, the quaternary carbon of C-1 at $\delta$ 95.4, confirming its attachment to two oxygens, and the C-1 methyl group at $\delta$ 24.7.

It was therefore proposed that the structure of (**B**) was 3-ethyl-1-methyl-2,9-dioxabicyclo-[3,3,1]-nonene, with an additional methyl group on the double-bond. The double-bond could be at either C-6,7 or at C-7,8.

A detailed analysis using two dimensional $^1$H nmr (2-D H,H COSY) allowed the double-bond position to be assigned to C-7,8 and the stereochemistry of the ethyl group to be demonstrated as equatorial as shown in structure (**6.3.7**). This left only the position of the methyl group to be determined.

In order to decide between the two possibilities, the corresponding dihydro-compounds (**6.3.8**) and (**6.3.9**) were synthesized and compared

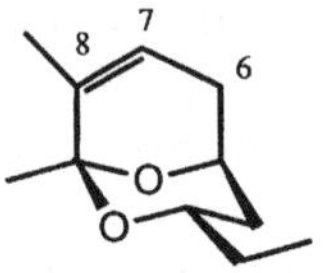

**(6.3.7)** 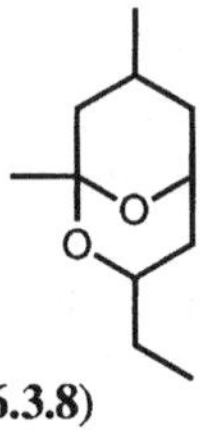 **(6.3.8)** 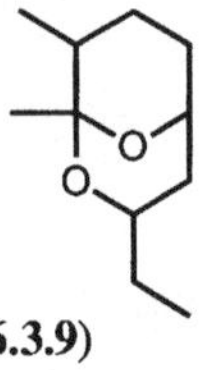 **(6.3.9)**

with dihydro-(**B**), which proved to have an MS identical to that of (**6.3.9**). This established the structure of compound (**B**) as (1*R*, 3*S*, 5*S*)-1,8-dimethyl-3-ethyl-2,9-dioxabicyclo-[3,3,1]-nonane – structure (**6.3.7**) or its enantiomer.

Compound (**C**) had HRMS molecular mass of 196.1149, corresponding to $C_{11}H_{16}O_3$. Reduction with $LiAlH_4$ gave a compound (**C′**) with molecular mass 198, showing the presence of an aldehyde or ketone group in (**C**). Catalytic hydrogenation of (**C**), as for (**B**), also gave a dihydro-compound, which indicated that (**C**) was an alkene with two rings and the carbonyl group.

Both the $^{13}C$ nmr and $^{1}H$ nmr spectra of (**C**) were similar to those of (**B**), suggesting similar skeletons, with one of the methylene groups replaced by a ketone. The signal due to the methyl group on the double-bond appears at $\delta$ 1.95 and the vinyl hdrogen at $\delta$ 6.15. The downfield shift of these two signals strongly indicates that the ketone group is at C-6, an inference supported by the coupling of the H at C-5 only to two other hydrogens (cf. in (**B**) where this is coupled to four others). Confirmation of these conclusions came from the catalytic hydrogenation of the alcohol (**C′**), which gave tetrahydro-(**C**), whose $^{1}H$ nmr spectrum showed significant new signals at $\delta$ 0.94 (3H, d) from the methyl at C-8, and $\delta$ 3.86 (1H, m) from the H at C-6 (> CH—OH), which was coupled to two hydrogens other than the one at C-5, i.e. those of the new $CH_2$ at C-7. Together the data confirm the structure of (**C**) as (1*R*, 3*S*, 5*R*)-1,8-dimethyl-3-ethyl-2,9-dioxabicyclo-[3,3,1]-non-7-en-6-one – structure (**6.3.10**) or its antipode.

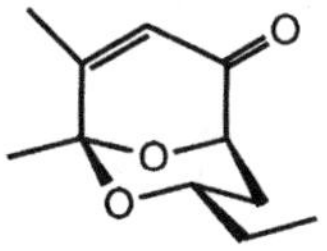

**(6.3.10)**

Five minor components, closely related to the three major compounds, were identified from subsequent GC–MS analysis of the crude scale-brush extract. These were the 2-ethyl homologue of (**A**), the 1-ethyl homologues of (**B**) and (**C**) and the compounds with a 3-methyl group instead of the ethyl group of (**B**) and (**C**).

## REFERENCES


Abley, P., McQuillan, F.J., Minnikin, D.E. *et al.* (1970) Location of olefinic links in long-chain esters by methoxymercuration–demercuration followed by gas chromatography–mass spectrometry. *J. Chem. Soc., Chem. Commun.*, 348–348.

Arn, H., Stadler, E., Rauscher, S. *et al.* (1980) Multicomponent sex pheromone in *Agrotis segetum*: preliminary analysis and field evaluation. *Z. Naturforsch.*, **35c** 986–989.

Arn, H., Guerin, P.M., Buser, H.R. *et al.* (1985) Sex pheromone blend of the codling moth, *Cydia pomonella*: evidence for a behavioural role of dodecanol. *Experientia*, **41**, 1482–1484.

Attygalle, A.B. and Morgan, E.D. (1988) Pheromones in nanogram quantities: structure determination by combined microchemical and gas chromatographic methods. *Angew. Chem. Int. Ed. Engl.*, **27**, 460–478.

Beroza, M and Bierl, B.A. (1967) Rapid determination of olefin position in organic compounds in microgram range by ozonolysis and gas chromatography. *Anal. Chem.*, **39**, 1131–1135.

Beroza, M and Sarmiento, R. (1966) Apparatus for reaction gas chromatography. *Anal. Chem.*, **38**, 1042–1047.

Bestmann, H.J., Erler, J. and Vostrosky, O. (1988) Determination of diel periodicity of sex pheromone release in three species of Lepidoptera by 'closed-loop-stripping'. *Experientia*, **44**, 797–799.

Bierl, B.A., Beroza, M., Staten, R. T. *et al* (1974) The pink bollworm moth sex attractant. *J. Econ. Entomol.*, **67**, 211–216.

Birch, A.J., Brown, W.V., Corrie, J.E.T. and Moore, B.P. (1972) Neocembrene-A, a termite trail pheromone. *J. Chem. Soc., Perkin Trans.* **1**, 2653–2658.

Braithwaite, A. and Smith, F.J. (1985) *Chromatographic Methods*, 4th edn. Chapman & Hall, London and New York.

Buser, H.R., Arn, H., Guerin, P. and Rauscher, S. (1983) Determination of double bond position in mono-unsaturated acetates by mass spectromety of dimethyldisulfide adducts. *Anal. Chem.*, **55**, 818–822.

Casey, M., Leonard, J., Lygo, B. and Procter, G. (1990) *Advanced Practical Organic Chemistry*, Blackie, Glasgow and London.

Francke, W., Mackenroth, W., Schröder, W. *et al.* (1985) Identification of cyclic enol ethers from insects: alkyldihydropyrans from bees and alkyldihydro-4H-pyran-4-ones from a male moth. *Z. Naturforsch.*, **40c**, 145–147.

Harwood, L.M. and Moody, C.J. (1989) *Experimental Organic Chemistry*, Blackwell Scientific Publications, Oxford.

Henderson, J.E., Warren, F.L., Augustin, O.P.H. *et al.* (1973) Isolation and structure of the sex pheromone of the moth *Nudaurelia cytherea*. *J. Insect Physiol.*, **19**, 1257–1264.

Kemp, W. (1991) *Organic Spectroscopy*, 3rd edn, Macmillan, London.

Klun, J.A., Plimmer, J.R., Bierl-Leonhardt, B.A. *et al.* (1979) Trace chemicals: the essence of sexual communication systems in *Heliothis* species. *Science*, **204**, 1328–1330.

Klun, J.A., Plimmer, J.R., Bierl-Leonhardt, B.A. *et al.* (1980) Sex pheromone chemistry of female corn earworm moths, *Heliothis zea*. *J. Chem. Ecol.*, **6**, 165–175.

McDonough, L.M., George, D.A., Butt B.A. *et al.* (1969) Isolation of a sex pheromone of the codling moth. *J. Econ. Entomol.*, **62**, 62–65.

Moore, B.P. (1966) Isolation of the scent-trail pheromone of an Australian termite. *Nature*, **211**, 746–747.

Moore, B.P. and Brown, W.V. (1971) Gas–liquid chromatographic identification of ozonolysis fragments as a basis for micro-scale structure determinations. *J. Chromatog.*, **60**, 157–166.

Moorhouse, J.E., Yeadon, R., Beevor, P.S. and Nesbitt, B.F. (1969) Methods for use in studies of insect communication. *Nature*, **223**, 1174–1175.

Morgan, E.D. (1990) Preparation of small-scale samples from insects for chromatography. *Analyt. Chim. Acta*, **236**, 227–235.

Morgan, E.D. and Wadhams, L.J. (1972) Gas chromatography of volatile compounds in small samples of biological materials. *J. Chromatogr. Sci.*, **10**, 528.

Pierce, H.D., Pierce, A.M., Millar, J.G. *et al.* (1984) Methodology for isolation and analysis of aggregation pheromones in the genera *Cryptolestes* and *Oryzaephilus* (Coleoptera: Cucjidae). *Proceedings of 3rd International Conference on Stored-product Entomology, Kansas State University, Manhattan, Kansas*, 121–137.

Ritter, F. J., Brüggemann-Rotgans, I.E.M., Verwiel, P.E.J. *et al.* (1977) Trail pheromone of pharaoh's ant, *Monomorium pharaonis*: isolation and identification of

faranal, a terpenoid related to Juvenile Hormone II. *Tetrahedron Letters*, **30**, 2617–2618.
Sakai, T., Makagawa, Y., Takahashi, J. *et al.* (1984) Isolation and identification of the male sex pheromone of the grape borer *Xylotrechus pyrrhoderus* Bates (Coleoptera: Cerambycidae). *Chem. Letters, 263–264.*
Sinnwell, V., Schulz, S., Francke, W. *et al.* (1985) Identification of pheromones from the male swift moth *Hepialus hecta L. Tetrahedron Letters*, **26**, 1707–1710.
Sower, J.L., Coffelt, J.A. and Vick, K.W. (1973) Sex pheromones: a simple method of obtaining relatively pure material from females of five species of moths. *J. Econ. Entomol.*, **66**, 1220–1222
Williams, D.H. and Fleming, I. (1989) *Spectroscopic Methods in Organic Chemistry* 4th edn, McGraw-Hill, Maidenhead.

# 7

# Synthesis of pheromones

Once the structure of a pheromonal constituent has been deduced, the next step requires the chemical synthesis of the compound so that it may be tested on the insect, both in the laboratory and in the field. Without such a step, the structure remains a chemical curiosity, of no value either to the biologist or to the chemist, or to the agriculturalist who may wish to control the population of the insect in question. Furthermore, the synthesis is the final step that is essential in the confirmation of the structural identification.

The aim of this chapter is to introduce the routes that have been used for certain classes of pheromonal compounds, to bring out the rationale behind these routes and to show how the methods of retro-synthetic analysis have been applied to the particular compounds discussed. It is not intended for the specialist synthetic chemist, but seeks to give an overview of the problems that are faced in any chemical synthesis and the methods available for tackling them. It is hoped that the comments will enable the biological specialists to gain some insight into the difficulties that need to be overcome in order that the synthetic chemist is able to present to an entomological colleague a vial containing the desired material.

The syntheses chosen are intended to be illustrative rather than exhaustive and could easily have been replaced with others equally valid.

## 7.1 GENERAL PROBLEMS OF SYNTHESIS

There is little doubt that every new compound presents a different challenge and that for any given compound a substantial number of different routes may lead to a successful synthesis. Indeed each person presented with the problem of preparing a given compound is likely to propose a pathway that differs in some degree from any other person's proposal. Hence the oft quoted truism that there is a great deal of 'Art in organic synthesis'.

The problems faced by the would-be synthesizer may be summarized as:

- assembling the correct carbon skeleton;
- ensuring that the required functional groups are in the correct places on the skeleton;
- making certain that any double-bonds have the correct geometry;
- devising a route that will provide the required diastereoisomer, exclusively for preference but, if this is not possible, then for ensuring that it can be separated out in a pure state;
- arranging that the desired optical isomer is isolated.

As well as these major points, the total number of steps needed to reach the target compound, the availability of starting materials and the overall cost will eventually also enter into the final choice of a synthetic scheme.

Lastly, but by no means least, is the 'Murphy's Law' factor, which will undoubtedly intervene at some stage in any new synthetic route and ensure that it will need to be modified while in progress.

All these points are most readily appreciated in relation to some real examples.

## 7.2 EXAMPLES OF SYNTHESIS

### 7.2.1 Straight chain compounds

In synthetic terms these represent the easiest targets. They also have widespread utility, because the majority of female lepidopteran sex pheromones fall into this class.

#### *(a) Monoene alcohols, acetates and aldehydes*

Z-9-Tetradecenyl acetate, a constituent of the pheromone of many *Adoxophyes* species, presents just two problems: the position of the double-bond in the chain, and the stereochemistry.

The most straightforward route to any monoene starts with the double-bond as an alkyne (formally acetylene) and builds up the chain on each side to the required length before finally reducing the triple-bond selectively to give the required geometry. The Scheme 7.1 illustrates this for Z-9-tetradecenyl acetate. Not counting the double-bond, we require eight carbons for the left-hand end of the molecule and four for the right. Thus from acetylene, the route might first add the four carbons on the right to give hex-1-yne. Happily this compound can be readily purchased and therefore forms a suitable starting material.

Scheme 7.1 Synthesis of (*Z*)-9-tetradecenyl acetate

(a) Disconnection

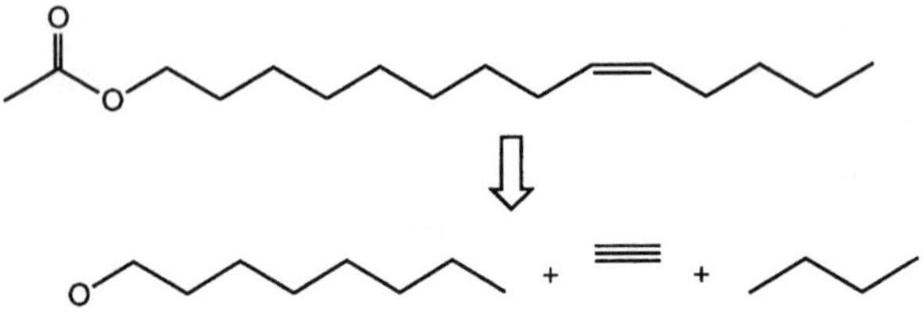

O

O

O + + 

(b) Forward synthesis

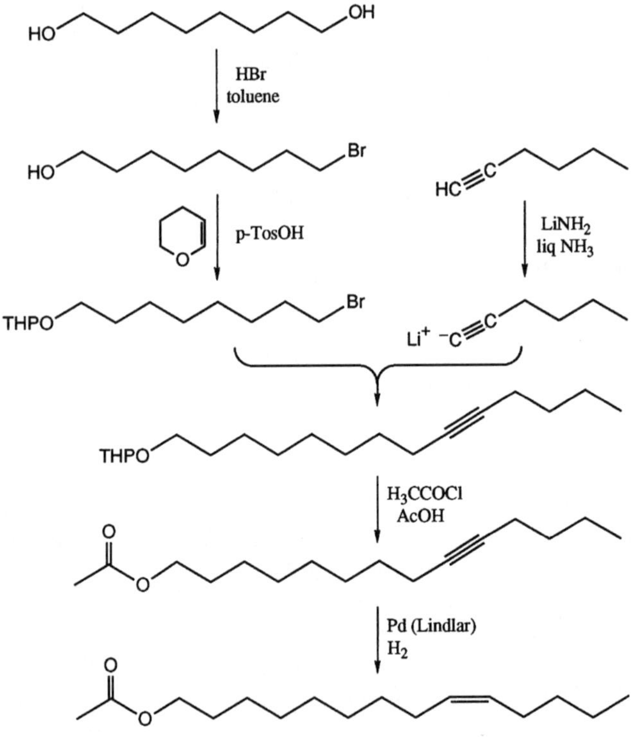

HO OH

HBr
toluene

HO Br

HC≡C

p-TosOH

LiNH$_2$
liq NH$_3$

THPO Br

Li$^+$ $^-$C≡C

THPO

H$_3$CCOCl
AcOH

O
O

Pd (Lindlar)
H$_2$

O
O

(*Z*)-9-tetradecenyl acetate

The eight carbons for the left-hand side need to have an oxygen atom at one end of the chain and a suitable functionality to link with the alkyne at the other. 8-Bromo-octan-1-ol suggests itself, because the bromine can be displaced by the sodium salt of the alkyne. However, this suffers from two drawbacks: it is expensive, and under the reaction conditions it can react suicidally with itself to generate polymers. This is a common problem found with any bi-functional compound and one must always be on guard about the possibility. The first drawback is overcome by starting with 1,8-octandiol (relatively cheap), which may be transformed into 8-bromo-octanol in a straightforward step, though some care is required, and the second by 'protecting' the alcohol group as a tetrahydropyranyl (THP) ether, which is an acetal. This group may be added and removed under acidic conditions but is perfectly stable to alkali.

Thus hex-1-yne is converted into its sodium (or lithium) salt using sodamide (lithamide) in liquid ammonia and then the 8-bromo-octanol THP ether, in tetrahydrofuran (THF) solution, is added. After a suitable work-up procedure, the crude tetradec-9-yn-1-ol THP ether is converted into the corresponding acetate. This may be done directly using acetyl chloride in acetic acid, which hydrolyses the THP ether and acetylates the resultant alcohol all in one pot. After removal of the excess reagents, the tetradec-9-yn-1-ol acetate may be semi-hydrogenated in pentane solution to the *Z*-alkene using a specific palladium catalyst (Lindlar's catalyst) in the presence of quinoline. Work-up and distillation then affords the desired *Z*-9-tetradecenyl acetate.

*Notes*

- The synthesis may be run on a scale sufficient to give up to 100 g of product using normal laboratory scale equipment. For large-scale work, the use of the lithium salt is preferable. The overall yield is from 70 to 75%; the isomer ratio ($Z/E$) is better than 99.8 : 0.2 and the overall purity greater than 98%.
- This basic synthetic route may be adapted to form any monoene alcohol, acetate or aldehyde.
- The position of the double-bond may be altered by using different alkynes and bromoalcohol THP ethers. Note that heptan-1,7-diol is expensive.
- The alkyne may be reduced to give the *E*-alkene with sodium metal in liquid ammonia.
- The THP ether group may be removed to give the alcohol by stirring with a trace of toluene-*p*-sulphonic acid in methanol, followed by neutralization and removal of solvent.
- The alcohol may be oxidized to the corresponding aldehyde by reaction with pyridinium dichromate (PDC) or pyridinium chloro-

chromate (PCC) in dichloromethane. For *E*-alkenes, this should only be done after reduction of the alkyne to the alkene.

### (b) Non-conjugated dienes

The two non-conjugated dienes *Z,Z*- and *Z,E*-7,11-hexadecadienyl acetate are the components of the female sex pheromone of the pink bollworm moth, *Pectinophora gossypiella* L. A similar strategy to that given above may be adopted, again using acetylene as the basis for each of the alkene bonds (Scheme 7.2).

Hex-1-yne is converted into oct-3-yn-1-ol exactly as for a monoene. Half of this alcohol is then reduced by each of the two routes: catalytic, to give the *Z*-3-octenol, and dissolving metal to give *E*-3-octenol. Conversion of the alcohol function into a suitable leaving group (bromide) must be carried out under conditions that do not affect the double-bond geometry. The use of triphenyl phosphine dibromide in the presence of pyridine in dioxan solution is mild and effective, affording the corresponding bromides in good yield. These may then be reacted with the lithium salt of oct-7-yn-1-ol THP ether in liquid ammonia–THF solution to give, from the *Z*-3-octenyl bromide, *Z*-11–hexadecen-7-yn-1-ol THP ether. Conversion to the corresponding acetates with acetyl chloride in acetic acid, followed by reduction using hydrogen and Lindlar's catalyst, completes the synthesis.

*Notes*

- The overall yields and purity are similar to those of the synthesis in section 7.2a. The isomer ratio is better than 99.5 : 0.5 for each of the components.
- Oct-7-yn-1-ol THP ether is prepared by reacting lithium acetylide (in ammonia) with 6-bromohexan-1-ol THP ether.

### (c) Conjugated dienes

The first lepidopteran pheromone to have its chemical structure elucidated is that from the silkworm moth, *Bombyx mori*. The major component is *Z,E*-10,12-hexadecadien-1-ol and there have been numerous syntheses reported. The one given here (Scheme 7.3; Negishi *et al.*, 1973) illustrates the use of boron chemistry to overcome the problem of linking two $sp^2$ centres, which is not possible by the route given above for monoenes. This apart, the synthesis shows strong similarities to those of other straight chain compounds.

The addition of borane to alkynes occurs stereospecifically so as to place the boron and the hydrogen on the same side of the new double-bond. Hence, addition of di(1,2-dimethylpropyl)borane to 1-pentyne in

Scheme 7.2 Synthesis of (*Z,Z*)- and (*Z,E*)-7,11-hexadecadienyl acetate

(i) $LiNH_2$, $NH_3$
(ii) $BrC_2H_4OTHP$

OTHP

p-TosOH
MeOH

OH

Pd (Lindlar)
$H_2$

Na
liq. $NH_3$

$C_4H_9$ OH

$C_4H_9$ OH

$Ph_3P$-$Br_2$
pyridine, dioxan

$Ph_3P$-$Br_2$
pyridine, dioxan

$C_4H_9$ Br

$C_4H_9$ Br

Li C≡C—$(CH_2)_6$-OTHP
liq. $NH_3$, THF

Li C≡C—$(CH_2)_6$-OTHP
liq. $NH_3$, THF

$C_4H_9$ $(CH_2)_6$-OTHP

$C_4H_9$ $(CH_2)_6$-OTHP

(i) AcCl, AcOH
(ii) Pd (Lindlar), $H_2$

(i) AcCl, AcOH
(ii) Pd (Lindlar), $H_2$

$C_4H_9$ $(CH_2)_6$-OAc

$C_4H_9$ $(CH_2)_6$-OAc

OAc

(*Z,Z*)-7,11-Hexadecadienyl acetate

OAc

(*Z,E*)-7,11-Hexadecadienyl acetate

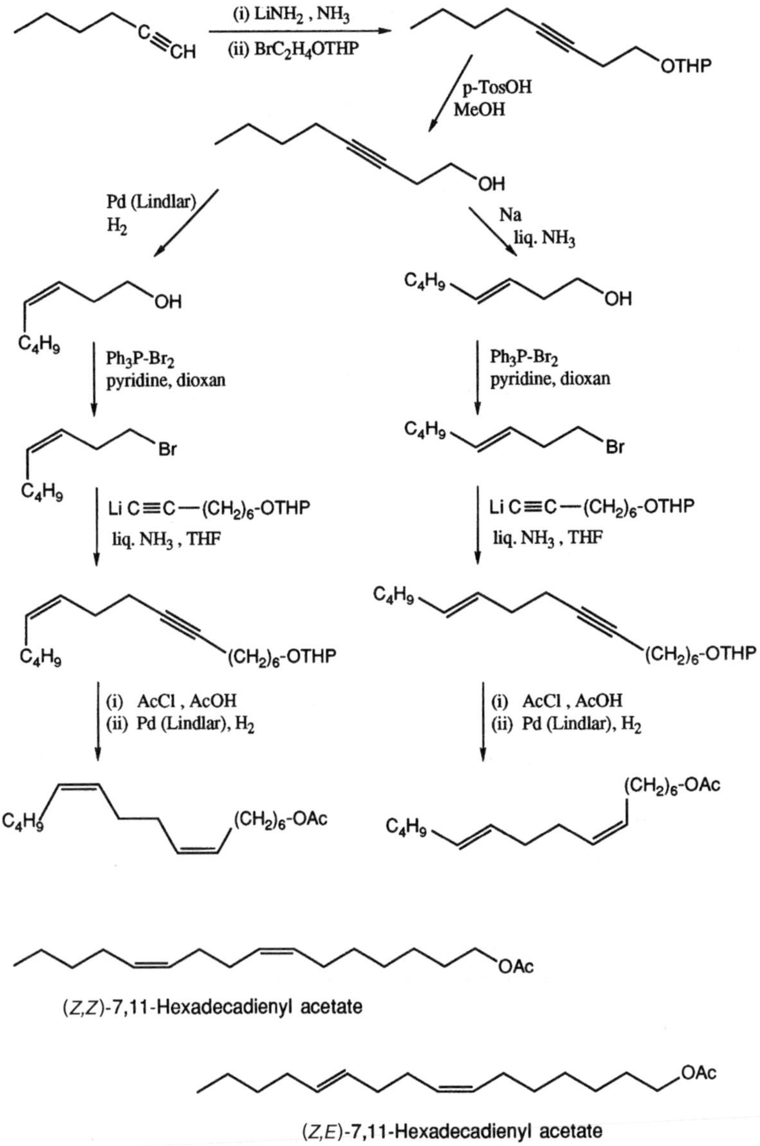

Scheme 7.3 Synthesis of (*Z*,*E*)-10,12-hexadecadien-1-ol (Negishi *et al.*, 1973)

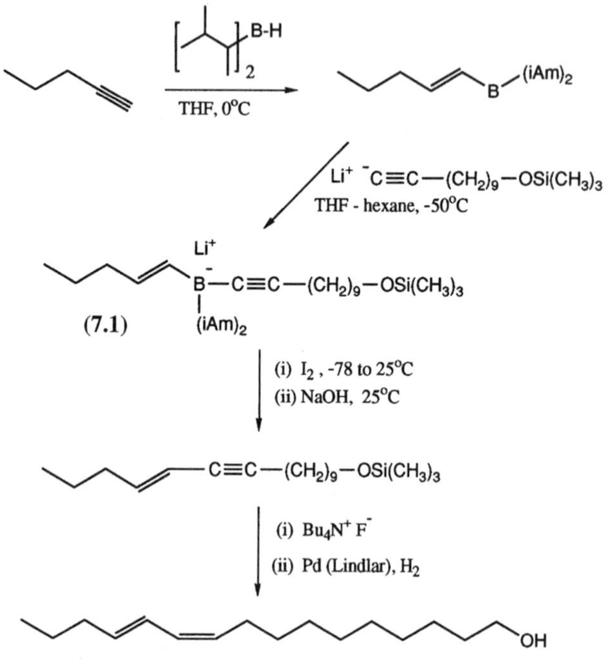

B-H 2

THF, 0°C

B (iAm)$_2$

Li$^+$ $^-$C≡C—$(CH_2)_9$—$OSi(CH_3)_3$

THF - hexane, -50°C

Li$^+$

B—C≡C—$(CH_2)_9$—$OSi(CH_3)_3$

(iAm)$_2$

**(7.1)**

(i) $I_2$, -78 to 25°C

(ii) NaOH, 25°C

C≡C—$(CH_2)_9$—$OSi(CH_3)_3$

(i) $Bu_4N^+ F^-$

(ii) Pd (Lindlar), $H_2$

OH

(*Z*,*E*)-10,12-hexadecadien-1-ol

THF solution at 0°C gives *E*-1-[di(1,2-dimethylpropyl)boryl]-pent-1-ene. Reaction of this with the lithium salt of undec-10-yn-1-ol trimethylsilyl ether gives the lithium tetra-alkyl borate (**7.1**) shown in Scheme 7.3. The key step is the reaction of this borate with iodine followed by dilute sodium hydroxide, which causes migration of the alkenyl residue on to the alkyne and elimination of the boron. The product is the enyne, *E*-12-hexadecen-10-yn-1-ol trimethylsilyl ether (63% yield). Reduction of the alkyne to the *Z*-alkene may be accomplished using Lindlar's catalyst and hydrogen, although the original workers used the addition of di(1,2-dimethylpropyl)borane, followed by protonolysis to accomplish the same transformation.

*Notes*

- The overall yield based on 10-undecynoic acid as starting material was about 45%. The isomeric purity was >99%.
- Undec-10-yn-1-ol trimethylsilyl ether was prepared from 10-undecy-

noic acid by reduction with $LiAlH_4$ and, after work-up, by treatment with chlorotrimethylsilane.

### 7.2.2 Branched chain alkenes

Trisubstituted double-bonds present a much greater challenge than disubstituted ones and a number of pathways for overcoming this problem have been used. When the desired stereochemistry is *E*, then the most straightforward manner for achieving the goal is to make use of the reaction discovered by Wittig. This involves the reaction of a ketone or aldehyde with a phosphonium ylid (equation 7.1).

$$R^1CH{=}O + Ph_3\overset{+}{P}{-}\overset{-}{C}H{-}R^2 \longrightarrow R^1{-}HC{=}CH{-}R^2$$

When this procedure is carried out on an aldehyde with a carbonyl-stabilized phosphonium ylid, then the major isomer formed is the *E*-alkene, even when a trisubstituted double-bond is created. The reaction is not totally stereospecific, but the greater the bulk of the alkyl group on the aldehyde, the more the preference for the formation of the *E*-alkene is enhanced; typical ratios range from 95:5 to greater than 99:1.

The greater grain borer, *Prostephanus truncatus* Horn, uses 1-methylethyl *E,E*-2,4-dimethylhepta-2,4-dienoate as the major component of the pheromone. This dienic ester contains two trisubstituted double-bonds conjugated to the ester group and is thus obviously suitable for a Wittig route. Because the Wittig reaction creates the double-bond, the retrosynthesis of the acid skeleton leads to three three-carbon fragments, two of propanal and one of propionic acid – Scheme 7.4a, where the stars indicate the required activation, and hence results in the forward route shown in Scheme 7.4b (Miles, 1989).

Propanal may be self-condensed in an aldol reaction to give the required *E*-2-methylpent-2-enal. The stabilized phosphonium ylid is prepared by reacting 1-methylethyl 2-bromopropionate with triphenylphosphine in toluene and then treating the phosphonium salt with aqueous sodium hydroxide. Reaction of the two components in dry dichloromethane leads to the desired target with a stereochemical purity of greater than 99.5%.

*Notes*

- The overall yield, based on the bromo ester, is 55 to 60% and batches of up to 150 g may be readily handled. The overall purity is >98% and the *E,E* isomer purity is better than 99.5:0.5.
- The 1-methylethyl 2-bromopropionate is prepared by reacting 2-bromopropionyl bromide with propan-2-ol. The reaction is almost quantitative.

Scheme 7.4 Synthesis of 1-methylethyl (*E,E*)-2,4-dimethylhepta-2,4-dienoate (Miles, 1989)

(a) Disconnection

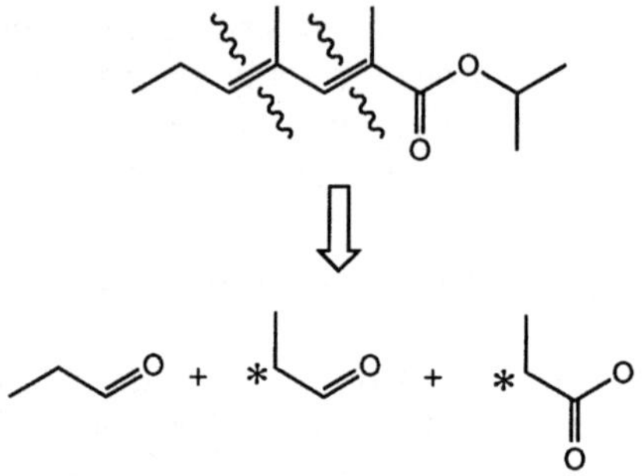

+ * + *

(b) Forward synthesis

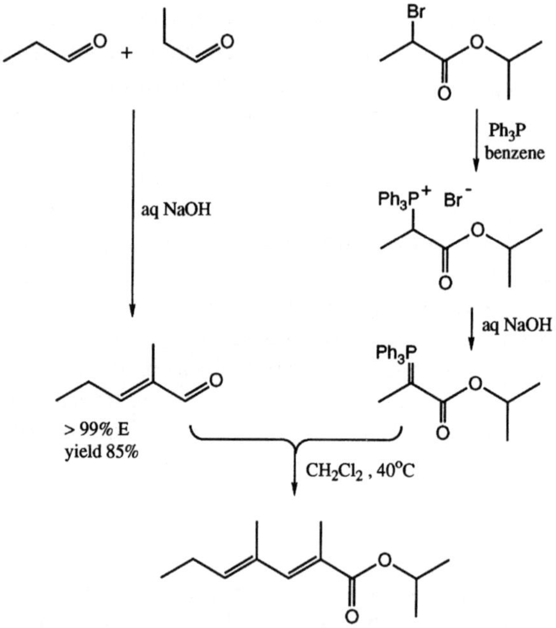

Br

+

$Ph_3P$
benzene

aq NaOH

$Ph_3P^+$ $Br^-$

aq NaOH

$Ph_3P$

> 99% E
yield 85%

$CH_2Cl_2$ , 40°C

- This procedure works well with non-conjugated aldehydes, but the $E/Z$ ratio is lower, typically 97 to 99 : 1.
- If the required double-bond is an isolated one, then the ester group may be reduced to give the allylic alcohol, which can be subsequently chain extended.

### 7.2.3 Cyclic compounds

In the synthesis of cyclic compounds there is the new problem of ensuring that the correct ring size is obtained. There are two general approaches to this problem: one is to use a starting material in which the required ring is already available and the other is to prepare an acyclic compound with the correct structural elements present and then to cyclize it. Each of these methods is exemplified.

#### *(a) Spirocyclic ketals*

The olive fly, *Dacus oleae*, has one major component (1,7-dioxaspiro[5,5]-undecane) and a number of minor constituents in the pheromone, amongst which is 4-hydroxy-1,7-dioxaspiro[5,5]undecane. The reactions of Scheme 7.5 show how each was synthesized from the lactone of 5-hydroxypentanoic acid (5-pentanolide.).

The major component (Fittig, 1890) results from the aldol type reaction of 5-pentanolide with itself to give the intermediate bicyclic compound (**7.2**), which is not isolated, but is heated in acid solution, which opens the lactone ring to produce the dihydroxy-keto-acid (**7.3**). This is not stable under the reaction conditions and decarboxylates immediately to form 5-oxononan-1,9-diol, which cyclizes spontaneously to the desired spiroketal target with loss of water.

Given the fact that 5-oxononan-1,9-diol cyclizes readily, the route to the minor component (Baker *et al.*, 1982) may be envisaged retrosynthetically as involving the triol (**7.4**), which could be formed from the unsaturated dihydroxyenone (**7.5**), itself prepared from 5-pentanolide.

The forward synthesis proceeded by the addition of the lithium salt of but-3-yn-1-ol THP ether to 5-pentanolide in liquid ammonia–THF solution. Work-up under buffered conditions gave the hemi-ketal (**7.6**), which was hydrogenated with Lindlar's catalyst to the alkenic hemi-ketal (**7.7**). When this was subjected to mild acid hydrolysis, the terminal hydroxy-group was liberated from the THP ether forming (**7.5**); water was added across the double-bond, assisted by the conjugated carbonyl group, and the molecule cyclized to give the desired 4-hydroxy-1,7-dioxaspiro[5,5]undecane.

Scheme 7.5 Synthesis of spirocyclic ketals

(a) 1,7-dioxaspiro[5,5]undecane (Fittig, 1890)

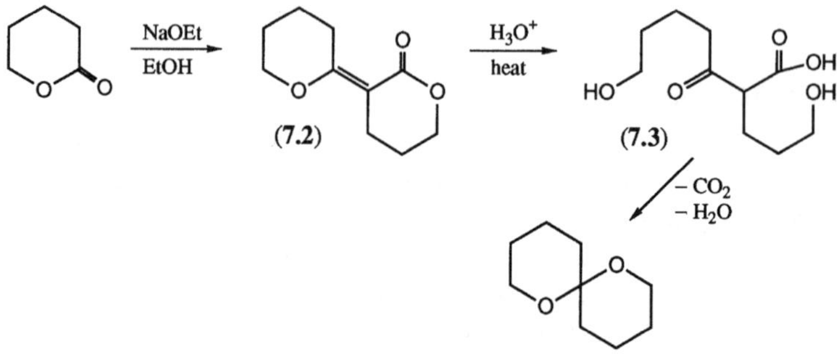

NaOEt / EtOH → (7.2) → $H_3O^+$ / heat → (7.3) → $-CO_2$, $-H_2O$

(b) 4-hydroxy-1,7-dioxaspiro[5,5]undecane (Baker, 1982)

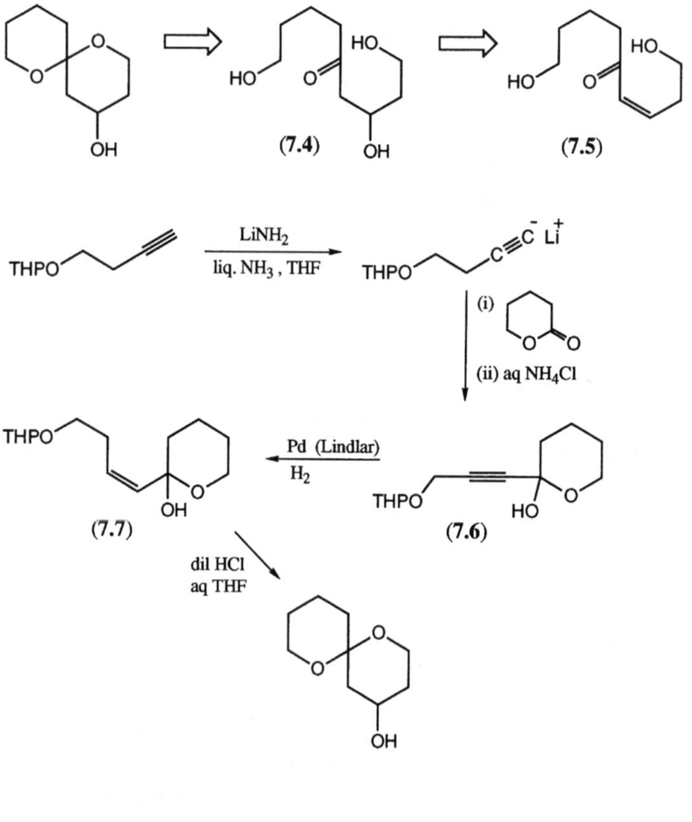

(7.4) (7.5)

$LiNH_2$ / liq. $NH_3$, THF

(i); (ii) aq $NH_4Cl$

Pd (Lindlar) / $H_2$

(7.7) (7.6)

dil HCl / aq THF

*Notes*

- The first of these two syntheses can be adapted to form any **symmetric** spiroketal ring compound, where the ring sizes are 5, 6 or 7.
- The second process is applicable to the formation of many **unsymmetric** spiroketals by using a lactone with a larger or smaller ring and changing the number of carbons in the alkyne. Again the ring sizes are limited to 5, 6 and 7 members.

### (b) *Large ring lactones*

The saw-toothed grain beetle, *Oryzaephilus surinamensis*, uses an aggregation pheromone consisting of the three macrocyclic lactones: Z,Z-3,6-dodecadien-11-olide, Z,Z-3,6-dodecadien-12-olide and Z,Z-5,8-tetradecadien-13-olide, structures (**7.8**), (**7.9**) and (**7.10**). The difficulty in all three is the 'skipped' diene unit: $-C{=}C-CH_2-C{=}C-$, and the large size of the ring. In addition, the first two have the skipped diene positioned one carbon away from the carbonyl group; the double-bonds would be more stable if they were conjugated with the ester carbonyl and they will readily isomerize into this position if incorrectly handled.

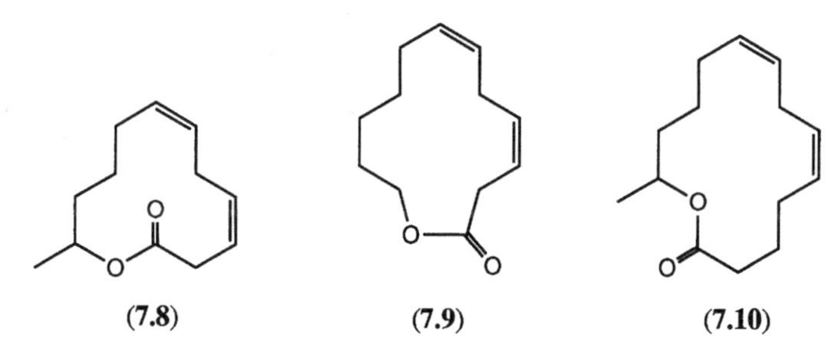


(**7.8**) (**7.9**) (**7.10**)

The formation of large rings requires that the two reactive ends of the molecule be brought together, while at the same time preventing the reaction of one molecule with another, which will lead initially to a dimer and thence to a polymer. Because this 'curling round' of linear molecules is most difficult for ring sizes of 10 to 12 members, the synthesis described is that of the compound with the smallest ring, (**7.8**) (Boden *et al.*, 1993).

As ring closure was to be the final step in the synthesis, the initial requirement was for the open chain compound (3Z,6Z)-11-hydroxydodeca-3,6-dienoic acid. This was obtained by using the Wittig synthesis of double-bonds, which with unstabilized phosphonium ylids gives Z-disubstituted alkenes with very high stereoselectivity (97 to 99.8%). The path used is shown in Scheme 7.6.

Reaction of the Grignard reagent (organomagnesium) from 1-bromo-

Scheme 7.6 Synthesis of (*Z,Z*)-3,6-dodecadien-11-olide (Boden, 1993)

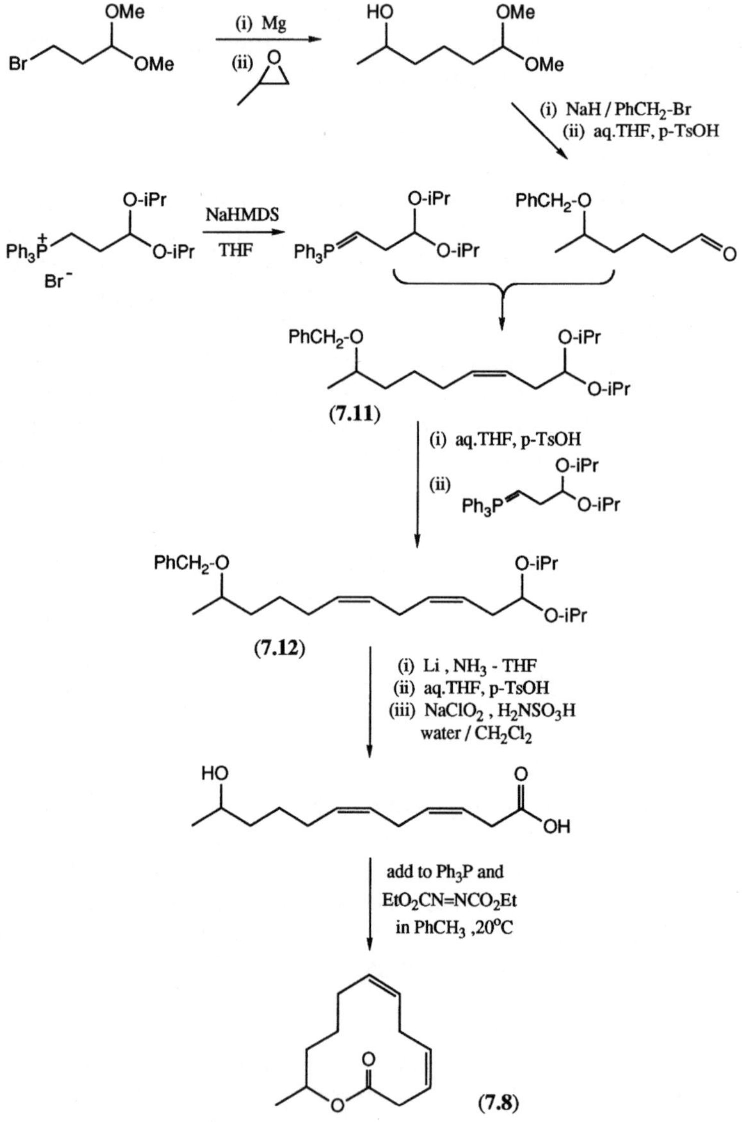
OMe
Br
OMe
(i) Mg
(ii)
O
HO
OMe
OMe
(i) NaH / PhCH$_2$-Br
(ii) aq.THF, p-TsOH
O-iPr
$Ph_3\overset{+}{P}$
O-iPr
Br$^-$
NaHMDS
THF
Ph$_3$P
O-iPr
O-iPr
PhCH$_2$-O
O
PhCH$_2$-O
O-iPr
O-iPr
(7.11)
(i) aq.THF, p-TsOH
(ii)
Ph$_3$P
O-iPr
O-iPr
PhCH$_2$-O
O-iPr
O-iPr
(7.12)
(i) Li , NH$_3$ - THF
(ii) aq.THF, p-TsOH
(iii) NaClO$_2$ , H$_2$NSO$_3$H water / CH$_2$Cl$_2$
HO
O
OH
add to Ph$_3$P and
EtO$_2$CN=NCO$_2$Et
in PhCH$_3$ ,20°C
O
O
(7.8)

3,3-dimethoxypropane with propylene oxide gives 6,6-dimethoxyhexan-2-ol, the six-carbon fragment needed for the left-hand end. The hydroxyl group is protected as the benzyl ether by reaction with sodium hydride followed by benzyl bromide, after which the aldehyde group is unmasked by mild acid hydrolysis of the methoxy acetal. Wittig reaction with the ylid derived from (3,3-di-isopropoxypropyl)triphenylphosphonium bromide gives the *Z*-3-alkene acetal, (**7.11**). The masked aldehyde may then be released, as in the previous step, and the Wittig reaction repeated to give the necessary carbon skeleton, compound (**7.12**). The benzyl ether protection is then removed by reaction with lithium metal in liquid ammonia, followed by hydrolysis of the acetal function. The liberated aldehyde is immediately oxidized in a two-phase system with sodium chlorite, which is mild enough not to permit any isomerization – a process that occurs readily with other oxidants. The resulting acid is then ready for the cyclization step. This is best achieved by activating the hydroxyl group and allowing reaction to occur under conditions of high dilution, which minimizes any intermolecular reaction. Very slow addition of the hydroxy-acid to a mixture of triphenylphosphine and diethyl azodicarboxylate in toluene at 20°C gives the macrolide in yields of better than 70%. The pure lactone is isolated by chromatography after completion of the reaction.

*Notes*

- The overall yield is 45% based on 1-bromo-3,3-dimethoxypropane.
- The basic method was used for the synthesis of all three of the lactones (**7.8**) to (**7.10**) and also for three others produced by *Cryptolestes* species.

### 7.2.4 Chiral compounds

The synthesis of compounds that are single enantiomers (homochiral) presents the final and probably most difficult challenge to the organic chemist. All chemical procedures that create chiral centres in a molecule will do so to give equal amounts of the two enantiomers, unless one of the reactants or reagents is itself homochiral, but this situation is not the norm. Hence, a straightforward synthesis, followed by separation of the two enantiomers, inevitably leads to the loss of at least 50% of the material that has been created – hardly an efficient procedure. Furthermore, separation of enantiomers is not always straightforward and sometimes not even possible.

In tackling the difficulty posed by this problem, there are three significant methods at the chemist's disposal. The first is to start with a suitable naturally available homochiral compound, the second is to make use of a chemical asymmetric reaction and the third is to employ an appropriate biologically mediated chemical transformation; this makes use of the fact that enzymes are homochiral reagents and will normally accept only one out of the two enantiomers as a substrate. Each of these procedures is illustrated.

### (a) *Citrus mealybug*

The citrus mealybug, *Planococcus citri*, Risso uses the cyclobutane derivative (1*R*)-*Z*-2,2-dimethyl-3-(1-methylethenyl)-cyclobutanemethanol acetate, structure (**7.13**), as the major component of the sexual attractant pheromone for the males. Field testing showed that a lure with 1 ng of the synthetic (1*R*) enantiomer was about five times as effective in trap captures as a virgin female (Ortu and Delrio, 1982).

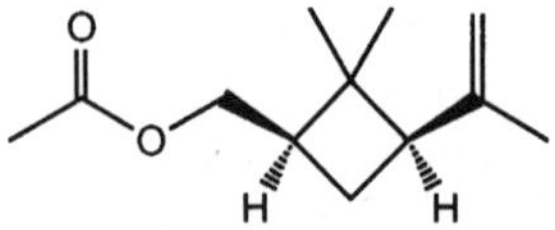

(**7.13**)  (**7.14**)

Scheme 7.7 Synthesis of citrus mealybug pheromone (Ferreira *et al*., 1987)

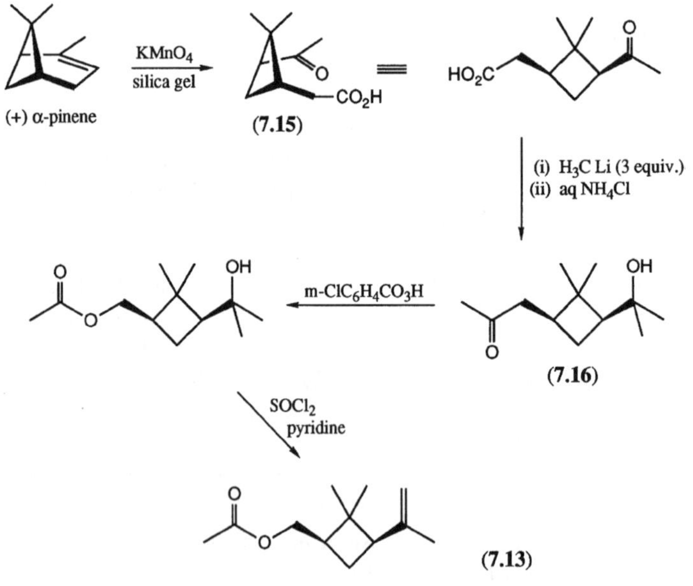

(+) α-pinene → $KMnO_4$, silica gel → (**7.15**) ≡ $HO_2C$…

(i) $H_3C$ Li (3 equiv.)
(ii) aq $NH_4Cl$

(**7.16**) → m-$ClC_6H_4CO_3H$ → OH

$SOCl_2$
pyridine

(**7.13**)

The pattern of methyl groups signals the terpenoid origin of the compound and suggests the use of *a*-pinene, structure (**7.14**), as an appropriate starting material, particularly as this is readily available in large quantities and in homochiral form from pine resin. The route used (Ferreira *et al.*, 1987) is shown in Scheme 7.7. Cleavage of the double-bond of (+) α-pinene with potassium permanganate on silica gel gives the (+) *cis*-pinonic acid (**7.15**), which after reaction with three molecular equivalents of methyl lithium gives the hydroxy-cyclobutyl-acetone (**7.16**).

Oxidation of this ketone with *m*-chloroperoxybenzoic acid converts the ketone into an acetyl group by migration of the cyclobutylmethyl residue and the synthesis is completed by the dehydration of the tertiary alcohol to the required alkene using thionyl chloride in pyridine.

### (b) *Cigarette beetle*

The cigarette beetle, *Lasioderma serricorne* Fabricius, has a multi-component sexual pheromone system, of which (4*S*,6*S*,7*S*)-4,6-dimethyl-7-hydroxynonan-3-one, (**7.17**) (serricornin), is the most abundant and most active component. The (4*S*,6*S*,7*R*)-isomer has been shown to inhibit the response of the natural compound.

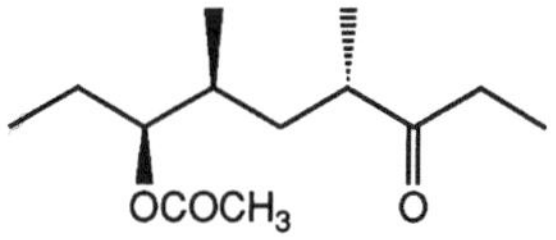

$OCOCH_3$ O

(**7.17**)

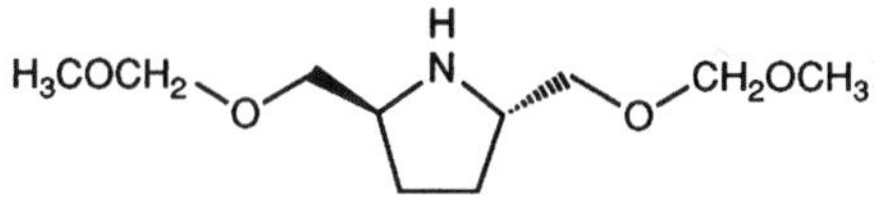

$H_3COCH_2$ O H N O $CH_2OCH_3$

(**7.18**)

The synthesis (Katsuki and Yamaguchi, 1987) is shown in Scheme 7.8 and starts with the chiral auxiliary (2*S*,5*S*)-2,5-bis(methoxymethoxy-methyl)pyrrolidine (**7.18**), which is converted to the *N*-propanoyl amide (**7.19**) with propanoic anhydride. This amide is converted to its zirconium enolate and then condensed with propanal. After hydrolysis with dilute hydrochloric acid, the homochiral (2*R*,3*S*) diastereomer of 2-methyl-3-hydroxypentanoic acid is obtained and the chiral auxiliary (**7.18**) is recovered. These two chiral centres will become carbons 6 and 7, respectively, of the final target. The diastereoselectivity of the reaction is imposed by the three-dimensional shape of the auxiliary (**7.18**), and the particular enantiomer used controls the enantioselectivity. Use of the antipode of (**7.18**), the (2*R*,5*R*) compound, would have given the same diastereomer, but the opposite enantiomer of the product, i.e. (2*S*,3*R*)-2-methyl-3-hydroxypentanoic acid.

The alcohol is converted into the methoxymethyl ether for protection, a process which also changes the acid into the methoxymethyl ester.

Scheme 7.8 Synthesis of (4*S*,6*S*,7*S*)-4,6-dimethyl-7-hydroxynonan-3-one (Katsuki, 1987)

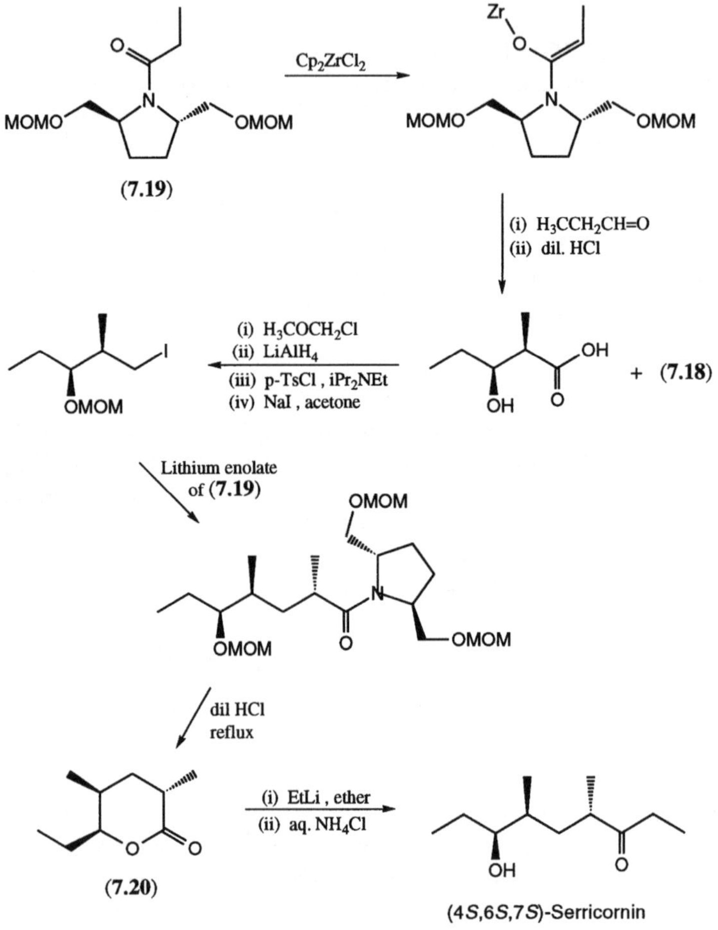

(**7.19**)

$Cp_2ZrCl_2$

(i) $H_3CCH_2CH{=}O$
(ii) dil. HCl

+ (**7.18**)

(i) $H_3COCH_2Cl$
(ii) $LiAlH_4$
(iii) p-TsCl, $iPr_2NEt$
(iv) NaI, acetone

Lithium enolate of (**7.19**)

dil HCl
reflux

(**7.20**)

(i) EtLi, ether
(ii) aq. $NH_4Cl$

(4*S*,6*S*,7*S*)-Serricornin

Reaction with $LiAlH_4$ reduces the ester to an alcohol, which is converted in two steps into the corresponding iodide. When this iodide is reacted with the lithium enolate of the amide (**7.19**), alkylation occurs at the activated carbon and the third chiral centre of serricornin is created with the correct diastereo- and enantio-selectivity. Hydrolysis of the amide bond gives the lactone (**7.20**), which still requires two more carbons. These are added using ethyl lithium in ether to give, after work-up, the desired target compound: (4*S*,6*S*,7*S*)-serricornin.

*(c) Yellow scale insect*

The major component of the sexual pheromone of the yellow scale, *Aonidiella citrina* Coquillett, is (3*S*)-3,6-dimethyl-6-(1-methylethyl)-*E*-5,8-decadien-1-ol acetate, (**7.21**). Neither the (3*R*) enantiomer nor the *Z*-5 isomer is biologically active. Although this, like the mealybug pheromone (**7.13**), is also a terpenoid, no suitable terpene precursor is available.

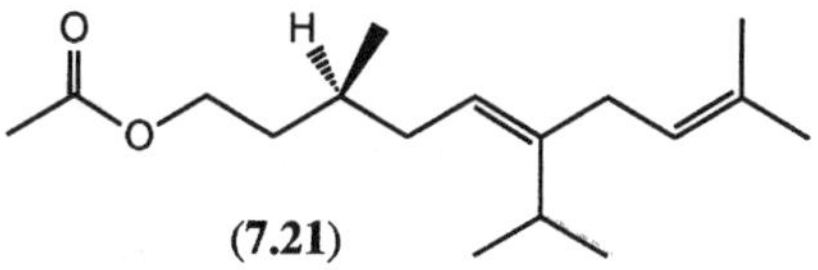

(**7.21**)

The synthetic route (Alvarez *et al.*, 1988) started with the creation of the required homochiral centre from the achiral diester, dimethyl 3-methylglutarate (**7.22**), by enzymatic hydrolysis with pig liver esterase, which hydrolyses selectively the ester group at the pro-*S* site to give the (3*R*) half ester (**7.23**). The significant point about this process is that, because of its plane of symmetry, **all** of the achiral starting material is converted into the **single** homochiral product. Selective reduction of the carboxylic acid group to a hydroxyl group, followed by lactonization, forms (3*R*)-3-methylpentan-5-olide, as shown in Scheme 7.9.

Reaction of this lactone with the lithium enolate of 1-(4-methylpent-3-enyl)sulphonylbenzene gives the β-sulphonyl hydroxyketone (**7.24**). The alcohol group is protected as the *t*-butyldimethylsilyl ether and the ketone is then reduced to the corresponding alcohol with sodium borohydride. Conversion of this alcohol group into an acetate, with acetic anhydride and triethylamine, followed by elimination of acetic acid, by heating with powdered sodium hydroxide in ether, affords the *E*-sulphonylalkene (**7.25**). The last three carbons of the skeleton are then added by displacing the benzenesulphonyl group, with retention of configuration, with the required 1-methylethyl group by reacting (**7.25**) with the Grignard reagent from 2-bromopropane in the presence of ferric chloride. This forms (**7.26**), together with a compound which has a hydrogen on the double-bond instead of the desired 1-methylethyl group. Without separation, the protection is removed from the terminal alcohol using tetrabutylammonium fluoride and the alcohol group is acetylated. The pheromone is finally purified by chromatography on silver nitrate impregnated silica gel to give (3*R*)-3,6-dimethyl-6-(1-methylethyl)-*E*-5,8-decadien-1-ol acetate.

*Notes*

- The overall yield is about 17% with an isomeric purity of 98%.

Scheme 7.9 Synthesis of yellow scale pheromone (Alvarez *et al.*, 1988)

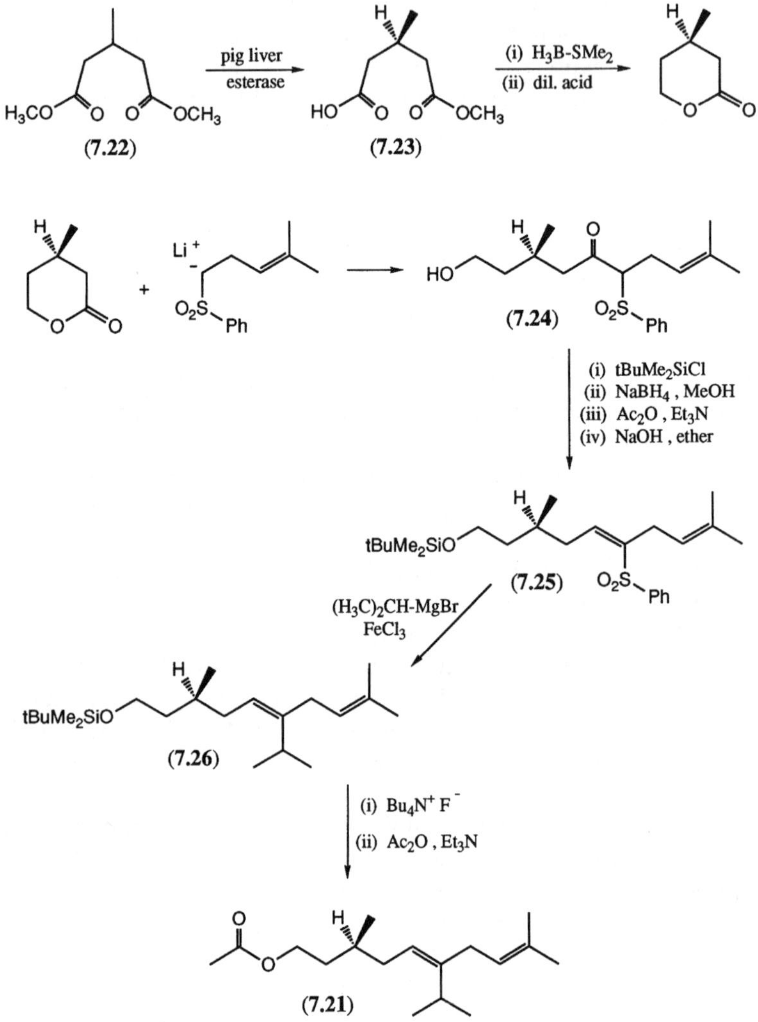

$H_3CO$ O O $OCH_3$ (7.22) — pig liver esterase → HO O O $OCH_3$ (7.23) — (i) $H_3B$-$SMe_2$ (ii) dil. acid → H O O

H O O + $Li^+$ $^-$ $O_2S$ Ph → HO H O $O_2S$ Ph (7.24)

(i) $tBuMe_2SiCl$
(ii) $NaBH_4$, MeOH
(iii) $Ac_2O$, $Et_3N$
(iv) NaOH, ether

$tBuMe_2SiO$ H (7.25) $O_2S$ Ph

$(H_3C)_2CH$-MgBr
$FeCl_3$

$tBuMe_2SiO$ H (7.26)

(i) $Bu_4N^+ F^-$
(ii) $Ac_2O$, $Et_3N$

O H O (7.21)

- The displacement of the benzenesulphonyl group by a Grignard reagent illustrates an alternative method for the stereoselective creation of a trisubstituted double-bond. The selectivity is better than 98% and is established in the elimination step forming the sulphonyl alkene (**7.25**).

## REFERENCES

Alvarez, E., Cuvigny, T., Hervé du Penhoat, C. and Julia, M. (1988) Syntheses with Sulfones XLIX; Stereo- and enantio-selective synthesis of (*S*) (-) 3,9-dimethyl-6-(1-methylethyl) (*E*) 5,8-decadien-1-ol acetate, sexual pheromone of the yellow scale. *Tetrahedron*, **44**, 119–126.

Baker, R., Herbert, R.H. and Parton, A.J. (1982) Isolation and synthesis of 3- and 4-hydroxy-1,7-dioxaspiro[5,5]undecanes from the Olive fly (*Dacus oleae*). *J. Chem. Soc., Chem. Commun.*, 601–603.

Boden, C.D.I., Chambers, J. and Stevens, I.D.R. (1993) A concise, efficient and flexible strategy for the synthesis of the pheromones of *Oryzaephilus* and *Cryptolestes* grain beetles. *Synthesis*, 411–420.

Ferreira, I.T.B., Cruz, W.O., Vieira, P.C. and Yonashiro, M. (1987) Carbon–carbon double bond cleavage using solid supported potassium permanganate on silica gel. *J. Org. Chem.*, **52**, 3698–3699, and personal communication.

Fittig, R. (1890) Ueber Lactonsauren, Lactone und ungesattige Sauren. *Annalen*, **256**, 50–62.

Katsuki, T. and Yamaguchi, M. (1987) Syntheses of optically active insect pheromones, (2*R*,5*S*)-2-methyl-5-hexanolide, (3*S*,11*S*)-3,11-dimethyl-2-nonacosanone and serricornin. *Tetrahedron Letters* **28**, 651–654.

Miles, M.V. (1989) Synthesis of an aggregation pheromone. 3rd Year Project Report, University of Southampton.

Negishi, E.-I., Lew, G. and Yoshida, T. (1973) Stereoselective syntheses of conjugated *trans*-enynes readily convertible into conjugated *cis, trans*-dienes and its application to the synthesis of the pheromone Bombykol. *J. Chem. Soc., Chem. Commun.*, 874–875.

Ortu, S. and Delrio, M.G. (1982) Field trials with synthetic sexual pheromone of *Planococcus citri*. *Redia* **65**, 341.

# 8

# Structure and species specificity

The early investigators of pheromones believed that each species investigated used a unique compound for signalling purposes and, when they had isolated and identified the component found in major amount, were content to use that compound for biological investigations. However, it soon became apparent that these single compounds when used in isolation were not as attractive as the virgin females nor, often, as the crude extracts obtained from them. This chapter aims to indicate some of the investigations that have been carried out to unravel the way that insects respond to variations in pheromone structure and composition. In practical terms, this means that the ideal material required for control purposes is often not a single compound, but will consist of a mixture whose composition needs to be established by cage and field testing. Not infrequently, individual components do not exhibit great biological activity in isolation and this is particularly true of minor components in a pheromone mixture. Hence, there is frequently the need to investigate further the minor components of a gland extract, once the identity of the major component has been established.

## 8.1 STRUCTURAL VARIATION AND PHEROMONAL ACTIVITY

The data presented in Fig. 8.1 summarize the results of a systematic study by Roelofs and Comeau (1971) on the effects of structural modification on attraction of male red-banded leaf rollers, *Argyrotaenia velutinana*, by compounds closely related to the major component of the pheromone, (Z)-11-tetradecenyl acetate – (Z)11–14:Ac). The following points may be noted.

- Keeping the relationship between the acetate and the alkene constant, the removal of one carbon from the end of the chain, or the addition of one carbon to that position, causes a very marked loss of activity.
- Keeping the chain length constant and moving the alkene group one

carbon closer to the acetate, (Z)10–14:Ac, results in total loss of activity. The *E*-isomer, (*E*)10–14:Ac, is also inactive.

- Hydrolysis of the acetate group to the alcohol, (Z)11–14:OH, or conversion into either the formate or the propanoate ester gives a compound which acts as an inhibitor of the attractant.
- Changing the stereochemistry of the double-bond from *Z* to *E* also produces an inhibitor of response.
- The saturated compound of two fewer carbons, dodecanyl acetate (12:Ac), acts as a synergist for the (Z)11–14:Ac, although it is not itself significantly attractive. Similar results are obtained with the two saturated compounds in which an ether oxygen has been placed at one end or the other of the position originally occupied by the double-bond (last two entries). The synergistic effect is to increase trap capture by a factor of from four- to six-fold.

These results demonstrate the sensitivity and discrimination of the semiochemical sensory apparatus of insects. The origin of this specificity at the electrophysiological level is the subject of much ongoing research; it is not strictly of relevance to the practical application of pheromones and will not be discussed here.

## 8.2 CHIRALITY AND ACTIVITY

In section 5.2.1, the concept of chirality and optical activity was introduced and information given on the specific enantiomers of structures which are used as pheromones. Among chiral pheromones, it has been observed that there are a number of different types of response by the insect to the particular homochiral compound presented. In a review on the synthesis of optically active pheromones, Mori (1989) has placed the types of response exhibited into eight different categories:

(A) Only **one** enantiomer is active and the antipode does not inhibit the action of the pheromone. A large number of chiral pheromones fall into this category, which may be further subdivided depending on whether the antipode also shows some attractivity or not. Examples are given in Fig. 8.2(a).

(B) Only **one** enantiomer is active but the antipode or a diastereomer **does inhibit** the action of the pheromone. Luckily, this is relatively rare and most of the known examples are given in Fig. 8.2(b).

(C) **All** the stereoisomers are biologically active. Here the insects do not discriminate among the stereoisomers; examples are in Fig. 8.2(c).

(D) In the same genus different species use different enantiomers. So far only ipsdienol, used by *Ips* species, falls into this category, although sulcatol, which is placed in the next category, shows some similarities (see below).

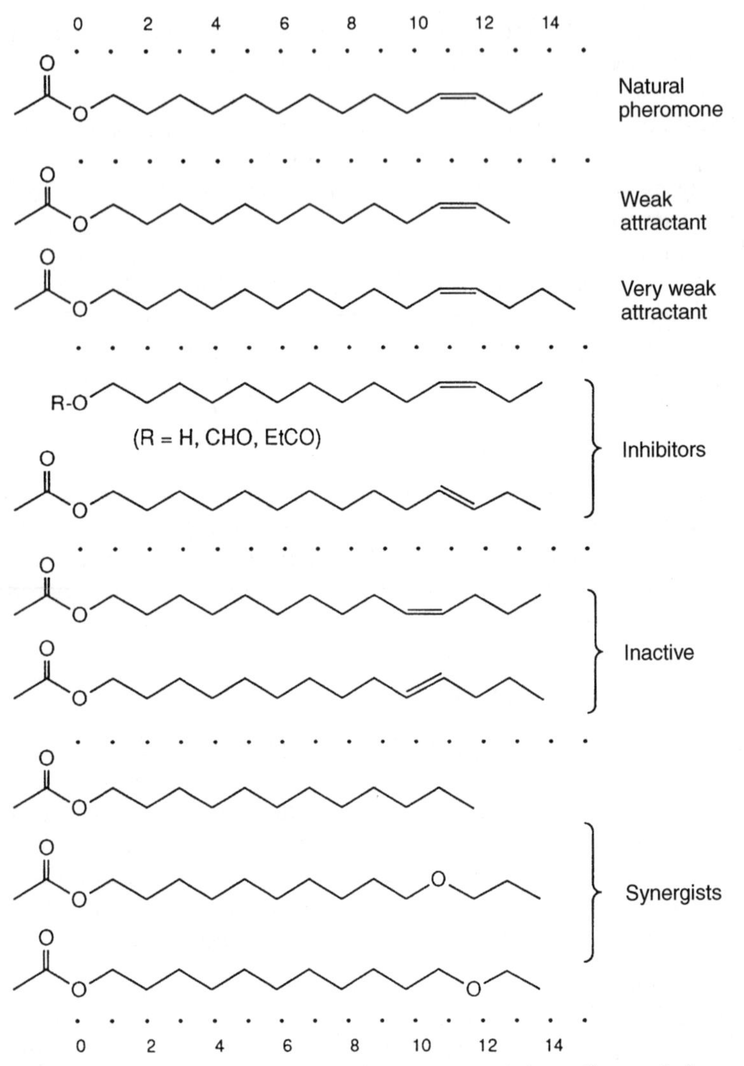


**Fig. 8.1** Structural modification and pheromone activity: effects of changes in structure and geometry on the attraction of the male red-banded leafroller (*Argyrotaenia velutinana*). (Data from Roelofs and Comeau, 1971.)

(E) **Both** the enantiomers are required for bioactivity. The ambrosia beetle, *Gnathotricus sulcatus*, is unresponsive to either pure enantiomer of sulcatol alone. Again this is the only example.

(F) Only **one** enantiomer is as active as the natural pheromone, but its activity can be enhanced by the addition of a less active stereoisomer. The red flour beetle, *Tribolium castaneum*, uses (4*R*,8*R*)-4,8-

dimethyldecanal as a sexual attractant and its activity is increased by the addition of the (4*R*,8*S*)-diastereomer.

(G) Male insects use one enantiomer and females use the antipode. The spiroketals used by the olive fly, *Dacus oleae*, fall into this category.

(H) Only the meso-isomer is active. Two examples are known: the tsetse fly, *Glossina pallidipes*, responds only to meso-13,23-dimethylpentatriacontane and the related *G. morsitans morsitans* only to meso-17,21-dimethylhentriacontane.

In category (A) one finds frontalin, (**8.2.1**), (Fig. 8.2.(a)), the major aggregation pheromone of the southern pine beetle, *Dendroctonus frontalis*). It is produced as an 85:15 mixture of the (+) and (–) enantiomers and both attract beetles. However, the (–) isomer is significantly more attractive than its optical antipode and EAG studies reveal that the insects possess antennal receptors responsive to each enantiomer (Payne *et al.*, 1982).

In category (B) one may cite particularly the case of the Japanese beetle, *Popillia japonica*, in which the inhibition to the pheromone (**8.2.2**), (Fig. 8.2(b)) caused by the (4*S*)-enantiomer is so marked that the racemate of (5Z)-tetradecen-4-olide completely lacks biological activity.

When there is more than one chiral centre in the molecule, then the situation often becomes quite complex. Here one of the biologically inactive diastereomers may show an inhibitory effect and this is illustrated in the data presented in Fig. 8.3, which shows the results of a study on the northern corn rootworm, *Diabrotica barberi*) (Guss *et al.*, 1985).

The four diastereomers of 8-methyldecan-2-ol propanoate were tested in a field trapping environment. The results in Fig. 8.3(c) show that only the (2*R*,8*R*)-isomer is attractive to males. The results in Fig. 8.3(b) show that the antipode of the pheromone, namely the (2*S*,8*S*)-isomer, acts as an inhibitor, although not very strongly, and the (2*R*,8*S*)-isomer has no effect. On the other hand, its antipode, the (2*S*,8*R*)-isomer, is sufficiently strong an inhibitor to act as an antagonist for the natural pheromone, completely suppressing trapping when present at the equivalent concentration.

The pheromone system of diprionid sawflies is based on 2,7-dimethylpentadecan-2-ol (diprionol) and contains three chiral centres. The active components are the acetate and propanoate esters of the alcohols, and the most widely used active diastereomer is the (2*S*,3*S*,7*S*) acetate, structure (**8.2.3**). In a very careful study, Löfqvist (1986) showed that for the European pine sawfly, *Neodiprion sertifer*, the attractant effect of the pheromone was completely inhibited by as little as 3% of the (2*S*,3*R*,7*R*/*S*) diastereomers, structures (**8.2.4**) and (**8.2.5**), and that this effect decreased with decreasing amounts of the inhibitor until the concentration reached 0.3%. Below this concentration, the mixture of

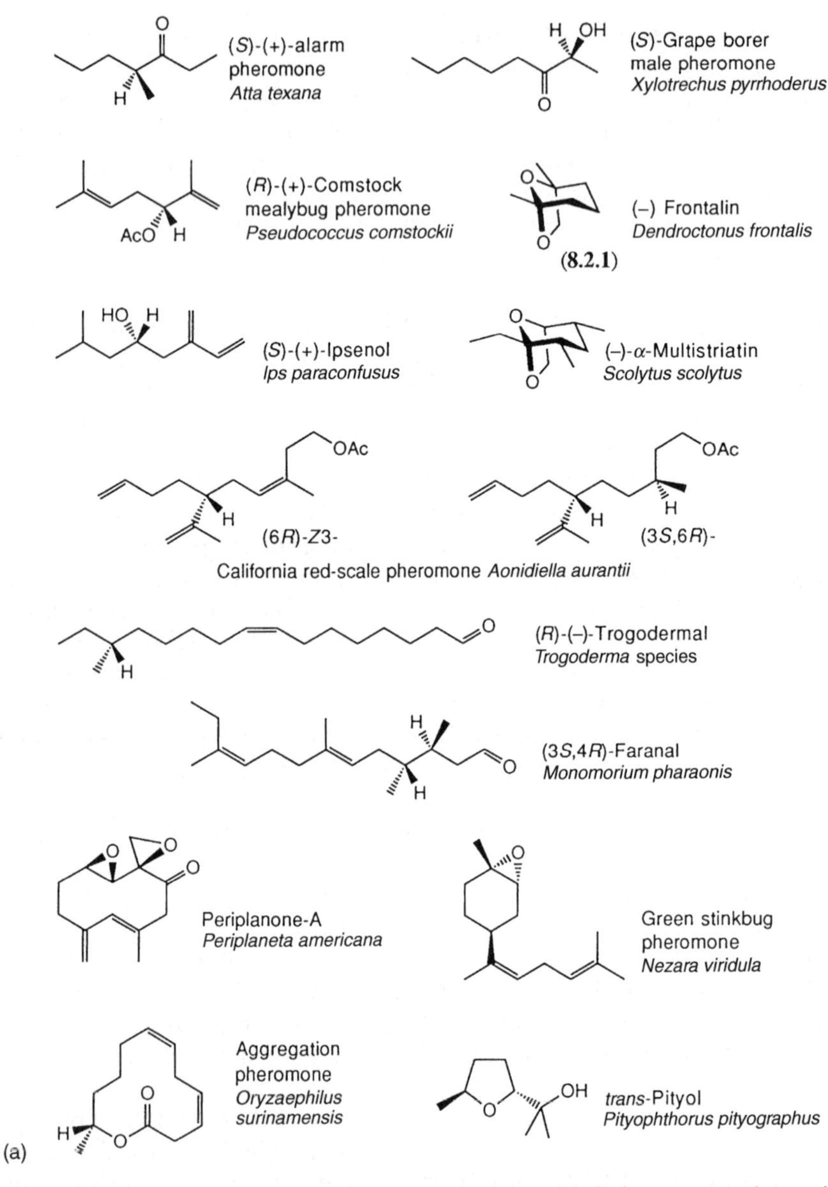


**Fig. 8.2** Optical isomerism and pheromone activity. (a) Only one enantiomer is active. The antipode does not inhibit the action. (b) Only one enantiomer is active. The antipode inhibits the action. (c) All the stereoisomers are bioactive. (d) Different species of the same genus use different enantiomers. (e) Both the enantiomers are required for bioactivity. (f) Only one enantiomer is as active as the natural pheromone, but its activity can be enhanced by the addition of a less active stereoisomer. (g) One enantiomer is active on male insects; the other is active on females. (h) Only the *meso*-isomer is active.

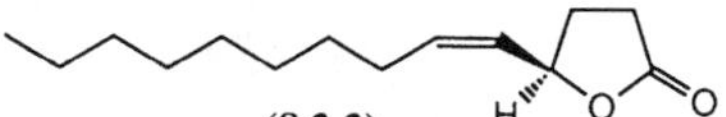

**(8.2.2)**

(*R*)-(*Z*)-Japonilure
*Popillia japonica*

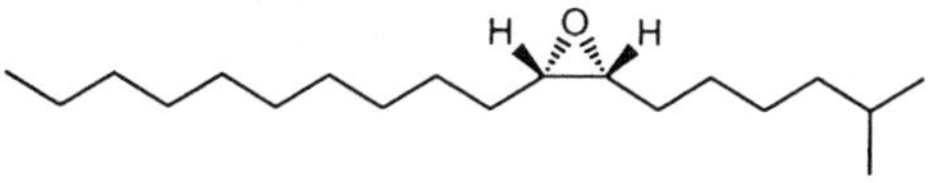

(7*R*,8*S*)-Disparlure
*Lymantria dispar*

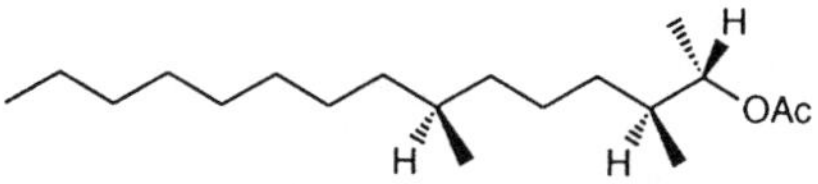

(2*S*,3*S*,7*S*)-Diprionyl acetate
*Neodiprion sertifer*

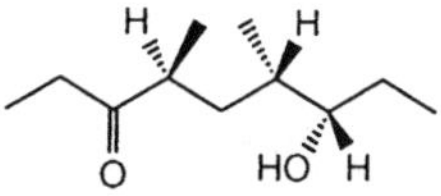

(4*S*,6*S*,7*S*)-Serricornin
*Lasioderma serricorne*

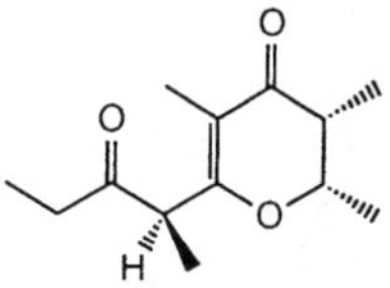

Stegobinone
*Stegobium paniceum*

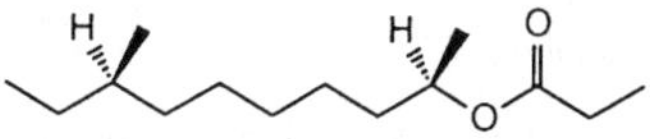

(2*R*,8*R*)-8-methyldecan-2-ol propanoate
*Diabrotica barberi*

(b)

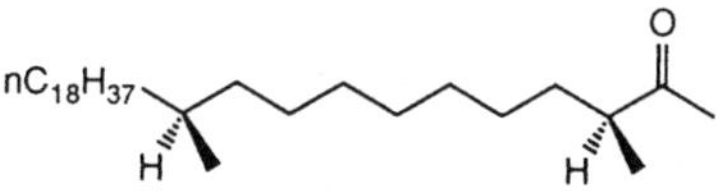

$nC_{18}H_{37}$

3,11-dimethylnonacosanone

German cockroach
*Blatella germanica*

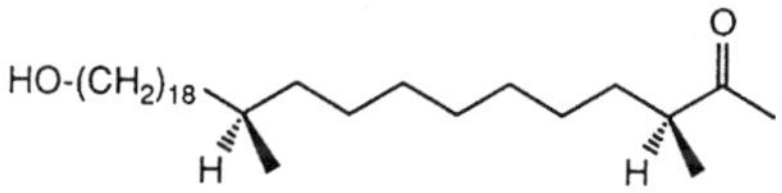

$HO\text{-}(CH_2)_{18}$

29-hydroxy-3,11-dimethylnonacosanone

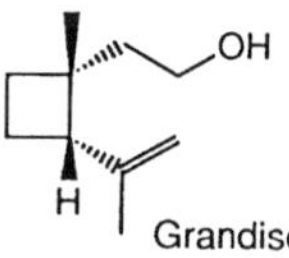

Grandisol
*Anthomis grandis*

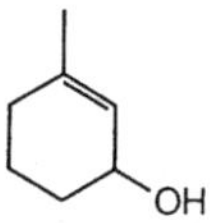

Seudenol
Douglas-fir beetle

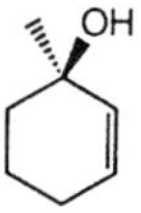

1-methylcyclohex-2-enol
*Dendroctonus pseudotsugae*

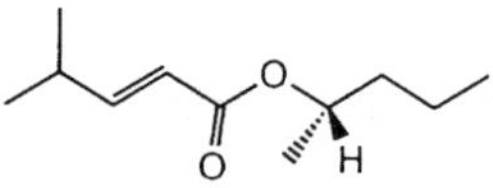

Dominicalure 2
*Rhyzopertha dominica*

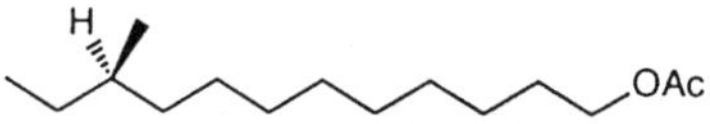

smaller tea-tortrix
*Adoxophyes* spp.

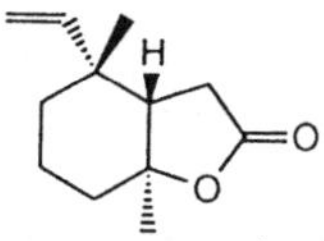

Anastrephin
*Anastrepha suspensa*

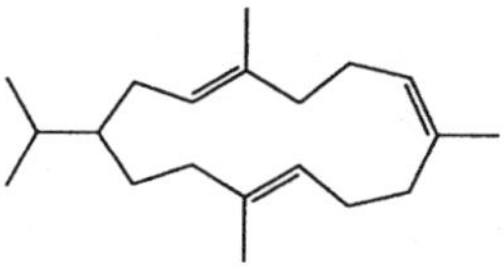

Neocembrene-A
*Nasutitermes exitiosus*

(c)

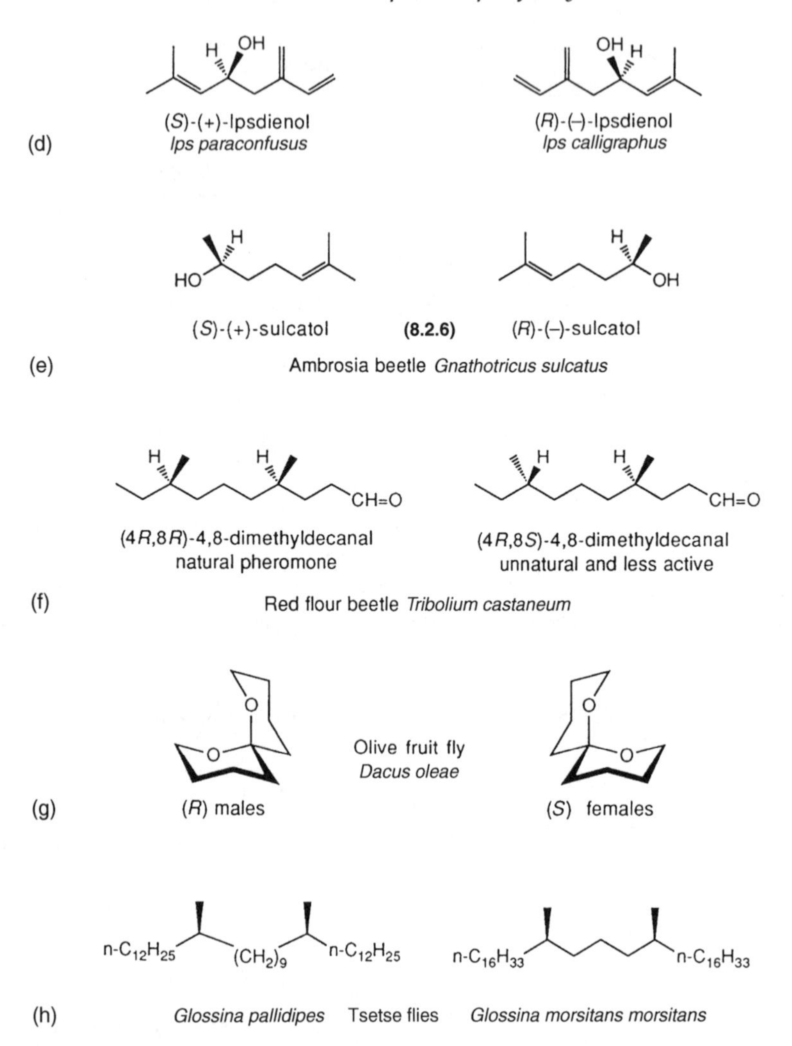


**(8.2.4)** and **(8.2.5)** acted as a **synergist**, increasing trap catches by as much as 55% (Fig. 8.4).

Sulcatol, **(8.2.6)**, which has been placed in category (*E*), is the pheromone produced by two species of ambrosia beetle: *Gnathotricus sulcatus* and *G. retusus*. Laboratory studies using the pure enantiomers and various mixtures of them gave the results shown in Table 8.1(a) while the results of field trapping experiments are shown in Table 8.1(b) (Borden *et al.*, 1980).

The data clearly demonstrate that for *G. sulcatus* both the enantiomers need to be present to elicit attraction and, though this is maximal for

| Structure | Isomer | Trap catch |
|---|---|---|
| 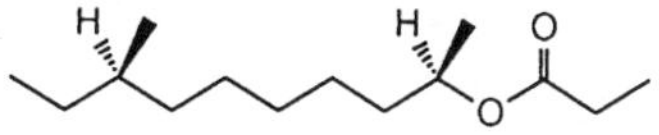 | (2*R*,8*R*) | 123.5 ± 48.2 |
| 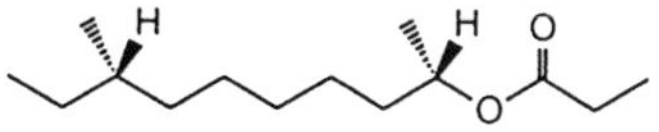 | (2*S*,8*S*) | 4.7 ± 3.0 |
| 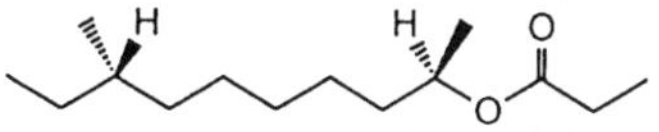 | (2*R*,8*S*) | 2.0 ± 1.4 |
| 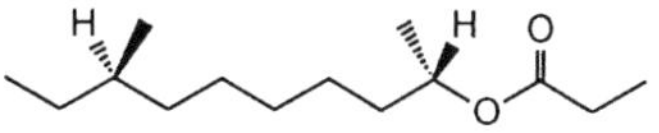 | (2*S*,8*R*) | 5.7 ± 3.6 |
| (a) | solvent blank | 5.7 ± 3.8 |

(b)

| Isomer mixture | Trap catch |
|---|---|
| (2*R*,8*R*) alone | 175.0 ± 21.6 |
| (2*R*,8*R*) + (2*S*,8*S*) | 68.7 ± 17.7 |
| (2*R*,8*R*) + (2*R*,8*S*) | 154.0 ± 44.3 |
| (2*R*,8*R*) + (2*S*,8*R*) | 7.0 ± 3.5 |
| solvent blank | 12.0 ± 3.2 |

**Fig. 8.3** Influence of diastereoisomerism on pheromone activity: response of males of northern corn rootworm (*Diabrotica barberi*) to diastereomers of 8-methyldecan-2-ol propanoate. (a) Trap catch for each pure isomer (1 μg/lure). (b) Trap catch for mixtures of (2*R*, 8*R*) isomer with added other isomer (at 1 μg + 1 μg/lure).

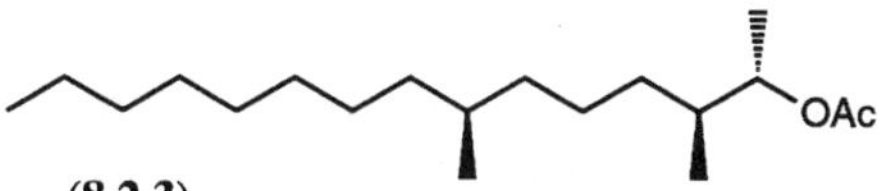

OAc

**(8.2.3)**

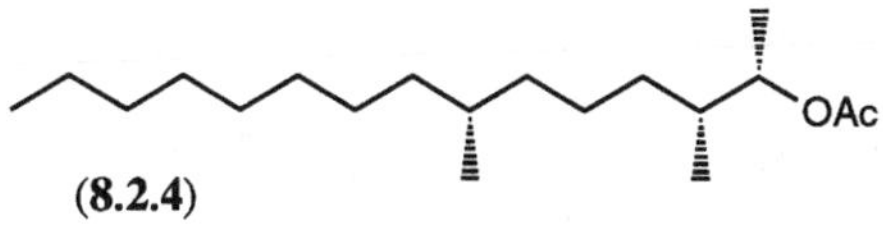

OAc

**(8.2.4)**

OAc

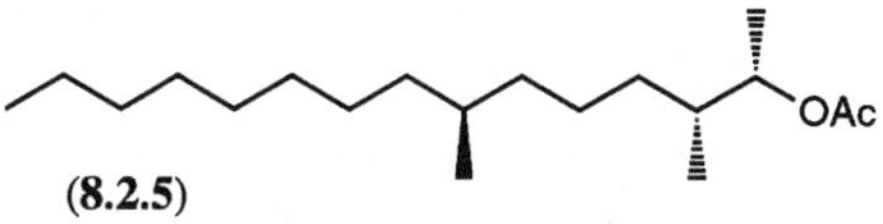

**(8.2.5)**

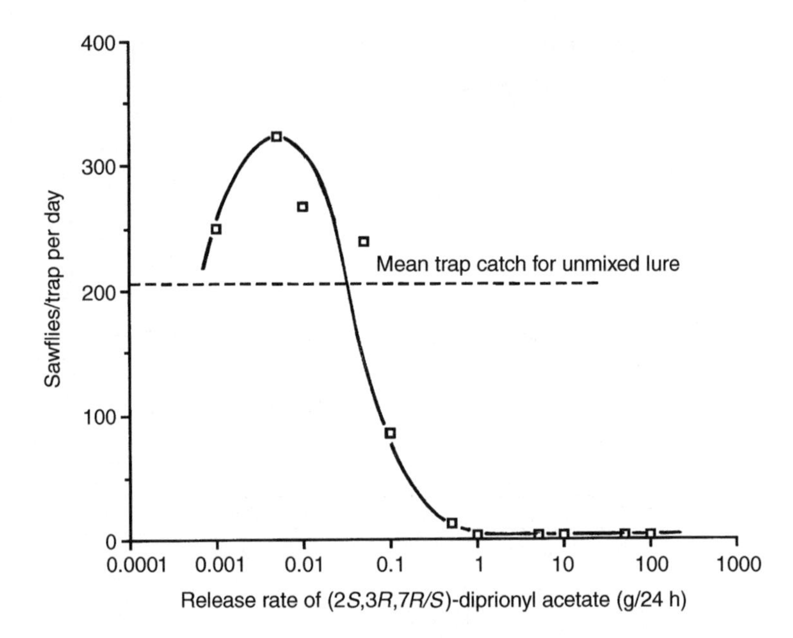


**Fig. 8.4** Field trapping of pine sawflies (*Neodiprion sertifer*). Trap catch vs release rate of added (2*S*, 3*R*, 7*R*/*S*)-diprionyl acetates in presence of lure releasing (2*S*, 3*S*, 7*S*)-diprionyl acetate at 10 µg/24 h. Redrawn from data from Löfqvist (1986).

the racemate, there is a broad window of possible mixtures. On the other hand, the response of *G. retusus* is clearly inhibited by even small amounts of the (*R*)-enantiomer and therefore, for this species, the compound falls into category (B). Finally one may note that even with *G. sulcatus* it has been suggested that the evidence appears to indicate that the (*R*)-enantiomer is inhibitory.

Lastly we may note that there is a substantial number of homochiral pheromones for which the investigations required to establish the category into which they fall have not yet been carried out. The structures may be found in Mori's review and these compounds represent a challenge for the future.

## 8.3 SYNERGISM: MULTI-COMPONENT PHEROMONES AND MIXTURE–ACTIVITY RELATIONS

From the evidence presented in section 8.1, it is clear that many insects, and particularly lepidopteran species, respond more strongly to mixtures of compounds than to the pure components in isolation. This synergistic effect may be related to questions of long-range and short-

**Table 8.1** Response by *Gnathotricus sulcatus* and *G. retusus* females to enantiomers of sulcatol (data from Borden *et al.*, 1980)

(*S*)-(+)-sulcatol **(8.2.6)** (*R*)-(-)-sulcatol

(a) Laboratory olfactometer, stimulus sulcatol (0.5 $\mu$g in 25 $\mu$l pentane

| Stimulus | Ratio of (*S*) : (*R*) | Response by 100 females | |
|---|---|---|---|
| | | *G. sulcatus* | *G. retusus* |
| Pentane (control | | 4a | 5a |
| Sulcatol | | | |
| | > 97 : < 3 | 13b | 44b |
| | 90 : 10 | 34cd | – |
| | 70 : 30 | 40cde | – |
| | 50 : 50 | 49e | 2a |
| | 20 : 80 | 46de | – |
| | 10 : 90 | 40cde | – |
| | < 3 : > 97 | 5a | 4a |

Percentages within a column followed by the same letter are not significantly different.

(b) Numbers of *G. sulcatus* and *G. retusus* females captured on sticky traps baited with mixtures of sulcatol enantiomers

| Stimulus | Ratio of (*S*) : (*R*) | Beetles captured | |
|---|---|---|---|
| | | *G. sulcatus* | *G. retusus* |
| Pentane (control) | | 6a | 19a |
| Sulcatol | | | |
| | 100 : 0 | 39ab | 188bc |
| | 99 : 1 | 143bc | 180bc |
| | 98 : 2 | 162c | 175c |
| | 97 : 3 | 120bc | 54abc |
| | 96 : 4 | 303cd | 33ab |
| | 90 : 10 | 663d | 76abc |
| | 75 : 25 | 786d | – |
| | 50 : 50 | 547d | – |
| | 25 : 75 | 1125d | – |
| | < 3 : > 97 | 4a | – |

Totals within a column followed by the same letter are not significantly different

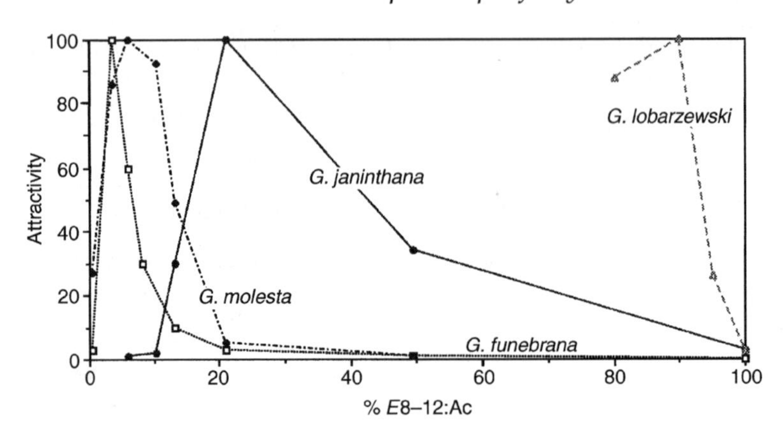


**Fig. 8.5** Attraction vs composition of pheromone lure for *Grapholita* species (percentage *E*-8-dodecenyl acetate in *Z*-8-dodecenyl acetate lure).

range attractivity, but clearly also serves to ensure species specificity in attraction.

A good example of the variation in attractancy with mixture composition is found in the Tortricid moths, which infest fruit orchards in Western Europe. The four species investigated all belong to the genus *Grapholita* and are the oriental fruit moth (*G. molesta*), the plum fruit moth (*G. funebrana*), the hawthorn leaf-roller (*G. janinthana*) and *G. lobarzewski*. All four species use dodec-8-en-1-ol acetate (8-12:Ac) as the major component, with the first three having a preponderance of the (*Z*)-isomer, while the last uses mainly the (*E*)-isomer.

The results are depicted graphically in Fig. 8.5, which shows a plot of the percentage attractivity of a lure, based on moth capture, against the percentage of *E*-isomer, (*E*)8–12:Ac, in the lure (Biwer, 1978). The three peaks of attractancy that are observed show that *G. janinthana* and *G. lobarzewski* are readily able to maintain the species purity without any problems of cross-attraction. However, the window of attraction for *G. molesta* clearly encompasses the whole of that for the plum fruit moth, *G. funebrana*, and indeed traps baited with a 95:5 ratio of (*Z*):(*E*)-8-dodecenyl acetate catch about equal numbers of both species. In this case, the integrity of the species is maintained by a time differential in mating flights, which occurs in the evening for the oriental fruit moth and in the morning for the plum fruit moth. There may, in addition, be other relatively minor components present in their pheromonal blends which add to the differentiatory mechanism (see below).

The case of the corn earworm, *Heliothis zea* (recently assigned to the genus *Helicoverpa*) and the tobacco budworm, *Heliothis virescens*, presents a good example where the composition of the two pheromone

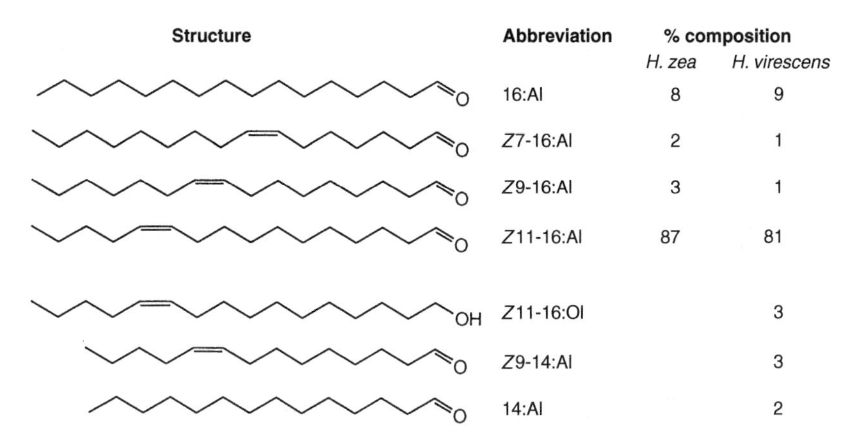

| Structure | Abbreviation | % composition | |
|---|---|---|---|
| | | H. zea | H. virescens |
| | 16:Al | 8 | 9 |
| | Z7-16:Al | 2 | 1 |
| | Z9-16:Al | 3 | 1 |
| | Z11-16:Al | 87 | 81 |
| | Z11-16:Ol | | 3 |
| | Z9-14:Al | | 3 |
| | 14:Al | | 2 |

**Fig. 8.6** Structures of compounds and composition of pheromone mixtures from *Heliothis zea* and *H. virescens*. (Data from Klun *et al.*, 1979.)

mixtures is more complex than that above. These two species are sympatric over wide areas of the United States. For each of the two species the major component was identified as (Z)-11-hexadecenal. However, lures baited with this compound alone were not nearly as attractive as those baited with virgin females. Further analysis of the pheromone gland and ovipositor for each species showed that that of the corn earworm contained four components, whose structures and ratio are given in Fig. 8.6, while that of the tobacco budworm contained these same four compounds plus three others, i.e. seven components in all; composition and ratio also shown in Fig. 8.6 (Klun *et al.*, 1979).

Traps baited with a four-component lure of the same composition as the corn earworm gland were found to catch *H. zea* exclusively, with no cross-trapping of the tobacco budworm. For *H. virescens*, however, a two-component lure containing (Z)11–16:Al and (Z)9–14:Al was sufficient to ensure that no cross-trapping of *H. zea* occurred. Each of the three compounds unique to *H. virescens* was examined for its influence on the four-component lure which was effective for the corn earworm, with the results shown graphically in Fig. 8.7, which shows percentage attractivity against percentage of the component added.

While the saturated tetradecanal has little effect, both of the other compounds contribute to the decrease in attraction, with the effect of the (Z)9–14 aldehyde being quite dramatic. The data in Table 8.2 show the effect on trap catch of the addition of (Z)9–14:Al to the *H. zea* four-component lure and the complete cross-over from the trapping of one species to that of the other that is observed in consequence (Klun *et al.*, 1980).

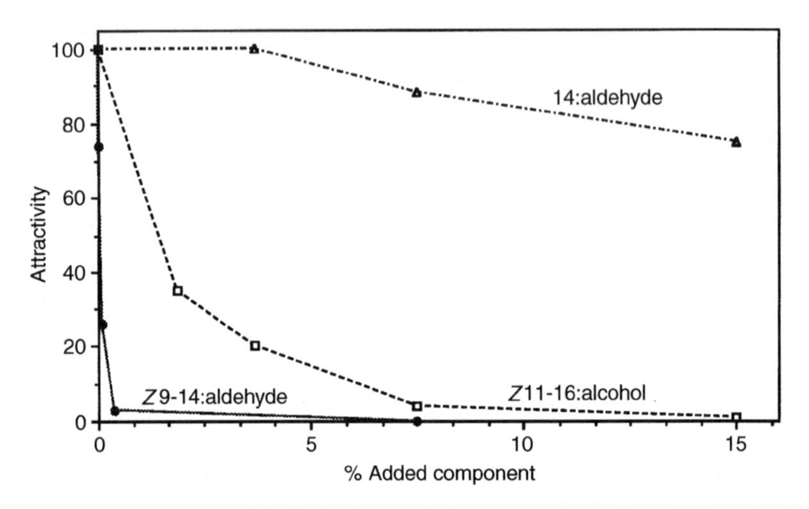


**Fig. 8.7** Field trapping of *Heliothis zea* with four-component lure. Effect of added components from *H. virescens* pheromone mixture. Trap catch vs percentage added component: tetradecanal (14-aldehyde), Z-11-hexadecenol (Z11-16:alcohol) and Z-9-tetradecenal (Z9-14:aldehyde).

Finally it is interesting to note that even if one restricts the examination to a single species, significant differences may be observed when a comparison is made between one population and another. Thus in the North American bark beetle *Ips pini*, populations from the eastern and western states produce different enantiomer ratios of ipsdienol and respond preferentially to the pheromone of their own population. Eastern *I. pini* produce and respond to a 65:35 ratio of (*S*)-(+):(*R*)-(–)-ipsdienol, whereas the western populations use solely the (*R*)-enantiomer. This attraction is not inhibited by the (*S*)-isomer and the

Table 8.2 Effect on trap catch of added (Z)-9-tetradecenal to four-component lure for *Heliothis zea* (2.5 mg) (data from Klun, 1980)

| Weight of (Z)-9-14Al ($\mu$g) | Numbers caught/trap *H. zea* | *H. virescens* |
|---|---|---|
| 0.0 | 25.4 | 0 |
| 0.25 | 21.0 | 0 |
| 2.5 | 16.5 | 0 |
| 25.0 | 1.4 | 6.6 |
| 250.0 | 0.1 | 34.6 |

**Table 8.3** Pheromone composition ot turnip moth (*Agrotis segetum*) ovipositor glands and nationality of population (normalized to (Z)7-12 : Ac = 100)

| Compound | Swedish | French | Hungarian | English | Armenian | Bulgarian |
|---|---|---|---|---|---|---|
| (Z)5-10:Ac | 8±2 | 116±39 | 6±1 | 19±5 | < 2 | < 2.5 |
| (Z)7-12:Ac | 100 | 100 | 100 | 100 | 100 | 100 |
| (Z)9-14:Ac | 85±11 | 32±5 | 139±13 | 103±22 | 90 | 136 |
| (Z)11-16:Ac | 23±8 | 4±1 | 37±6 | 29±7 | 35 | 68 |
| (Z)5-14:Ac | 33±10 | 11±2 | 25±5 | 17±5 | – | – |

– no data reported.
Swedish, French, Hungarian and English data from Löfstedt et al. (1986); Armenian and Bulgarian data from Hansson *et al.* (1990).

result is that the eastern populations respond more strongly to the racemate than do the western ones. Similar differences have been observed between European and South American populations of the green stinkbug, *Nezara viridula*.

An extensive study of the European populations of the turnip moth, *Agrotis segetum*, has demonstrated that although females from all the countries examined produce in their ovipositor pheromone glands the same 17 compounds, the ratio is different for every country, though these differences are statistically significant only for the population from France (Löfstedt *et al.*, 1986; Hansson *et al.*, 1990). Table 8.3 gives the composition of the pheromone with respect to the important major components, which fall into a homologous series in which the double-bond is five carbons from the methyl end of the chain: (Z)5–10:Ac, (Z)7–12:Ac, (Z)9–14:Ac and (Z)11–16:Ac. As well as these four compounds, (Z)5–14:Ac is the other major ester. Alcohols corresponding to the lowest two members of the homologous series are also present.

From this information it may be seen that the detail of pheromone communication in *A. segetum* varies from country to country, with each population having its own 'dialect'. It has been suggested that the origin of this variation may be an adaptive shift caused by interspecific interactions with other noctuids in a particular geographical area, these other species reacting to similar blends of the three major acetates. On the other hand they might arise from purely random shifts in the communication system. In either case, it is clear that the investigator has to take regional differences of these types into account when seeking an effective lure, or mating disruption formulation, for any given pest species.

These variations provide a significant word of warning for those wishing to control insect populations by the use of pheromone-based methods: population pressure will almost certainly lead to adaptive changes in the pheromonal communication system used by the species concerned and hence to the need to modify the control techniques along with the insect's development.

## REFERENCES

Biwer, G. (1978) Les phéromones facteurs d'isolement sexuel. *La Recherche*, **9**, 799–801.

Borden, J.H., Handley, J.R., McLean, J.A. *et al* (1980) Enantiomer-based specificity in pheromone communication by two sympatric *Gnathotricus* species. *J. Chem. Ecol.*, **6**, 445–455.

Guss, P.L., Sonnet, P.E., Carney, R.L. *et al* (1985) Response of northern corn rootworm, *Diabrotica barberi* Smith and Lawrence, to stereoisomers of 8-methyl-2-decyl-propanoate. *J. Chem. Ecol.*, **11**, 21–26.

Hansson, B.S., Toth, M.I., Löfstedt, C. *et al.* (1990) Pheromone variation among eastern European and a western Asian population of the turnip moth, *Agrotis segetum. J. Chem. Ecol.*, **16**, 1611–1622.

Klun, J.A., Plimmer, J.R., Bierl-Leonhardt, B.A. *et al.* (1979) Trace chemicals: the essence of sexual communication systems in *Heliothis* species. *Science*, **204**, 1328–1330.

Klun, J.A., Bierl-Leonhardt, B.A., Plimmer, J.R. *et al* (1980) Sex pheromone chemistry of the female tobacco budworm moth, *Heliothis virescens. J. Chem. Ecol.*, **6**, 177–183.

Löfqvist, J. (1986) Species specificity in response to pheromone substances in diprionid sawflies. In *Mechanisms in Insect Olfaction* (eds T.L. Payne, M.C. Birch and C.E.J. Kennedy), Clarendon Press, Oxford, pp. 123–129.

Löfstedt, C., Löfqvist, J., Lanne, B.S. *et al* (1986) Pheromone dialects in European turnip moths, *Agrotis segetum. Oikos*, **46**, 250–257.

Mori, K. (1989) Synthesis of optically active pheromones. *Tetrahedron*, **45**, 3233–3298.

Payne, T.L., Dickens, J.C., West J.R. *et al* (1982) Southern pine beetle: olfaction receptor and behaviour discrimination of enantiomers of the attractant pheromone frontalin. *J. Chem. Ecol.*, **8**, 873–881.

Roelofs, W.L. and Comeau, A. (1971) Sex pheromone perception: synergists and inhibitors for the red-banded leaf roller attractant. *J. Insect Physiol.*, **17**, 435–448.

# *Part Three*

# Practical Applications of Pheromones and Other Semiochemicals

*O.T. Jones*

## INTRODUCTION

Since the 1940s insect pest control has been achieved primarily through the use of conventional pesticides which are essentially man-made molecules with little resemblance in most cases to anything found in nature. The first group of insecticides to be developed, the organochlorines, had a tremendous impact on insect pest control through their high degree of efficiency and their persistence in the environment. However, by the late 1950s environmental and health concerns were already emerging relating to their use and with the publication of *Silent Spring* (Carson, 1962) the issue became highly visible. These concerns receded somewhat in the late 1960s and early 1970s with the advent of new compounds with less persistence, but by the mid 1980s issues such as pesticide residues in food (Foschi, 1989), groundwater contamination by pesticides (Roberts, 1989) and the increasing appearance of resistance in insects to pesticides (Georghiou and Saito, 1983) raised the concerns relating to pesticides to new heights (Dinham, 1993).

With this background of concern relating to conventional pesticides, much research effort by academics and government researchers, and more recently by industrial researchers, has been directed at developing environmentally friendly alternatives to conventional pesticides. Several lines of investigation have been explored, most notably:

- microbial pesticides (bacteria, fungi, viruses, etc.);
- beneficial insects (predators and parasites);
- hormonal pesticides (juvenile hormones, insect growth regulators);

- naturally occurring insecticides (pyrethrum, derris, neem, etc.);
- semiochemicals (pheromones, kairomones, allomones, etc.).

Many of these lines of research have led to useful products and techniques for controlling pests and form the basis of a very large database of literature all of which, with the exception of semiochemicals, are beyond the scope of this book. However, Arnason *et al.* (1989), DeBach and Rosen (1991), Franz (1986), Lisansky and Coombes (1994), Ravensberg (1994) and Starnes *et al.* (1993) provide useful overviews of the other fields of study.

Integrated pest management (IPM) is a term that is often used these days in the context of pest control. It is defined by the International Organization for Biological Control (IOBC) as a 'pest management strategy employing all methods consistent with economic, ecological and toxicological requirements to maintain pests below the economic threshold while giving priority to natural limiting factors' (in Katsoyannos, 1992). Unlike the use of broad spectrum conventional pesticides, IPM strategies require a full understanding of the biology and life history of a pest and its natural enemies within any ecosystem (Dent, 1991). Any product that is used to reduce pest populations should be as selective as possible and present a minimal risk to humans and the environment. Many of the products and techniques mentioned above, including those based on semiochemicals, are therefore highly suited for inclusion in any IPM strategy. Similarly, continuous monitoring of pests and natural enemies is central to any IPM programme, and the role of semiochemicals in this context forms the subject of Chapter 9.

# 9

# Pest monitoring

## 9.1 GENERAL PRINCIPLES

When developing an insect pest monitoring system based on semiochemicals it is important, right from the start, to determine the purpose for which that system is to be used. The end use in most cases will determine what form the monitoring system will take (Wall, 1990). Both scientific and commercial factors should be taken into account when defining such objectives (Wall, 1989).

### 9.1.1 Scientific factors

It is important in every case to identify and define the problem. Is it one of detecting a particular pest, timing of control measures, or assessing the risk posed by some particular pest? Detection requires only a sensitive trapping method that provides qualitative (present or absent) information. Timing, on the other hand, requires quantitative information from the trap catches and often requires additional biological and meteorological data with which to predict the occurrence of the susceptible stage in the life cycle. For risk assessment, the relationship between trap catch and population density or subsequent damage needs to be established.

Table 9.1 shows the main uses currently made of semiochemical-based monitoring traps and shows the complex nature of the information that is obtained through using such systems in their various applications. Detection of presence or absence is all that is required for early warning of emergence, for warning of arrival or departure of a pest within a crop, and for survey and quarantine work. The timing of control measures, on the other hand, usually requires the establishment of a threshold catch before further action takes place. The threshold can indicate the emergence or arrival of a pest within a crop in sufficient numbers to warrant a spray application. The decision to spray, however, may depend on further information regarding the stage of

**Table 9.1** Main uses of pheromone-monitoring traps

| *Information from trap catch* | *Application* |
|---|---|
| Detection | Early warning |
| | Survey |
| | Quarantine |
| 'Threshold' | Timing of treatments |
| | Timing of other sampling methods |
| | Risk assessment |
| Density estimation | Population trends |
| | Dispersion |
| | Risk assessment |
| | Effects of control measures |

development of the host crop and additional information on the pest itself from other sampling data obtained on the number of eggs or larvae, or degree of parasitism, together with meteorological information relating to the pest's development. A threshold catch may also be used to establish the level of risk that the pest poses to the crop. This is usually done by comparing relative numbers of pests from year to year and relies on the operator's experience with the system. Such assessments of risk are not as dependable as those in which a relationship has been established between trap catch and population density. These relationships are often difficult to establish and become increasingly less reliable with greater time intervals between the trap catch data and the data relating to the phenological stage of the pest being measured. Thus it is often easier to establish relationships between trap catches and oviposition by adults from the same generation as that being caught in the trap than relationships between trap catch and subsequent larval, pupal or adult populations of the following generation. Similarly, correlations between catch trap and subsequent damage caused by immature stages of the following generation are often difficult to establish because of many factors, including predation and parasitism, which can influence damage levels in the interim. Where such correlations are established, however, they can be used to study long-term population trends, and the patterns of pest dispersion within a cropping area, and also to establish the effectiveness or otherwise of control measures.

### 9.1.2 Practical and commercial factors

One of the major considerations relating to the design of a semiochemical-based monitoring system is the intended end-user. If the user is an advisory or extension service professional, they will be able to operate a sophisticated system and integrate it with other information and procedures. If the system is to be operated by people with only minimal training and understanding of the system's capabilities, then the degree of sophistication introduced into the system has to be restricted.

By the same token, the end-user's needs and expectations of the system may not be sophisticated even though the system would have great capability. If the end-user wants to know only about the presence or absence of a particular pest there is no point in over-engineering the system to do more. Similarly, it is important to establish the frequency with which the operator is prepared to visit the traps. It is often better to opt for a slightly lower frequency of trap counts which the operator is prepared to adopt than to risk a system made inefficient by failure to examine traps at the correct times. The siting of traps may also have to be a compromise between science and practicality. Traps may well catch more insects at the canopy of trees 10 m high but not every operator has the inclination, time or physical ability to climb up to them to count the insects.

Finally, for a monitoring system to be commercialized successfully, it has to fulfil the manufacturer's requirements. The monitoring system has to have a sufficiently large market and produce a sufficiently large profit margin in that market for the manufacturer to cover the costs of development, promotion and manufacture. Problems sometimes arise with pests of minor crops where the market opportunity is not very great but where the end-users have a real need for a system. Under these circumstances, public sector financial support for the system's development and commercialization can be a great help.

## 9.2 BASIC COMPONENTS OF A MONITORING SYSTEM

The basic components of a semiochemical-based monitoring system are the attractant source, the trap and where to place it, and sufficient knowledge of the pest biology to interpret the catches.

### 9.2.1 Attractant source of lure

The attractant source consists basically of two components: the active ingredient and its controlled release device.

#### *(a) Active ingredients*

A lot has already been said in the book about pheromone components and their role in attracting insects. Only those aspects relating to their role as attractant sources will be mentioned here. It is often the case that the active ingredients used in an attractant source need not be identical to those used by the insects in the wild. Indeed many minor components isolated from pheromone glands are not required to be included in a lure for it to work adequately as an attractant source in a trap. *Adoxophyes orana* (Fisher von Roslerstamm), for instance, has been monitored for many years using a 9 : 1 mixture of (Z)-9-14:Ac and (Z)-11-14:Ac (Minks and Voerman, 1973; Minks and de Jong, 1975; Alford *et al.*, 1979). More recent work has shown that the pheromone gland of this insect contains a complex mixture of at least 14 components (Den Otter and Klijnstra, 1980; Charmillot, 1981; Guerin *et al.*, 1986). In the case of the pea moth, *Cydia nigricana*, whose natural pheromone consists of (*E*,*E*)-8,10-12:Ac as a major component, Greenway and Wall (1981) have shown that it is better to use the structurally similar compound (*E*)-10-12:Ac in monitoring traps since it is a weaker attractant – the real major component attracts too many moths and causes problems of trap saturation. In the case of the Mediterranean fruit fly, *Ceratitis capitata*, the active chemical components of the male-produced pheromone have still to be fully elucidated, but the insect can be adequately attracted into monitoring traps using a parapheromone– Trimedlure. This chemical is structurally related to components of angelica seed oil, which is known to be attractive to males of this species (Beroza *et al.*, 1961; Cunningham, 1989).

Whatever attractants are used, the objective is to sample a representative number of insects from the population at any given time. It is also desirable that the system samples selectively for any given species. Minor components may therefore be added to the lure not to enhance the catch of the target species but to inhibit the capture of non-target species. This has been possible in the case of the two closely related species, *Prays oleae* and *Prays citri*, which share a common major component, (Z)-7-14:Ald. The addition of 10% (Z)-9-14:Ald will inhibit the attraction of *P. oleae* so that in a mixed population of the two species *P. citri* can be selectively captured (Campion *et al.*, 1979; Renou *et al.*, 1979). The purity of the substances used can also influence the degree of selectivity imbued to the lure. Purity in this context can refer to chemical purity in general, isomeric purity, or chiral (enantiomeric) purity if the chemical is optically active. It has been shown, for instance, that the (+) form of Disparlure is up to 10 times more attractive in traps for *Lymantria dispar* than the racemic mixture (Vité *et al.*, 1976; Cardé *et al.*, 1977a,b; Plimmer *et al.*, 1977).

*(b) Controlled release devices or dispensers*

Most pheromone components are volatile substances and would volatilize very quickly unless they are formulated in some sort of controlled release device. The controlled release technologies used to date have been many and varied. They rank from relatively simple substrates such as cotton dental wicks, rubber septa and polyethylthene vials to more complex proprietorial technologies of laminates (Hercon Flakes), hollow fibres (Scentry™), membranes (Biolures™) and polymeric systems (Polytrap™ and Selibate™). The objective in most cases is to slow down the release of the pheromone so that the dispenser covers at least the flight period of the insect – usually 4 to 8 weeks. Most of the simpler devices do not release the pheromone at a constant rate during their period of operation in the field; they tend to release more attractant when first placed in the field compared with the amount they release at the end of their useful life. This is referred to as a 'first order' release profile (Fig. 9.1). The ideal release profile, on the other hand, is one in which the amount of attractant released per unit time remains constant throughout the life time of the lure ('zero order' release profile). The more sophisticated devices are capable of doing this in many cases and are therefore better at reflecting population changes with time because they are sampling more consistently from the population.

The controlled release device is also responsible in many cases for protecting the pheromone components from degradation by ultraviolet light (UV) and oxidation. This can sometimes be accomplished by physical means whereby the active ingredient is entrapped in the controlled release device or, as is more often the case, through the addition of antioxidants or UV blockers. This is especially important in the case of pheromones with conjugated dienes and aldehydic functional groups, because these compounds tend to suffer much more from degradation than others. Jones *et al.* (1989) showed that, using

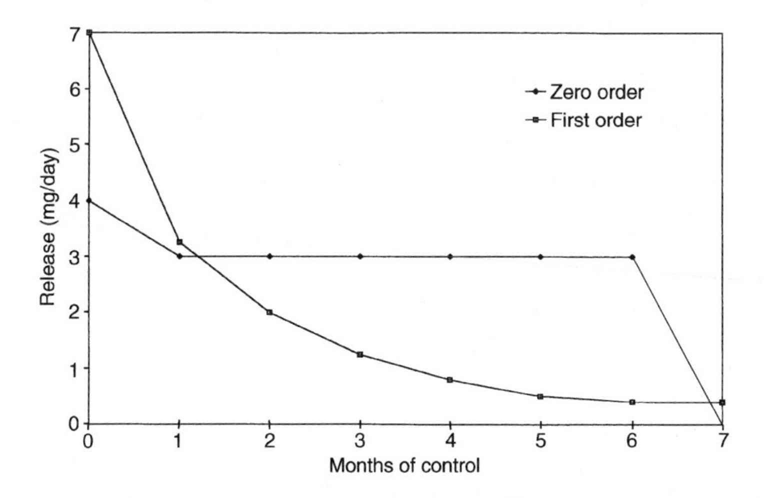


**Fig. 9.1** Release rate curves for zero- and first-order kinetics.

PVC polymers, it was possible to overcome such problems through the addition of suitable antioxidants and UV blockers.

Selection of an appropriate dose for monitoring purposes will determine in most cases the area sampled by a trap. The latter is determined by the length of the plume of odour emanating from the trap, which is in turn proportional to the concentration of the attractant emanating from the source (Wall and Perry, 1987). Ideally, the trap should only sample from the crop that it monitors. However, there are many examples in the literature of traps catching large numbers of insects over distances far beyond the boundaries of the crop (Charmillot, 1980; Henneberry and Clayton, 1982). Lowering the dose of attractant in such cases is one way of reducing the attractive range (Daterman, 1982).

### 9.2.2 Trap design

The variety of trap designs used in conjunction with semiochemicals for insect pest monitoring is extensive and a selection is shown in Fig. 9.2. Most designs are the result of empirical selection and very few systematic studies of insect behaviour and pheromone plume formation with relation to trap design have been undertaken. Lewis and Macaulay (1976) showed that, in the case of *Cydia nigricana*, trap design had a profound influence on plume shape, which in turn influenced greatly the number of moths attracted to, and caught in, the trap. Indeed the process of catching insects in a semiochemical-baited trap can be seen as two stages – the attraction stage and the entrapment stage. The trap design will have some influence on attraction efficiency in that it can influence the shape of the plume that emanates from it. Clearly, an omnidirectional trap such as a funnel trap will have a more consistent plume form, independent of wind direction, compared with a delta trap which will produce a different plume form depending on the angle with which the wind impinges on it. Similarly, the degree of resistance that the trap presents to the wind will determine the turbulence around the trap and hence the cohesion, form and length of the plume that emanates from it.

Entrapment systems tend to fall into two groups: those that involve a non-drying adhesive or water to retain the insects caught and those that have a one-way entrance system with no exit. The latter may in some cases involve a toxicant to immobilize the insect once it has passed through to the interior of the trap. This may not be to prevent the escape of the insects while the trap is in use but to prevent escape of live insects when the traps are opened for inspection and counting.

The size of the trap is often related to the size and number of insects that it is intended to catch. A sticky trap, for instance, is often not

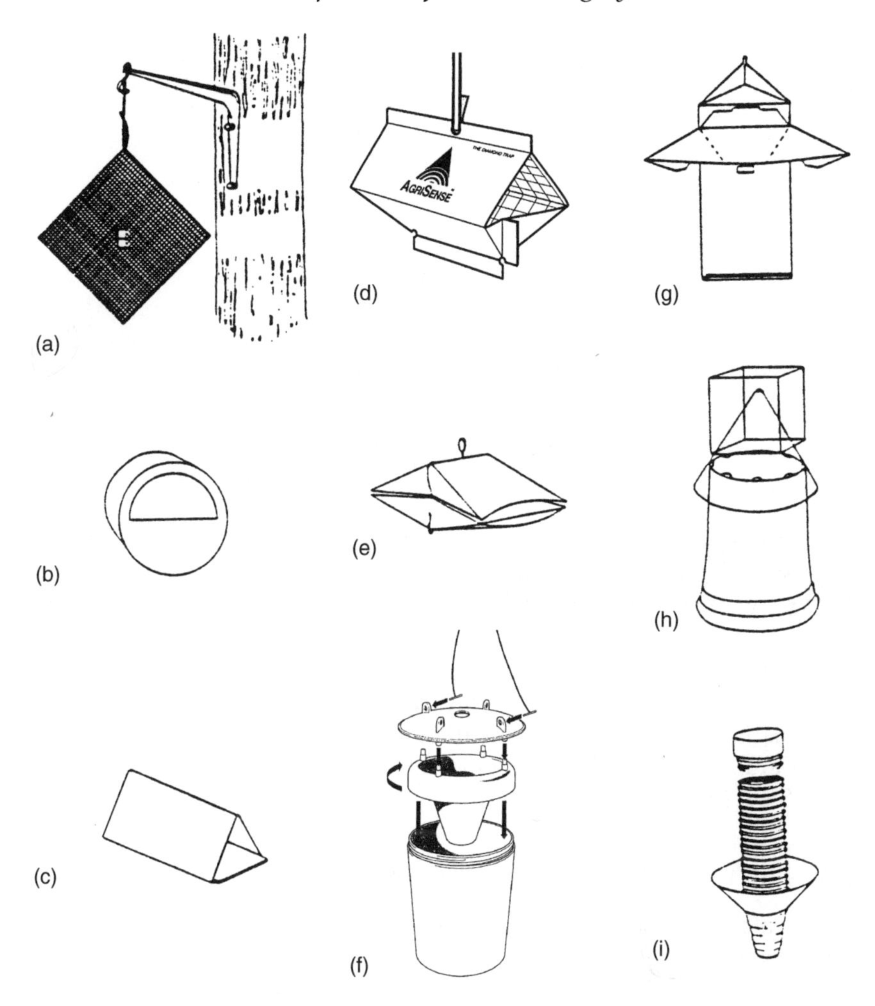


**Fig. 9.2** Examples of pheromone trap designs. (a) Vertical sticky plate for *Scolytus multistriatus*; (b) IOBC trap for orchard Lepidoptera; (c) delta trap for *Cydia nigricana*; (d) tent trap; (e) wing trap for many species of Lepidotpera; (f) funnel trap for larger moths; (g) milk carton trap for *Lymantria dispar*; (h) Hardee trap for *Anthonomus grandis grandis*; (i) cylinder trap for *Ips typographus*. (Adapted from Wall, 1989.)

suitable for large moths such as noctuids because of the risk of trap catch saturation. It was very well demonstrated by Sanders (1981) that even with a small tortricid moth (*Choristoneura fumiferana*) vastly differing catches can be obtained, depending on the frequency with which the sticky bases of Pherocon 1CP traps were changed (Fig. 9.3).

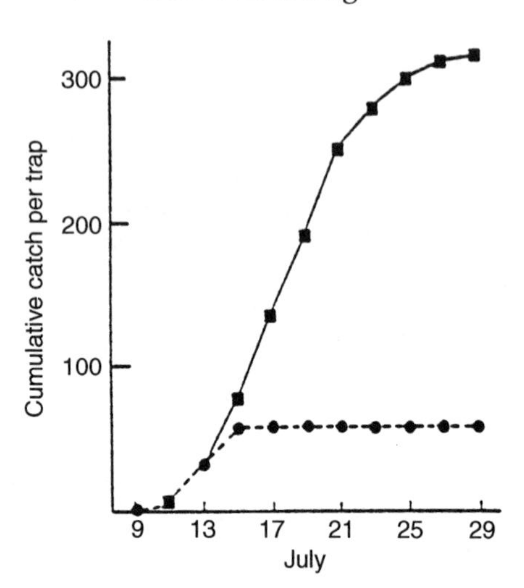


**Fig. 9.3** Catches of male *C. fumiferana* moths in Pherocon 1CP traps, solid line, trap bottoms changed every two days; broken line, trap bottoms unchanged throughout. (Redrawn from Sanders, 1981.)

Trap colour will sometimes influence the number of insects caught, but this is generally more pronounced with beetles (Leggett and Cross, 1978) and fruit flies (Katsoyannos, 1989) than with nocturnal Lepidoptera (Sanders, 1978).

## 9.3 FACTORS THAT INFLUENCE TRAP CATCH

Trap catch, in addition to being influenced by the design of the trap and attractant source, is affected by two further factors: trap placement and the biology of the pest.

### 9.3.1 Trap placement

Three important elements relating to trap placement are trap height, position with respect to vegetation and trap density. Ideally, traps should be placed at the optimum height for catching insects, though this may not always be practical in plantation crops. In broad-acre field crops, it is often better to relate the height of the trap to the height of the vegetation (Riedl *et al.*, 1979) and make adjustments as the latter grows (Lewis and Macaulay, 1976). The orientation of the trap with respect to the prevailing wind direction is also important (Lewis and Macaulay, 1976); as a rough rule of thumb, the trap should be positioned such that the plume which emerges from it passes downwind over the crop area being monitored.

Trap density is much more complex. Two questions need to be considered: what density of traps is required to sample the pest population adequately, and at what density of traps do individual traps start to interfere with each other? Data collected from single traps at any location are always less reliable than data that are the average of more than one trap at any given location. Commercial trapping systems are usually sold as a kit of two or three traps and deployed as a battery of traps adequately spaced to monitor any given area. The question then arises as to how far apart they should be so that they do not interfere with each other. Field studies on various tortricid moths have shown that traps can influence each other at distances of 300 m or more, but again the question of practicality comes into play and traps are often placed at 30 to 50 m apart. Although such traps will have some influence on each other, this distance is regarded as sufficient to prevent one trap from 'poaching' moths from the plume of another trap.

### 9.3.2 Influence of pest biology on trap catch interpretation

When using semiochemical-based trapping systems it should be remembered that they measure insect behaviour and are not direct measures of populations, as are sweep netting or 'sondage' sampling. Insects have to be in the right physiological stage to respond to the pheromone; there have been instances when insects have been present in the field but have not been caught in traps for a whole series of reasons, mostly related to their behavioural physiology.

Trap catch interpretation can be effective only if the pest's biology is also known in detail. For instance, it is important to know the difference between male and female emergence patterns. As in most Lepidoptera it is the male which is attracted to the traps, it is important to know if the females emerge at the same time or in advance of the males, or at some time after the males. This information will have a profound effect on spray-timing decisions based on the trapping system. It is also important to know whether males, females or both sexes go through a post-emergence dispersal phase, or even a prolonged period of migration. In the latter case, they may not respond to the pheromone until they have gone through such a phase of obligatory migration. Population density can also affect the proportion of insects sampled from the population. This may be due to the competition from virgin females in the field and, as their numbers increase with every subsequent generation in a multi-voltine species, the competition between them and the traps gets greater and the traps in turn sample a smaller proportion of the male population.

## 9.4 USE OF MONITORING SYSTEMS IN PEST DETECTION

### 9.4.1 Early warning of pest incidence

A good example of the use of pheromone traps to detect the arrival of dispersing or migrating insect pests can be found in East Africa, where a network of traps is used over the whole of that region to monitor movements of the African armyworm, *Spodoptera exempta*. Information on pest appearance is collected from traps and fed back to a coordination centre at the Desert Locust Control Organization for East Africa, from where timely warnings are transmitted to areas about to experience an attack from the pest. The system allows enough time for control measures to be organized and applied before the larvae from the invading population have done too much damage (McVeigh *et al.*, 1990).

### 9.4.2 Survey to define infested areas

Perhaps the best-known example of the use of pheromone traps for this purpose is in monitoring the spread of gypsy moth, *Lymantria dispar*, an introduced pest of broadleaf forests in North America (Schwalbe, 1981). Some 250 000 traps baited with the pheromone for this pest, (+) disparlure, are now used annually by the US Department of Agriculture to survey the areas infested by this insect.

### 9.4.3 Arrival of quarantine pests in pest-free areas

The pheromone-based detection system described above is now also being used to monitor the potentially very damaging race of *L. dispar*, the Asian gypsy moth, in those countries where it is regarded as a major quarantine pest. Traps are deployed around major ports or railway depots where the pest could be introduced with timber that is being imported from areas in which the pest is endemic, such as Eastern Europe or North East Asia.

The Mediterranean fruit fly, *Ceratitis capitata*, is a serious pest of many fruit species and has so far been kept out of the US mainland. It has been introduced into the country on a number of occasions but has always been eliminated. The Inspection Services in the United States are constantly vigilant with respect to any re-introductions of this pest and use significant numbers of sticky traps baited with Trimedlure, the para-pheromone for this insect, as monitoring tools. Countries wishing to export fruit to the United States or to other countries that do not have the pest, such as Japan, have established 'fly-free zones' in their fruit-growing areas where the fly is either known not to exist or where it has been annihilated through intensive spraying or sterile insect techniques (SIT). Such zones can be monitored using the same trapping systems as those described above for the US mainland and act as proof that the areas are truly fly-free.

## 9.5 USE OF MONITORING SYSTEMS IN TIMING OF CONTROL TREATMENTS

### 9.5.1 Timing of spray treatments

This approach can be applied to both immigrant pest populations of insects such as the pea moth, *Cydia nigricana* (Macaulay *et al.*, 1985) and emergent but resident populations such as those of the codling moth, *Cydia pomonella* (Glen and Brain, 1982). A threshold catch is usually established for any trapping system before any spray treatment is justified. If that threshold catch is superseded, then a period of time is allowed to pass before the spray application is made. This period will vary with temperatures and a 'heat-sum' is often calculated in order to determine the precise time for spray applications. During the period between the threshold date and the spray date, males and females mate, the females lay their eggs and the spray is applied as those eggs begin to hatch. The spray application is therefore targeted at the neonate larvae before they penetrate the fruit or pod, within which they would be protected and therefore difficult to control with contact pesticides.

### 9.5.2 Timing of other sampling methods

Trap catch information is often not sufficient by itself to take pest management decisions. In the case of the olive fly *Bactrocera oleae*, reaching a threshold catch in pheromone-baited traps triggers the time to take a sample of females from traps baited with food lure such as a protein hydrolysate. This sample of females is then examined under a binocular microscope to check the state of maturity of the eggs in the ovaries. If they are at the point of laying, a spray application is made. If they are still not mature, spraying is delayed for a further week when another sample of females is inspected (Montiel Bueno, 1986). In bollworm pests of cotton, a threshold catch in a pheromone trap triggers the taking of a sample of cotton bolls for inspection for eggs or young larvae. If this is shown to be above a critical level, then a spray application is made (Lopez *et al.*, 1990; Campion, 1994).

## 9.6 USE OF MONITORING SYSTEMS IN RISK ASSESSMENT AND POPULATION DENSITY ESTIMATES

### 9.6.1 Risk assessment

The olive fly, *Bactrocera oleae*, is often present in olive groves during the early summer months as shown by catches of both males and females in food-baited McPhail traps. The flies are usually not sexually active because the diameters of the young fruits have not reached a sufficient size for oviposition to be possible. As the summer progresses, the fruitlets grow and there comes a point at which oviposition is possible and sexual activity is triggered in the adult flies. This can be observed by comparing catches of male flies in vertical yellow sticky traps, baited with the pheromone, with catches in unbaited traps. During the early summer, when there is no sexual activity, catches are more or less equal in the baited and unbaited traps. With the initiation of mating, the catches in the baited traps suddenly become up to 10 times higher. This signals the need to take a look at the state of development of the ovaries in the females. The two factors are then taken together as an indication of risk of infestation to the olive fruits in that particular grove (Montiel Bueno, 1989).

### 9.6.2 Population trends

One of the objectives of monitoring *Choristoneura fumiferana* is to identify population trends over many generations. Sanders (1990) has shown that trap catches obtained since 1973 do broadly reflect trends in population density during that period, and as this pest tends to be cyclical in importance (outbreaks lasting 5–10 years in cycles of 35–40 years) then the pheromone-baited trapping system can be used to provide a very good indication of population trends over a long period.

### 9.6.3 Population density correlations

Obtaining quantitative information about pest populations from trap catch data has proved difficult in many cases, especially in strong-flying, highly mobile insects such as *Heliothis* spp. (Srivastava *et al.*, 1992). In many cases, species have been trapped successfully in the field, but their presence in traps has not correlated well with subsequent sampling of eggs or larvae or with damage. This may be due to the fact that sex pheromones, in Lepidoptera at least, usually attract the males and a correlation between male and female numbers in the field does not always exist. Indeed, the females may not be present, or are in a migratory phase and not laying eggs, or have already oviposited elsewhere before the males moved to the site being monitored. Conversely, there have been examples of pest damage in the field when insects have not been caught in traps. In such cases, only mated females have flown into the crop.

Such problems have not occurred with enough frequency to dissuade researchers from continuing to seek quantitative data from trapping systems and many useful correlations have been established, especially with the smaller, less dispersive moth species. The olive moth, *Prays oleae*, serves as a good example of what is sometimes required to establish such correlations (Ramos *et al.*, 1989). It took 10 years of studies before researchers were able to establish a suitably predictive correlation between male trap catch in pheromone traps and subsequent infestation of fruit. The degree of synchrony between pest emergence and the olive tree phenology was crucial in determining the subsequent infestation level; a high degree of asynchrony meant that only the late-emerging moths could attack fruits. Once this factor was included in the equations, correlations between trap catch and subsequent infestations were established which were valid over many years.

### 9.6.4 Effects of control measures

Figure 9.4 shows catches of *Ephestia kuhniella* in funnel traps in a flour mill over a period of 8 months and clearly demonstrates the effects of

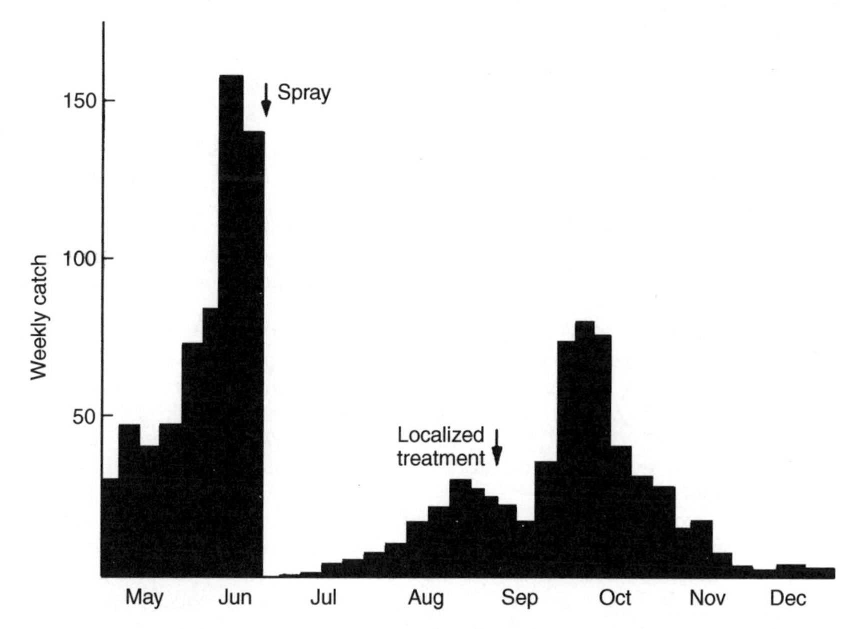


**Fig. 9.4** Weekly catches of *Ephestia kühniella* in funnel traps in a flour mill. (From Spinelli and Arsura, 1984)

total fumigation of the mill in June of that year. The trap catches are reduced to zero for a period of 2 weeks. Populations subsequently recover and a need is indicated for some localized treatments to maintain the population below the threshold, which would require a total fumigation once more. This localized treatment is again seen to be effective in maintaining the population at a controllable level and as October is reached, the cool autumnal temperatures have the effect of naturally reducing the mill population (Spinelli and Arsura, 1984).

To conclude, it is quite clear that the semiochemical-based trapping systems have become very well established in terms of population monitoring in all its various forms. Their successful use has led in many cases to substantial reductions in the overuse of insecticides. Documentary evidence for this is often circumstantial but it has been estimated that in Ontario, Canada, over 90% of apple growers use integrated methods of pest control, which has brought about a 25% reduction in pesticide use. This reduction is due, in the main, to the widespread acceptance among growers of insect monitoring systems. Similarly, some 50% of apple producers in British Columbia (Canada) use monitoring systems for insect pests and corresponding reductions in pesticide use have also taken place in that province (Canadian Government statistics: Agriculture Canada, 1986). As concerns about insecticide use increase, the future for semiochemical-based monitoring systems looks bright and their use will inevitably increase.

## REFERENCES

Alford, D.V., Carden, P.W., Dennis, E.B., *et al.*, (1979) Monitoring codling and tortrix moths in United Kingdom apple orchards using pheromone traps. *Ann. Appl. Biol.*, **91**, 165–178.

Arnason, J.T., Philogene, B.J.R. and Morand, P. (1989) *Insecticides of plant origin. Proceedings of a symposium held in Toronto, Ontario, Canada, June 5–11, 1988.* ACS Symposium Series No. 387, American Chemical Society, Washington, DC.

Beroza, M., Green, N., Gertler, S.I., *et al.* (1961) New attractants for the Mediterranean fruit fly. *Agricultural and Food Chemistry*, **9**, 361–365.

Campion, D.G. (1994) Pheromones for the control of cotton pests. In *Insect Pests of Cotton* (ed. G.A. Matthews), CAB International, Wallingford, pp.505–534.

Campion, D.G., McVeigh, L.J. Polyrakis, J. *et al.* (1979) Laboratory and field studies of the female sex pheromone of the olive moth, *Prays oleae. Experientia*, **35**, 1146–1147.

Cardé, R.T., Doane, C.C., Baker, T.C., *et al.* (1977a), Attractancy of optically active pheromone for male gypsy moths. *Environ. Entomol.*, **6**, 768–772.

Cardé, R.T., Doane, C.C., Granett, J., *et al.* (1977b) Attractancy of racemic disparlure and certain analogues to male gypsy moths and the effect of trap placement. *Environ. Entomol.*, **6**, 765–767.

Carson, R. (1962) *Silent Spring*, Houghton Mifflin Co., Boston, MA.

Charmillot, P.J. (1980) Développement d'un système de prévision et de lutte contre le carpocapse (*Laspeyresia pomonella* L.) en Suisse romande: role du service regional d'avertissement et de l'arboriculture. *Bull. OEPP*, **10**, 231–239.

Charmillot, P.J. (1981) Technique de confusion contre la tordeuse de la pelure *Adoxophyesorana* F. v. R. (Lep: Tortricidae): II. Deux ans de'essais de lutte en vergers. *Mitt. Schweiz. Ent. Ges.*, **54**, 191–204.

Cunningham, R.T. (1989) Parapheromones. In *Fruit Flies: Their Biology and Control*, Volume 3A, *World Crop Pests* (eds A.S. Robinson and G. Hooper), Elsevier, Amsterdam, pp. 221–230.

Daterman, G.E. (1982) Monitoring insects with pheromones: trapping objectives and bait formulations. In *Insect Suppression with Controlled Release Pheromone Systems* (Vol. 1) (eds A.F. Kydonieus and M. Beroza), CRC Press, Boca Raton, Florida, pp. 195–212.

DeBach, P. and Rosen, D. (1991) *Biological Control with Natural Enemies*, Cambridge University Press, Cambridge, 456 pp.

Den Otter, C.J. and Klijnstra, J.W. (1980) Behaviour of male summer fruit tortrix moths, *Adoxophyes orana* (Lepidoptera: Torticidae), to synthetic and natural female sex pheromone. *Ent. Exp. Appl.*, **28**, 15–21.

Dent, D.R. (1991) *Insect Pest Management*, CAB International, Wallingford, 604 pp.

Dinham, B. (1993) *The Pesticide Hazard: a Global Health and Environment Audit*, The Pesticides Trust, London.

Foschi, S. (1989) Pesticide residues in food. *Difesa delle Piante*, **12**, 41–64.

Franz, J.M. (ed.) (1986) *Biological Plant and Health Protection (Progress in Zoology* **32**), Gustav Fischer Verlag, Stuttgart.

Georghiou, G.P. and Saito, T. (eds) (1983) *Pest Resistance to Pesticides*, Plenum Press, New York.

Glen, D.M. and Brain, P. (1982) Pheromone-trap catch in relation to the phenology of codling moth (*Cydia pomonella*). *Ann. Appl. Biol.*, **101**, 429–440.

Greenway, A.R. and Wall, C. (1981) Attractant lures for males of the pea moth, *Cydia nigricana* (F.), containing (*E*)-10-dodecen-1-yl acetate and (*E,E*)-8,10-dodecadien-1-yl acetate. *J. Chem. Ecol.*,**7**, 563–573.

Guerin, P.M., Arn, H., Buser, H.R. and Charmillot, P.J. (1986) Sex pheromone of *Adoxophyes orana*: additional components and variability of (Z)-9- and (Z)-11-tetradecenyl acetate. *J. Chem. Ecol.*, **12**, 763–772.

Henneberry, T.J. and Clayton, T.E. (1982) Pink bollworm of cotton (*Pectinophora gossypiella* (Saunders)): male moth catches in gossyplure-baited traps and relationships to oviposition, boll infestation and moth emergence. *Crop Protection*, **1**, 497–504.

Jones, O.T., Hall, D.R. and Smith, J.L. (1989) A new controlled release formulation which protects pheromones from isomerisation and degradation under field conditions. In *The Use of Pheromones and Other Semiochemicals in Integrated Control* (eds H. Arn and R. Bues), Proceedings of a working group meeting held on 20–22 September, 1988, Avignon, France. IOBC-WPRS Bulletin XII, pp. 151–152.)

Katsoyannos, B.I. (1989) Responses to shape, size and colour. In *Fruit Flies: Their Biology and Control*, Vol. 3A, *World Crop Pests* (eds A.S. Robinson and G. Hooper), Elsevier, Amsterdam, pp. 307–324.

Katsoyannos, P. (1992) *Olive Pest Problems and their Control in the Near East*. FAO Plant Production and Protection Paper XXX, FAO, Rome.

Leggett, J.E. and Cross, W.H. (1978) Boll weevils: the relative importance of colour and pheromone in orientation and attraction to traps. *Environmental Entomology*, **7**, 4–6.

Lewis, T. and Macaulay, E.D.M. (1976) Design and elevation of sex-attractant traps for the pea-moth, *Cydia nigricana* (Steph.), and the effect of plume shape on catches. *Ecol. Entomol.*, **1**, 175–187.

Lisansky, S.G. and Coombs, J. (1994) Developments in the market for biopesticides. In *Proceedings of the Brighton Crop Protection Conference – Pests and Diseases–1994*, Volume 3, British Crop Protection Council, Farnham, pp. 1049–1054.

Lopez, J.D., Shaver, T.N. and Dickerson, W.A. (1990) Population monitoring of *Heliothis* spp. using pheromones. In *Behaviour-Modifying Chemicals for Insect Management*, (eds R.L. Ridgway, R.M. Silverstein and M.A. Inscoe), Marcel Dekker, New York, pp. 473–496.

Macaulay, E.D.M., Etheridge, P., Garthwaite, D.G. *et al.* (1985) Prediction of optimum spraying dates against pea moth, (*Cydia nigricana* (F.)), using pheromone traps and temperature measurements. *Crop Protection*, **4**, 85–98.

McVeigh, L.J., Campion, D.G. and Critchley, B.R. (1990) The use of pheromones for the control of cotton bollworms and *Spodoptera* spp. in Africa and Asia. In *Behaviour-Modifying Chemicals for Insect Management*, (eds R.L. Ridgway, R.M. Silverstein and M.A. Inscoe), Marcel Dekker, New York, pp. 407–415.

Minks, A.K. and de Jong, D.J. (1975) Determination of spraying dates for *Adoxophyes orana* by sex pheromone traps and temperature recordings. *J. Econ. Entomol.*, **68**, 729–732.

Minks, A.K. and Voerman, S. (1973) Sex pheromones of the summer fruit tortrix moth, *Adoxophyes orana*: trapping performance in the field. *Ent. Exp. Appl.*, **16**, 541–549.

Montiel Bueno, A. (1986) The use of sex pheromone for monitoring and control of the fruit fly. In *Proceedings of the Second International Symposium on Fruit Flies*, held on 16–21 Sept, 1986, Colymbari, Crete, Greece, (ed. A.P. Economopoulos), Elsevier Science Publishers, Amsterdam, pp. 483–494.

Montiel Bueno, A. (1989) Control of the olive fly by means of its sex pheromone. In *Proceedings of the CEC/IOBC International Symposium on Fruit Flies of Economic Importance*, held on 7–10 April, 1987 in Rome, Italy (ed. R. Cavalloro), A.A. Balkema, Rotterdam, pp. 443–453.

Plimmer, J.R., Schwalbe, C.P., Paszek, E.C. *et al.* (1977) Contrasting effectiveness of (+) and (–) enantiomers of disparlure for trapping native populations of the gypsy moth in Massachusetts. *Environ. Entomol.*, **6**, 518–522.

Ramos, P., Campos, M., Ramos, J.M. and Jones, O.T. (1989) Nine years of studies on the relationship between captures of male olive moths, *Prays oleae* Bern. (Lepidoptera: Hyponomeutidae) in sex pheromone baited traps and fruit infestation by the subsequent larval generation (1979–1987). *Tropical Pest Management*, **35**, 201–204.

Ravensburg, W.J. (1994) Biological control of pests: current trends and future prospects. *Proceedings of the Brighton Crop Protection Conference – Pests and Diseases – 1994*, Volume 3, British Crop Protection Council, Farnham, pp. 591–600.

Renou, M., Descoins, C., Priesner, E. *et al.* (1979) Le tétradécene-7 Z al-1, constituant principal de la sécrétion phéromonale de la Teigne de l'Olivier: *Prays oleae* Bern. (Lépidoptere Hyponomeutidae). *C.R. Acad. Sc. Paris*, **288**, 1559–1562.

Riedl, H., Hoying, S.A., Barnett, W.W. and Detar, J.E. (1979) Relationship of within-tree placement of the pheromone trap to codling moth catches. *Environ. Entomol.*, **8**, 765–769.

Roberts, T. (1989) Pesticides in drinking water. *Shell Agriculture*, **3**, 18–20.

Sanders, C.J. (1978) Evaluation of sex attractant traps for monitoring spruce budworm populations (Lepidoptera: Tortricidae). *Can. Entomol.*, **110**, 43–50.

Sanders, C.J. (1981) Sex attractant traps: their role in management of spruce budworm. In *Management of Insect Pests with Semiochemicals* (ed. E.R. Mitchell), Plenum, New York, pp. 75–91.

Sanders, C.J. (1990) Practical use of insect pheromones to manage coniferous tree pests in Eastern Canada. In *Behaviour-Modifying Chemicals for Insect Management*, (eds R.L. Ridgway, R.M. Silverstein and M.N. Inscoe), Marcel Dekker Inc., New York, pp. 345–361.

Schwalbe, C.P. (1981) Disparlure-baited traps for survey and detection. In *The Gypsy Moth: Research Toward Integrated Pest Management*, eds C.C. Doane and M.L. McManus, *USDA Tech. Bull., No. 1584, pp. 542–548.*

Spinelli, P. and Arsura, E. (1984) Il feromone sessuale nella lotta alle tignole delle derrate. *Disinfestazione*, 1984 Vol. 4, pp. 4–7.

Srivastava, C.P., Pimbert, M.P. and Reed, W. (1992) Monitoring of *Helicoverpa (=Heliothis) armigera* (Hubner) moths with light and pheromone traps in India. *Insect Science and its Application*, **13**, 205–210.

Starnes, R.L., Liu, C.L. and Marone, P.G. (1993) History, use and future of microbial insecticides. *American Entomologist*, **39**, 83–91.

Vité, J.P., Klimetzek, D., Loskant, G. *et al.* (1976) Chirality of insect pheromones: response interruption by inactive antipodes. *Naturwissenschaften*, **68**, 582–583.

Wall, C. (1989) Monitoring and spray timing. In *Insect Pheromones in Plant Protection* (eds A.R. Jutsum and R.F.S. Gordon), John Wiley & Sons Ltd, pp. 39–66.

Wall, C. (1990) Principles of monitoring. In *Behaviour-Modifying Chemicals For Insect Management* (eds R.L. Ridgway, R.M. Silverstein and M.N. Inscoe), Marcel Dekker Inc., New York, pp. 9–23.

Wall, C. and Perry, J.N. (1987) Range of action of moth sex-attractant sources. *Entomol. Exp. Appl.*, **44**, 5–14.

# 10

# Mass trapping

## 10.1 GENERAL PRINCIPLES

Mass trapping in concept would appear to be simple: place a high density of traps in the crop to be protected and achieve a measure of protection through removal of a sufficiently high proportion of individuals from the population by trapping. In practice, it is rarely that simple. Several factors make the technique non-viable on a large scale. These include:

- lack of attraction of females by the attractant source used;
- lack of highly efficient traps;
- problem of high insect populations and trap saturation;
- need for a high density of traps per unit of surface area, which in turn renders the technique too costly.

### 10.1.1 Attractant source

In moth species, it is generally the females which attract the males through the use of sex pheromones and as a consequence the use of pheromones for mass trapping of these insects has not, for the most part, been very successful. A male moth can fertilize more than one female and as a result a very high proportion of males would have to be removed from the population before the fecundity of the females starts to be affected. Roelofs *et al.* (1970) have calculated that, for certain moth species, a ratio of 5 traps to every 'calling' female is needed to trap sufficient males to obtain a 95% reduction in female fecundity.

It is clear that the attractant source which is used should catch females. A reduction in females as well as males can impact the population numbers significantly. There are some insects in which the sexual attractant pheromone is produced by the male and then often attracts both the females and other males, as in the case of the boll weevil (*Anthonomus grandis*) discussed later. If a pheromone is indeed used for mass trapping it is often an aggregation pheromone which is attractive to both sexes.

Some researchers have turned to sources other than pheromones to attract females. These have included food and oviposition attractants. They consist of olfactory cues which are emitted by food sources or oviposition sites that the females use in the wild. Tephritid fruit flies have been captured for many years using either protein hydrolysates or ammonium salts. The attraction of both males and females by such materials is thought to mimic their attraction in the field to sources of protein that the females, in particular, require for egg production. Many of these tephritid fruit flies are also long-lived and require a highly developed olfactory sense to find food and survive during long dry periods when there are no fruits available for oviposition.

Host plant volatiles have also been used as olfactory attractants either alone or in combination with pheromones. The attraction of certain weevils has been found to be greatly enhanced when the sex or aggregation pheromone is used in combination with host plant volatiles for mass trapping. The Japanese beetle, *Popillia japonica* (Ladd and Klein, 1986) and palm weevils, *Rhynchophorus* spp. (Chinchilla *et al.*, 1993; Oehlschlager *et al.*, 1993) provide good examples of this phenomenon.

### 10.1.2 Trapping system

Many of the trapping systems used to date are inherently inefficient, with capture efficiencies as low as 0.4% and 8.7% quoted for some trapping experiments with *Heliothis virescens* (Lingren *et al.*, 1978). However, experiments carried out on gypsy moth, *Lymantria dispar*, by Elkinton and Childs (1983) have shown that, whereas trapping efficiency may be low for most trap designs, the efficiency of recruitment of insects to the plumes emanating from such traps, with subsequent orientation up the plume to within 0.5 m of the source, is at least 95%. The technique of mass trapping will therefore be most effective where a powerful attractant draws insects from a distance to within 0.5 m of a trap and then good trap designs, coupled possibly with secondary visual and olfactory cues, make for efficient entrapment of the insects from that point on. Recent work by Quartey and Coaker (1992) has shown the importance of visual cues in entrapment of *Ephestia cautella* in funnel traps. They showed that vertical stripes 7.5 mm wide on the outside of funnel traps caught 90% and 80% of moths released in a wind tunnel in moving and still air, respectively, compared with 70% and 35% for the best available commercial traps.

A lot of development work therefore needs to be carried out on the design of trapping systems for many species before mass trapping becomes a viable option for insect pest suppression.

### 10.1.3 High populations and trap saturation

The problem of trap saturation was mentioned in the previous chapter in the context of monitoring traps but a lot of the points made there apply equally well when referring to the use of traps for population suppression through mass trapping. Sticky traps quickly become saturated with insects if the insects are large and the trap service interval is long. Similarly, trap saturation may arise as a result of high densities of pests. This is probably one of the major factors limiting the development of this technique.

The boll weevil, *Anthonomus grandis*, is a serious pest of cotton in the United States. Cotton with a value of over $300 million is lost annually through boll weevil attacks (Hardee, 1982). With the identification of the male-produced pheromone, 'Grandlure' (Tumlinson *et al.*, 1969), a new tool was made available for controlling this pest.

Grandlure is a mixture of four components and acts both as a sex pheromone attracting females and as an aggregation pheromone in early and late season, when it attracts both sexes. Numerous attempts at mass trapping of boll weevil were conducted during the 1970s with various formulations and traps (Hardee, 1982) but the low trap capacities and low trapping efficiencies meant that supplemental insecticide applications also had to be used and mass trapping was almost abandoned as a means of combating this pest (Hardee, 1984, personal communication). Later in the same decade, however, an area-wide boll weevil eradication project was initiated in a number of the southern US states. The system used involved area-wide insecticide treatments to reduce boll weevil populations together with pheromone traps for monitoring and suppression of very low density populations. This programme has been highly successful in both the southeastern and

**Table 10.1** Boll weevil captures in fields in North Carolina (NC) and eastern South Carolina (SC), USA (from Ridgway *et al.*, 1990)

| *Location by year* | *Hectares* | *% of fields with number of weevils captured* | | |
|---|---|---|---|---|
| | | *0* | *1–5* | *> 5* |
| *1983* | | | | |
| NC | 6600 | 0 | 1 | 99 |
| SC | 21 240 | 2 | 5 | 93 |
| *1986* | | | | |
| NC | 8600 | > 99.9 | < 0.1 | 0.0 |
| SC | 34 400 | > 97.3 | 2.3 | < 0.4 |

**Table 10.2** Benefits associated with boll weevil eradication in Virginia (VA), North Carolina (NC) and South Carolina (SC) USA (modified from Carlson *et al.*, 1989).

| | Dollars $ha^{-1}$ | |
|---|---|---|
| *Benefit* | *Original eradication zones (VA, NC)* | *Expanded eradication zones (NC, SC)* |
| Net reduced pesticide | 72.18 | 75.08 |
| Acreage expansion | 33.20 | 34.50 |
| Yield effect | 86.25 | 86.25 |
| Total monetary benefit | 191.63 | 195.83 |

southwestern United States (Ridgway *et al.*, 1990). Table 10.1 shows the measure of success achieved in North and South Carolina between the years 1983 and 1986. Even more striking are the economic benefits which resulted from the programme (Table 10.2). In 1987, the south-eastern boll weevil eradication programme was expanded to Florida, most of Georgia and a major portion of Alabama and similar successes were recorded. In the southwestern programme, which started in 1985, the boll weevil had been eradicated from California, southwestern Arizona and adjoining areas of Mexico by 1987 through a similar system of trapping and insecticide applications. The expanded south-western programme was completed in 1990 and a buffer zone or containment programme is now in operation to prevent re-infestation (Ridgway and Inscoe, 1992). It is clear, therefore, that mass trapping can have a role where problems of high populations and trap saturation are overcome through supplementary spraying with insecticides and particularly where an area-wide programme is undertaken.

## 10.2 BARK BEETLES

Semiochemicals play a very important role in the chemical ecology of bark beetles (Chapter 2 and Fig. 10.1). Attractants that are released partly from the host tree, but mainly from pioneer beetles, lead to an aggregation of beetles on to that tree. Because of their large numbers, the bark beetles quickly overcome the tree's attempts at self-defence, i.e. the production of resins in the beetle galleries, and the tree eventually dies. Most species also introduce pathogenic blue-stain fungi that invade the sapwood of the tree, thus further weakening the tree to facilitate colonization. This process of colonization is known as mass attack and is a highly tuned process controlled for the most part through semiochemicals. The pioneer beetles which search for trees suitable for

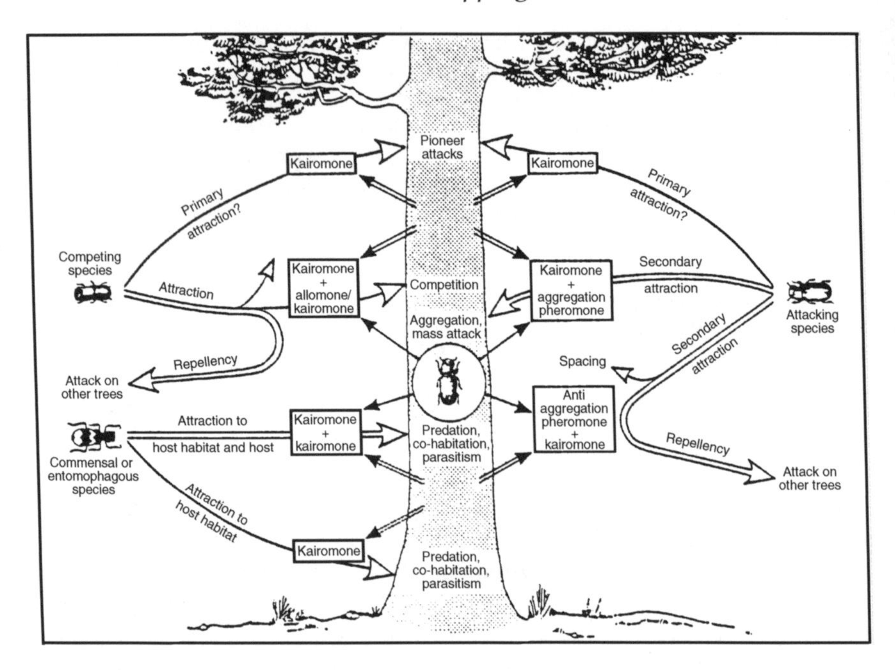


**Fig. 10.1** Semiochemical interactions of an attacking species of bark beetle with its host and other insect species. (Borden, 1989.)

mass attack depend solely on host stimuli during these initial phases. Later, when a suitable host is found and the pioneer beetles release their aggregation pheromones, the importance of the host odour diminishes, though in artificial aggregation pheromone lures (as we will see later) the addition of host plant secondary metabolites will usually enhance attraction of beetles.

Aggregation pheromones have been identified in several species of bark beetles (Borden, 1982). The sex that initiates gallery construction is the main pheromone producer. In monogamous bark beetle species this is the female; in polygamous species it is the male. In other species, both sexes contribute to the blend of components that make up the aggregation pheromone. The pheromones are generally multi-component, often with host plant terpenes as precursors (Vite and Francke, 1976).

The aggregation behaviour of bark beetles was exploited for their control long before the pheromones were isolated and identified. Mass trapping of beetles by means of trap trees has been practised in Europe for over 200 years (Gmelin, 1787). Living, mature spruce trees were felled in early spring and left in the forest during the main flight period. Such felled trees would prove very attractive to pioneer beetles, which would then induce mass aggregation of beetles. Once colonized by the attracted beetles, the logs would be removed from the forest with the beetles still in the bark. The bark would then be removed and burnt.

Trap trees are still used in North America as a means of controlling the spruce beetle, *Dendroctonus rufipennis* (Nagel *et al.*, 1957). Trees can even be baited with kairomones and pheromones to attract beetles to living trees (Gray and Borden, 1989; Borden, 1990). Such attractants usually induce only the first phases of the mass attack. The aggregated beetles eventually release more pheromones, which override the original tree bait. Again the attracted beetles are controlled by destroying the attacked trees through felling and burning or felling and removal.

Tree baits containing monoterpenoid and bicyclic ketal semiochemicals (Fig. 10.2) have been developed for four *Dendroctonus* bark beetle species (Table 10.3). These baits are used primarily to contain and concentrate known infestations of beetles until selective or clear-cut logging can be carried out to remove infested trees. Log processing then kills resident beetle colonies by destroying the cambial tissues where they feed and breed. One consequence of this technique is that the area that needs to be harvested is effectively reduced because the tree baits cause infestations to intensify in a limited area, rather than to expand.

As synthetic aggregation pheromones became available, experiments were initiated to replace trap trees with artificial traps baited with dispensers containing synthetic pheromones and host tree odours. For some species, it became possible to trap beetles in sufficient numbers to reduce their populations and achieve an acceptable level of control.

In North America it is predominantly beetles of the genus *Dendroctonus* that are responsible for mass attacks on living trees, while in central and northern Europe and in northeastern Asia it is the spruce bark beetle, *Ips typographus*, that causes the greatest amount of timber loss by the same process of attack. These beetles have much in common in terms of their attack strategies, and many of the control strategies that have been developed for them based on semiochemicals are similar. As an example, the development and use of a mass trapping system for *I. typographus* only will be described here.

The main flight period for *I. typographus* occurs between April and the end of May, resulting in one generation of larvae in northern areas, with the beetles overwintering as adults. In southern areas there may be two or even three generations a year. The male insects initiate the attack by carrying out the first borings. They attract up to four females by releasing an aggregation pheromone. At this point they transfer several species of blue-stain fungi into the tree, some of which can kill even healthy trees. Z-verbenol and methyl butenol are then released, together with ipsdienol as a minor component of the aggregation pheromone (Bakke *et al.*, 1977). Dispensers based on these three components are

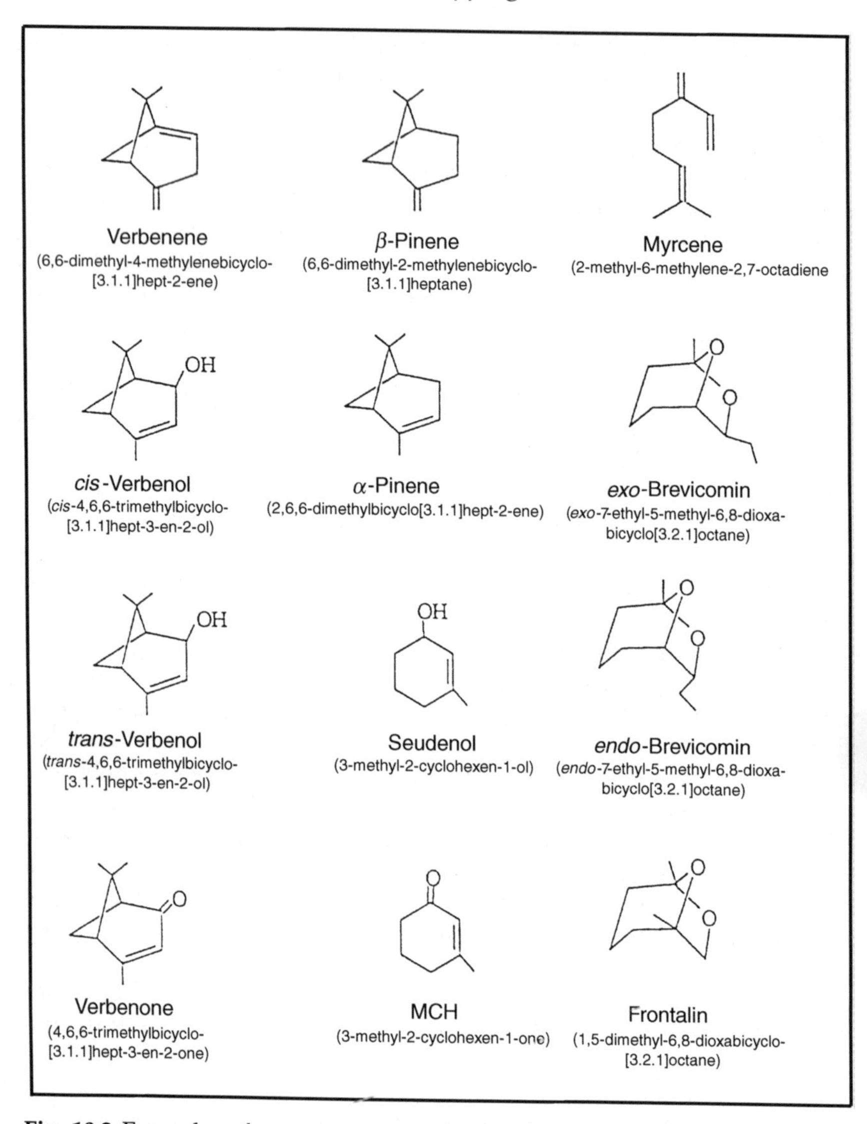


**Fig. 10.2** Examples of monoterpenes and related compounds having activity as semiochemicals for forest Coleoptera. (Burke, 1992.)

placed in one of several designs of trap (Fig. 10.3) in early May and then continue to catch for two to three months. Such lures were approved for use in drainpipe traps (Fig. 10.3) by the Pesticide Boards of the Ministry of Agriculture in Norway and Sweden for an IPM programme against *I. typographus* in those two countries (Bakke, 1981; Bakke *et al.*, 1983; Eidmann, 1983). During the 1970s there were severe

**Table 10.3** Semiochemicals in tree-bait products for four forest coleopteran pests

| *Pest species* | *Active ingredients in tree baits* |
|---|---|
| Douglas-fir beetle *Dendroctonus pseudotsugae* | frontalin, α-pinene, camphene |
| Mountain pine beetle *Dendroctonus ponderosae* | *trans*-verbenol, *exo*-brevicomin, myrcene |
| Spruce beetle *Dendroctonus rufipennis* | frontalin, α-pinene |
| Western balsam bark beetle *Dryocoetes confusus* | *exo*-brevicomin |

outbreaks of this pest in Scandinavia, with significant losses of good timber. Mass trapping was initiated in 1979 in Norway as part of an IPM package that involved changes to cultural practices, which also disfavoured the beetle. In addition to mass trapping, felling and removal of beetle-infested trees was an important component of the IPM approach.

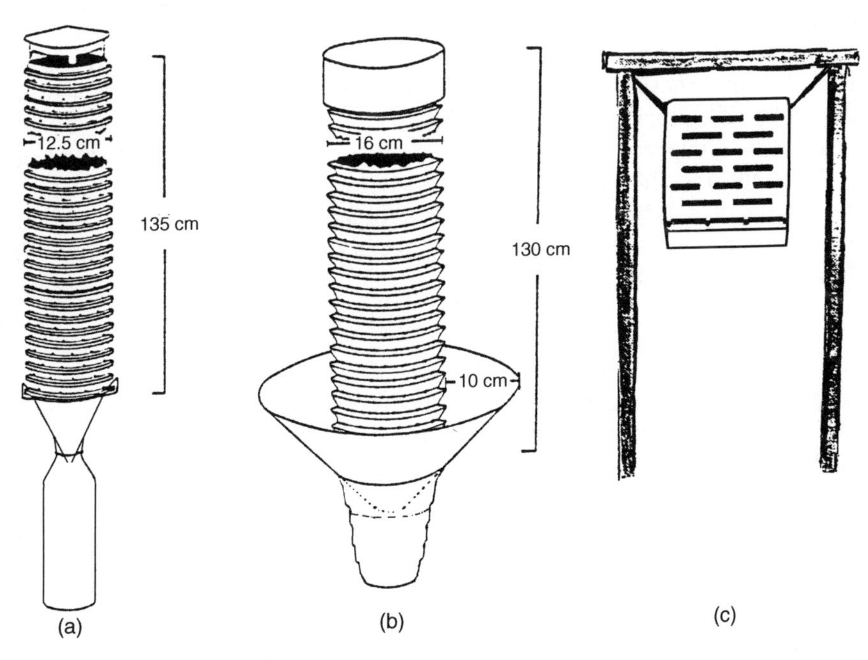


**Fig. 10.3** Drainpipe trap models used in control programme against spruce bark beetle in Norway: (a) 1979 model; (b) 1980 model. (c) Slit trap. (In part from Bakke and Lie, 1989.)

**Table 10.4** Number of traps deployed in Norway during the bark beetle control programme and estimated number of beetles caught in traps

| *Year* | *No. of traps* | *Estimated total capture of beetles (millions)* |
|---|---|---|
| 1979 | 600 000 | 2900 |
| 1980 | 590 000 | 4500 |
| 1981 | 530 000 | 2100 |

Table 10.4 shows the number of traps used during 1979–81 together with the estimated number of beetles caught. Trap catches peaked in 1980 and by 1982 the incidence of tree death had reduced dramatically. More than 99% of the insects trapped were of the genus *Ips*. A small percentage of a predatory clerid of the genus *Thanasimus* was also attracted by Z-verbenol and ipsdienol (Bakke and Kvamme, 1981), but, through selecting the right size of entry hole for the beetles to enter the trap, it was possible to exclude the vast majority of predators, thus increasing the ratio of predators to bark beetles in the forest as trapping proceeded.

As the IPM measures taken against *I. typographus* were many and varied, it is difficult to know what contribution to the overall effect was made by the mass trapping efforts. The incidence of the beetles in Scandinavia has not risen to epidemic levels since the early 1980s but indications are that this might happen again in the not too distant future and it will be interesting to see whether the authorities concerned will again mount such an ambitious programme: last time it covered 2.0 to 2.5 million hectares of forest.

## 10.3 MOTH PESTS

The polyphagous Egyptian cotton leafworm, *Spodoptera littoralis*, is a serious pest of many high value crops in the Mediterranean basin and further south into Africa. The major female sex pheromone component of this species was identified by Nesbitt *et al.* (1973) as (*Z,E*)-9,11-tetradecadienyl acetate. Because of the large size of the adult insects, sticky traps are not often used for this insect and funnel traps in various forms are preferred. A funnel connected to a plastic bag, which acts as a collecting device, is most commonly used. The funnel usually has a cover at about 3 cm above it to keep out dirt and rain water; the pheromone lure is usually located below this cover and hangs centrally between the lid and the broad end of the funnel (Fig. 10.4).

Large-scale trapping using funnel traps was started in Israel in 1975

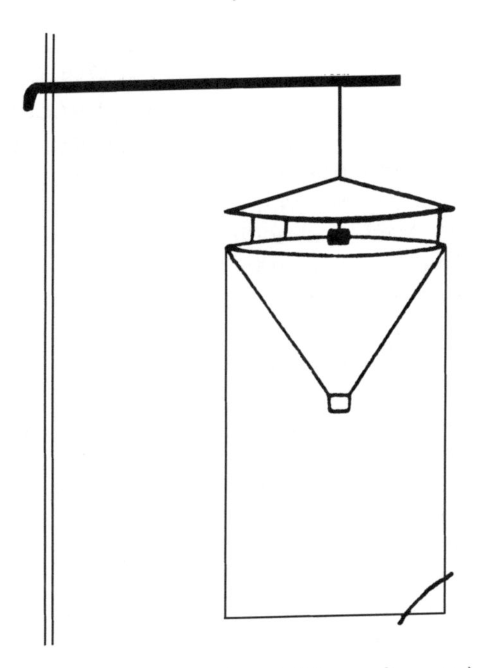

**Fig. 10.4** Funnel trap: a yellow plastic funnel (17 cm diameter) with a hood of the same material at a clearance of 2.5–3.0 cm. The pheromone source is suspended below the hood at the centre, and moths are caught in the polythene bag wired below the funnel.

(Teich *et al.*, 1979) using one trap per 1.7 ha. At this density they demonstrated a reduction in egg clusters of 40–50% and a corresponding reduction in insecticide treatments of 20–25% (Shani, 1982). The areas thus treated by mass trapping amounted to 2500 ha in 1975, 12 000 ha in 1980 and 20 000 ha in 1981. With a density of one trap per 0.5–0.6 ha, the results from Israel looked promising. Similar attempts to use the mass trapping technique on Crete and in Egypt were not always so successful (Campion and Nesbitt, 1981). These authors tried densities of traps ranging from three to 27 traps $ha^{-1}$ but found that the trap catch did not increase significantly with more than nine traps $ha^{-1}$. With a density of three to nine traps $ha^{-1}$, it was possible to reduce the number of egg masses in some areas while in others no significant reductions could be detected.

Some of the variable results obtained by Campion and Nesbitt in Egypt and Crete could be due to varying population density, which, as we have seen, is one of the key factors limiting the use of mass trapping as a technique. In those species that, by the nature of their life cycles and ecology, exist naturally at low numbers, the technique of mass trapping stands a much better chance of success. This is well illustrated by the work done on the cocoa pod borer moth, *Conopomorpha cramerella*, in Malaysia during the mid 1980s (Beevor *et al.*, 1993).

Cocoa pod borer is a relatively low density pest, with limited dispersal ability. It has virtually no alternative host in commercial cocoa plantations and earlier observations with catches of this insect in sex pheromone-baited traps had indicated a fall-off in catch with time, suggesting that males were being trapped out in the vicinity of the trap. As control methods for this pest are not always effective and tend to be labour intensive (Mumford and Ho, 1988), control through mass trapping in sex pheromone-baited traps was attempted. A 200 ha site on Sabah was mass trapped over a four-year period (1984–88) with maximum trap densities of 16 traps ha$^{-1}$. The moth catches declined significantly over two weeks following the start of the experiment; the male : female ratio changing from 1 : 1 to 1 : 5.8 and trap catches remained low thereafter. Damage to pods in the mass trapped area decreased steadily over the period of the trial while damage to pods in an untreated area remained high or showed some slight reduction.

The cost of pheromone trapping on an estate basis was about £35/ha per annum in 1990 but with the collapse of cocoa prices in the late 1980s and early 1990s the technique was not thought cost effective. However, with cocoa prices now recovering it will be interesting to see if pheromone-based mass trapping will once again be used for integrated pest management of the cocoa pod borer.

Another factor limiting both mass trapping and mating disruption (Chapter 12) is the immigration of mated females from outside the area being treated. This problem reduces in importance if the technique of mass trapping or mating disruption is carried out in an area where movement of individuals into, and out of, the population is substantially reduced. Stored product moths such as the Mediterranean flour moth, *Ephestia kühniella*, provide good examples of pest management techniques with pheromones in enclosed environments.

The pheromone of *E. kuhniella* was identified in the early 1980s as a blend of (Z,*E*)-9,12-14:Ac, (Z,*E*)-9,12-14:Alc and (Z)-9-14:Ac (Brady *et al.*, 1971a,b; Brady, 1973; Kuwahara and Casida, 1973; Sower *et al.*, 1974. However, for most trapping work on this species the major component, (Z,*E*)-9,12-14:Ac, alone is used. Indications that this species could be controlled by mass trapping appeared in the early 1980s (Levinson and Buchelos, 1981) but since that time considerable efforts have been made to perfect the technique.

Trematerra and Battaini (1987) demonstrated that an IPM approach for controlling *E. kuhniella* in flour mills was possible through a combination of mass trapping and a limited use of insecticides. Funnel traps were used, baited with 2 mg of the major component. In a subsequent trial, Trematerra (1988) showed that similarly baited funnel traps were capable of maintaining a population of *E. kuhniella* in a large flour mill at below economic levels for one year. A trap density of one per 260–280 $m^3$ was used in this trial. Additional traps were placed outside the mill to reduce the chances of re-infestation. A population reduction of 95–97% was achieved though the mill still carried out one major fumigation and some localized treatments with insecticides during the year. However, the trapping was continued at the same location for a further two years and a progressive decline was obtained in populations of 26% and 16% when compared with the first-year moth densities (Trematerra, 1990) (Fig. 10.5). Trematerra noted, however, that the technique did not eradicate *E. kuhniella* from the flour mill concerned and that the level of population reduction was achieved at least in part by more fastidious cleaning of the mill and its machinery, the latter being the harbourages that allow the insect to survive and re-infest. Other secondary measures of the success of the treatment included a reduction in damaged products in all departments of the mill and a reduction in insect fragments as measured by filth-tests on milled products.

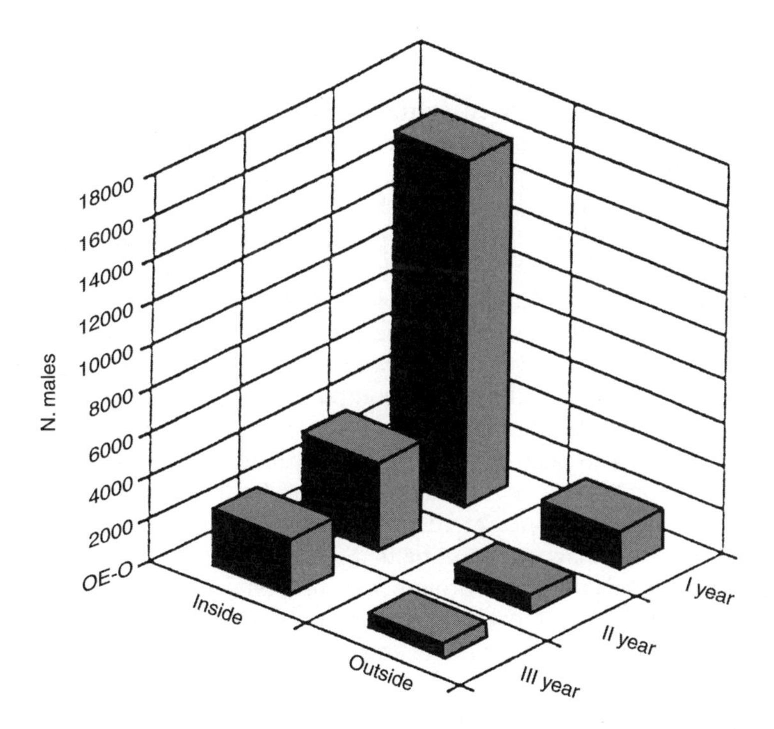


**Fig. 10.5** Dynamic population of *Ephestia kuehniella* Zeller males inside and outside flour mill during three years of mass trapping.

## 10.4 TEPHRITID FRUIT FLIES

The tephritid fruit flies constitute a major group of phytophagous Diptera which cause widespread damage and economic loss to fruit crops in the tropics and subtropics. They all belong to the family *Tephritidae* (also called *Tripetidae* in earlier literature) and number at least 3500 species, with countless subspecies and local variations in form. Fewer than 1% of these species account for 98% of the damage caused to fruit crops. Notable amongst the list of noxious species are the Mediterranean fruit fly, *Ceratitis capitata*, the olive fly, *Bactrocera* (= *Dacus*) *oleae*, the oriental fruit fly, *Bactrocera* (= *Dacus*) *dorsalis*, and many New World fruit flies of the genus *Anastrepha*. The adults of all these species damage the fruit when they oviposit, causing blemishes and discolorations of the fruit surface. This may also result in biochemical or physiological changes in the fruit at the oviposition site which can produce callus tissue to varying degrees. On hatching, the larvae bore into the fruit tissue and open the way for bacterial and fungal pathogens as secondary invaders. These in turn cause rapid and extensive necrosis and complete loss of the fruit.

### 10.4.1 Attractants

Fruit flies have always proved extremely difficult to control effectively, whilst maintaining environmental integrity using insecticides, because of the need to use large quantities of insecticides targeted mostly at the adult stages. Consequently, the search for effective attractants for these adult flies has long been given high priority amongst researchers. Over the years, this has led to the discovery and deployment of a range of parapheromones (Payne *et al.*,1973) Most of the parapheromones in use today are male attractants and include trimedlure, methyleugenol and 'Cue-lure' for *Ceratitis capitata, Bactrocera dorsalis* and *B. cucurbitae,* respectively (Fig. 10.6). All of these compounds are used extensively in detection and monitoring traps for these insects (Chapter 9) and in some cases have also been used in direct population suppression.

Parapheromones have not been developed for temperate zone fruit flies such as *Rhagoletis* spp. or the neotropical *Anastrepha* complex. For these species, proteinaceous food baits (Steiner, 1953) have to be used or compounds associated with protein degradation, such as ammonium salts and urea. These substances attract both males and females, are generally short range in their effect and sometimes also attract beneficial insects.

Although sex pheromones are known to exist in many tephritid fruit fly pests, only in the olive fruit fly, *Bactrocera* (= *Dacus*) *oleae,* have the chemicals involved been characterized to a sufficient degree that they

**Fig. 10.6** Male lures for tephritid fruit flies: (a) methyl eugenol; (b) cue-lure; (c) trimedlure.

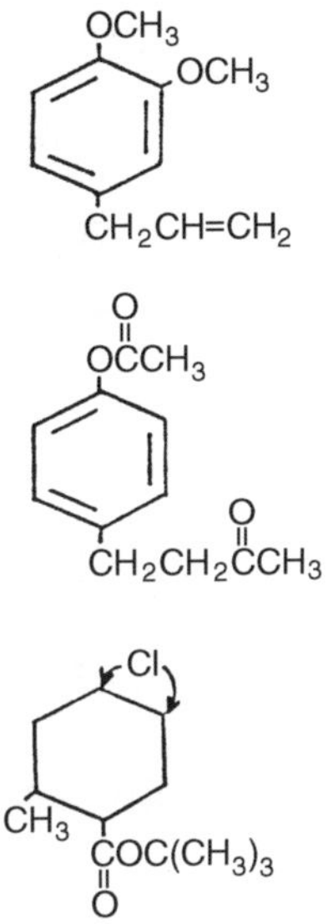

can be used in the field for monitoring and control purposes (Baker *et al.*, 1980, Jones *et al.*, 1983; Mazomenos and Haniotakis, 1985).

### 10.4.2 Traps

Traps have always been important tools in fruit fly management and the literature on the subject is extensive (Chambers, 1977, among others, for reviews). Boller (1982) neatly summarized the different characteristics of traps used for monitoring and control of fruit fly populations (Table 10.5). In essence, he states that trapping systems should be developed in accordance with their intended uses – exploiting, in doing so,

**Table 10.5** Characteristics of fruit fly traps used for monitoring and control

| *Trap type* | *Purpose* | *Characteristics* | *Example* |
|---|---|---|---|
| Monitor | Ecological studies | Weak; spot trapping | Visual traps |
| | Detection | Strong; wide-range | Olfactory traps |
| | Economic thresholds | Intermediate power | |
| Control | Male or female annihilation | Strong | Combination of all stimuli in super-trapping systems |
| | Suppression | Strong | |

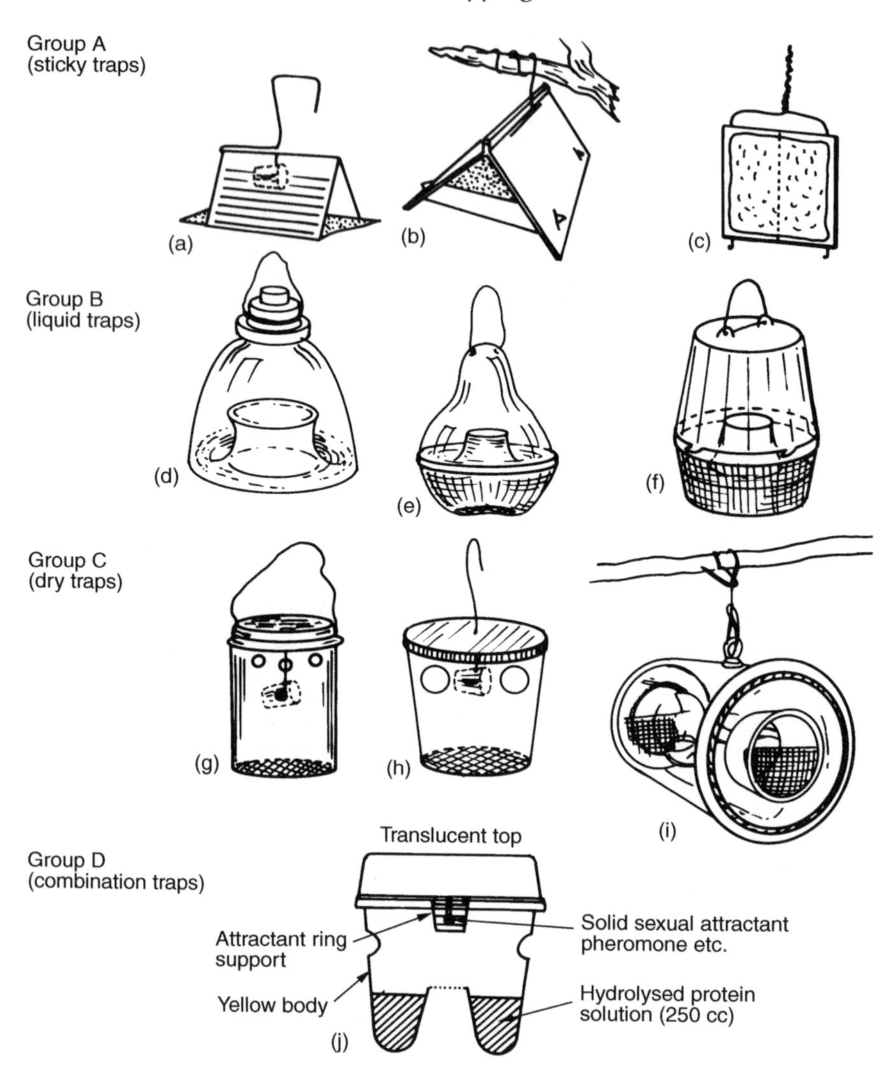


**Fig. 10.7** Various trap designs used to monitor and mass-trap tephritid fruit flies. Group A–sticky traps: (a) Jackson trap, with sticky insert and Trimedlure Plug bait; (b) Delta trap, usually yellow with a replaceable sticky base and baited with a Trimedlure Plug in a basket as in the Jackson trap; (c) vertical yellow sticky trap which can be baited as in the case of *Bactrocera oleae* with a sex pheromone. Group B–liquid traps, all usually baited with a protein hydrolysate or an ammonium salt solution: (d) glass McPhail trap; (e) plastic McPhail trap where the base is coloured yellow; (f) plastic McPhail trap where the whole trap is transparent. Group C–dry traps, all baited with Trimedlure Plugs and DDVP as a killing agent: (g) Nadel trap, Spanish version; (h) Nadel trap, Hawaiian version; (i) Steiner trap. Group D–combined liquid and Trimedlure trap: (j) Tephri trap.

particular properties of the trap design or attractant used. For instance, visual traps with their limited range of attraction are highly suited for ecological studies where the objective is spot-sampling of populations and not an area-wide collection of individuals.

Attractant-baited traps with potent lures, on the other hand, are suitable both for detection and male/female annihilation. Figure 10.7 shows a number of trap designs that have been used to date for both monitoring and, in some cases, control purposes of tephritid fruit flies.

It has long been observed that tephritid fruit flies are attracted to the colour yellow and detailed studies of the exact spectral sensitivity of *Ceratitis capitata, Bactrocera oleae* and *Rhagoletis cerasi,* such as those carried out by Agee *et al.* (1982), can reveal the exact shade of yellow that provides the optimal reflectance characteristics for any particular species of fruit fly. The above three named species showed a major sensitivity peak at 485–500 nm (green region) and a secondary peak at 365 nm (ultraviolet region). Only *R. cerasi* was shown to discriminate strongly between different shades of yellow with different reflectance characteristics when trapping experiments were conducted in the field.

### 10.4.3 Mass trapping

Optimization of visual trapping systems in terms of their spectral reflectance led researchers to evaluate them as a means of population suppression. Table 10.6 shows data from Switzerland on the area-wide use of yellow sticky traps to control *Rhagoletis cerasi* in commercial cherry orchards. As can be seen from the data, a high density of traps (one to five traps per tree) had to be used in order to achieve any population suppression.

In the olive fly, *B. oleae,* where olfactory attractants have been described for both males and females, the possibility of combining the visual attraction to yellow sticky traps with the response of the males to the female-produced sex pheromone, 1,7 dioxaspiro (5,5) undecane, and the female's response to ammonium salts has been investigated as a population suppression mechanism (Broumas *et al.,* 1982). Positive

**Table 10.6** Suppression of cherry fruit fly populations in Switzerland by mass-trapping with yellow visual traps (from Boller, 1982)

| *Year* | *Township* | *Cherry trees in trap area* | *No. of traps applied* | *Infestation level* | *Estimated effectiveness* |
|---|---|---|---|---|---|
| 1970 | Hersberg | 397 | 992 | 3.6% | 75% |
| 1971 | Nuglar | 1200 | 1680 | 0.0% | 100% |
| 1976 | Murenberg | 850 | 3500 | 0.24% | > 90% |

results were obtained in these early experiments in areas with 10 000 olive trees, but the system suffered from a number of the limitations noted at the beginning of this chapter: namely, the need for a high density of traps, their limited catching capacity (especially in areas with high populations) and increasing concerns about the non-selectivity of the traps when it comes to beneficial insects, such as hymenopteran parasites and egg predators, e.g. *Chrysoperla carnea* (Neuenschwander, 1982).

Many of these problems were overcome by abandoning sticky traps and turning instead to 'target' devices which were treated with insecticides to kill flies once they made contact with the device. In this way, problems of trap saturation are overcome because the insect, having picked up a lethal dose of insecticide from the target device, then flies or walks away from it until the toxic effects of the insecticide manifest themselves. Similarly, the use of targets with grey, green or brown colours reduces the attraction of beneficial insects and thus conserves their populations in the olive grove. This technique of lure and kill is dealt with in greater detail in Chapter 11.

## REFERENCES

Agee, H.R., Boller, E.F., Remund,U. *et al.* (1982) Spectral sensitivities and visual attractant studies on the Mediterranean fruit fly, *Ceratitis capitata* (Wiedemann), olive fly, *Dacus oleae* (Gmelin), and the European cherry fruit fly, *Rhagoletis cerasi* (L.). *Z. ang. Ent.*, **93**, 403–412.

Baker, R., Herbert, R., Howse, P.E. *et al.* (1980) Identification and synthesis of the major sex pheromone of the olive fly (*Dacus oleae*). *J. Chem. Soc. Chem. Commun. No. 1106*, 52–53.

Bakke, A. (1981) The utilisation of aggregation pheromone for the control of the spruce bark beetle. In *Insect Pheromone Technology; Chemistry and Applications* (eds B.A. Leonhardt and M. Beroza), ACS Symposium Series no. 190, Washington, DC, pp. 210–229.

Bakke, A. and Kvamme, T. (1981) Kairomone response in *Thanasimus* predators in pheromone components of *Ips typographus*. *J. Chem. Ecol.*, **7**, 303–312.

Bakke, A. and Lie, R. (1989) Mass trapping, In *Insect Pheromones in Plant Protection* (eds A.R. Jutsum and R.F.S. Gordon, John Wiley & Sons, Chichester, pp. 67–87.

Bakke, A., Froyen, P. and Skattebol, L. (1977) Field response to a new pheromonal compound isolated from *Ips typographus*. *Naturwissenschaften*, **64**, 98.

Bakke, A., Sether, T. and Kvamme, T. (1983) Mass-trapping of the spruce bark beetle *Ips typographus*. Pheromone and trap technology. *Medd. Nor. Inst. Skogforsk.*, **38**, 1–35.

Beevor, O.S., Mumford, J.D., Shah, S. *et al.* (1993) Observations on pheromone-baited masstrapping for control of cocoa pod borer, *Conopomorpha cramerella*, in Sabah, East Malaysia. *Crop Protection*, **12**, 134–140.

Boller, E.F. (1982) Biotechnical methods for the management of fruit fly populations. In *Fruit Flies of Economic Importance* (ed. R. Cavalloro), A.A. Balkema, Rotterdam, pp. 342–352.

Borden, J.H. (1982) Aggregation pheromones. In *Bark Beetles in North American Conifers: a System for the Study of Evolutionary Biology* (eds J.B. Milton and K.E. Sturgeon, University of Texas Press, Austin, pp. 74–139

Borden, J.H. (1989) Semiochemicals and bark beetle populations: exploitation of natural phenomena by pest management strategists. *Holarctic Ecology,* **12,** 501–510.

Borden, J.H. (1990) Use of semiochemicals to manage coniferous tree pests in western Canada. In *Behaviour-Modifying Chemicals for Insect Management: Applications of Pheromones and Other Attractants.* (eds R.L. Ridgeway, R.M. Silverstein and M.N. Inscoe, Marcel Dekker Inc., New York, pp. 281–315.

Brady, U.E. (1973) Isolation, identification and stimulatory activity of the second component of the sex pheromone system (complex) of female almond moth, *Cadra cautella. Life Sci.,* **13,** 227–235.

Brady, U.E., Nordlund, D.A. and Daley, R.C. (1971a) The sex stimulant of the Mediterranean flour moth *Anagasta kuehniella. J. Ga. Entomol. Soc.,* **6(4),** 215–217.

Brady, U.E., Tumlinson, J.H., Brownlee, R.B. and Silverstein, R.M. (1971b) Sex stimulant and attractant of the Almond Moth. *Science,* **171,** 802–804.

Broumas, Th., Katsoyannos, P., Yamvrais, C. *et al.* (1982) Control of the olive fruit fly in a pest management trial in olive culture. In *Fruit Flies of Economic Importance* (ed. R. Cavalloro), A.A. Balkema, Rotterdam, pp. 584–591.

Burke, S. (1992) Use patterns and non-target effects of forest coleopteran semiochemicals. In *Insect Pheromones and other Behaviour-modifying Chemicals: Applications and Regulation. Proceedings of a Symposium held at the Brighton Crop Protection Conference–Pests and Diseases, 19th November, 1990* (eds R.L. Ridgway, M. Inscoe and H. Arn, BCPC Monograph No. 51, pp. 60–78.

Campion, D.G. and Nesbitt, F. (1981) Recent advances in the use of pheromones in developing countries with particular reference to mass-trapping for the control of the Egyptian cotton leaf worm *Spodoptera littoralis* and mating disruption for the control of pink bollworm *Pectinophora gossypiella.* In *Les Mediateurs Chimiques Agissant sur le Comportement des Insectes,* Institut National de la Recherche Agronomique, Paris, pp. 335–342.

Chambers, D.L. (1977) Attractants for fruit fly survey and control. In *Chemical Control of Insect Behaviour: Theory and Application,* (eds H.H. Shorey and J.J. McKelvey, Wiley-Interscience, New York, pp. 327–344.

Chinchilla, C.M., Oehlschlager, A.C. and Gonzalez, L.M. (1993) Management of red ring disease in oil palm through pheromone-based trapping of *Rhynchophorous palmarum* (L.). *PORIM International Palm Oil Congress, 20–25 September, 1993, Kuala Lumpur, Malaysia.*

Eidmann, H.H. (1983) Management of the spruce bark beetle *Ips typographus* in Scandinavia using pheromones. *Proc. 10th Int. Congress Plant Protection,* **3,** 1042–1050.

Gmelin, J.F. (1787). *Abhandlung uber die Wurmtrocknis,* Verlag d. Crusiusschen Buchhandlung, Leipzig, 176 pp.

Gray, D.R. and Borden, J.H. (1989) Containment and concentration of mountain pine beetle (Coleoptera: Scolytidae) infestations with semiochemicals: validation by sampling of baited and surrounding zones. *J. Econ. Ent.,* **82,** 1399–1405.

Hardee, D.D. (1982) Mass trapping and trap cropping of the boll weevil *Anthonomus grandis* Boheman. In *Insect Suppression with Controlled Release Pheromone System,* Vol.II. (eds A.F.Kydonius and M. Beroza, CRC Press, Boca Raton, Florida, pp. 65–72.

Jones O.T., Lisk, J.C., Longhurst, C. *et al.* (1983) Development of a monitoring trap for the olive fly, *Dacus oleae* (Gmelin)(Diptera, Tephritidae) using a component of its sex pheromone as lure. *Bull. Ent Res.*, **73**, 97–106.

Kuwahara, Y. and Casida, J.E. (1973) Quantitative analysis of the sex pheromone of several phycitid moths by electron capture gas chromatography. *Agric. Biol. Chem.*, **37**, 681–684.

Ladd, T.L. and Klein, M.G. (1986) Japanese beetle (Coleoptera: Scarabididae) response to color traps with phenethyl propionate + eugenol + geraniol (3 : 7 : 3) and japonilure. *J. Econ. Entomol.*, **79**, pp. 84–86.

Levinson, H.Z. and Buchelos, C.Th. (1981) Surveillance of storage moth species (Pyralidae, Gelechiidae)) in a flour mill by adhesive traps with notes on the pheromone-mediated flight behaviour of male moths. *Z. and Ent.*, **92**, 233–251.

Lingren, P.D., Sparks, A.N., Raulson, J.R. and Wolf, W.W. (1978) Applications of nocturnal studies of insects. *Bull. Ent. Soc. Amer.*, **24**, 206–212.

Mazomenos, B.E. and Haniotakis, G.E. (1985) Male olive fruit fly attraction to synthetic sex pheromone components in laboratory and field tests. *J. Chem. Ecol.*, **11**, 397–405.

Mumford, J. D. and Ho, S.H. (1988) Control of the cocoa pod borer (*Conopomorpha cramerella*). *Cocoa Grower's Bull.*, **40**, 19–29.

Nagel, R.H., McComb, D. and Knight, F.B. (1957). Trap tree method for controlling the Englemann spruce beetle in Colorado. *J. For.*, **55**, 894–898.

Nesbitt, B.F., Beevor, P.S., Cole, R.A. *et al.* (1973) Sex pheromones of two noctuid moths. *Nature New Biol.*, **244**, 208–209.

Neuenschwander, P. (1982) Beneficial insects caught by yellow traps in mass trapping olive fly, *Dacus oleae. Ent. Exp. et Appl.*, **32**, 286–296.

Oehlschlager, A.C., Chinchilla, C.M., Gonzalez, L.M. *et al.* (1993) Development of a pheromone based trapping system for the American palm weevil, *Rhynchophorous palmarum. J. Econ. Ent.*, **86**, 1381–1392.

Payne, T.L., Shorey, H.H. and Gaston, L.K. (1973) Sex pheromones of Lepidoptera XXXVIII. Electroantennogram responses in *Autographa californica* to *cis*-7-dodecenyl acetate and related components. *Annals of the Entomological Society of America*, **66**, 703–704.

Quartey, G.K. and Coaker, T.C. (1992) The development of an improved model trap for monitoring *Ephestia cautella. Entomol. Exp. Appl.*, **64**, 293–301.

Ridgway, R.K. and Inscoe, M.N. (1992) Insect behaviour-modifying chemicals: practical applications in the United States. In *Insect Pheromones and Other Behavior-Modifying Chemicals: Applications and Regulation. Proceedings of a Symposium held at the Brighton Crop Protection Conference – Pests and Diseases, 19th November, 1990* (eds R.L. Ridgway, M. Inscoe and H. Arn), BCPC Monograph No. 51, pp. 19–28.

Ridgway, R.L., Inscoe, M.N. and Dickerson, W.A. (1990) Role of the boll weevil pheromone in pest management. In *Behaviour-Modifying Chemicals for Insect Management: Applications of Pheromones and Other Attractants* (eds R.L. Ridgway, R.M. Silverstein and M.N. Inscoe), Marcel Dekker, Inc., New York, pp. 437–471.

Roelofs, W.L., Glass, E.H., Tette, J. and Comeau, A. (1970) Sex pheromone trapping for red-banded leaf roller control: theoretical and actual. *J. Econ. Ent.*, **63**, 1162–1167.

Shani, A. (1982) Field studies and pheromone application in Israel. Paper presented at the 3rd Israeli meeting on pheromone research, May 4, 1982, pp. 18–22.

Sower, L.L., Vick, K.W. and Tumlinson, J.H. (1974) (*Z*,*E*)-9,12-tetradecadien-1-ol: a chemical released by female *Plodia interpunctella* that inhibits the sex pheromone response of male *Cadra cautella*. *Environ. Entomol.*, **3**, 120–122.

Steiner, L.F. (1953) Fruit fly control in Hawaii with poison-bait strays containing protein hydrolysate. *J. Econ. Entomol.*, **45**, 838–843.

Teich, I., Neumark, S., Jacobson, M. *et al.* (1979) Mass trapping of males of Egyptian cotton leafworm (*Spodoptera littoralis*) and large-scale synthesis of prodlure. In *Chemical Ecology: Odour Communication in Animals*, (ed. F.J. Ritter), Elsevier, Amsterdam, pp. 343–350.

Trematerra, P. (1988) Suppression of *Ephestia kuehniella* Zeller by using a mass-trapping method. *Tecnica molitoria*, **18**, 865–860

Trematerra, P. (1990) Population dynamic of *Ephestia kuehniella* Zeller in flour mill: three years of mass-trapping. *Proc. 5th Int. Working Conf. stored-product prot.*, Vol III, Bordeaux, pp. 1435–1443.

Trematerra, P. and Battaini, F. (1987) Control of *Ephestia kuehniella* Zeller by mass-trapping. *J. Appl. Ent.*, **104**, 336–340.

Tumlinson, H.H., Hardee, D.D., Gueldner, R.C. *et al.* (1969) Sex pheromones produced by male boll weevil: isolation, identification, and synthesis. *Science*, **166**, 1010–1012.

Vite, J.P. and Francke, W. (1976) The aggregation pheromones of bark beetles: progress and problems. *Naturwissenschaften*, **63**, 550–555.

# 11
# Lure and kill

The last chapter discussed the use of semiochemical-based traps for population control through mass trapping. This theme will now be continued but 'lure and kill' techniques differ in one respect: the insect, once attracted by the semiochemical lure, is not 'entrapped' at the source of the attractant by adhesive, water or any other physical device. Instead, the insect is subjected to a killing or sterilizing agent, which effectively eliminates it from the population. This technique has been variously descried as 'lure and kill', 'attracticide' or 'attraction–annihilation' (Lanier, 1990), though the latter concept would also include mass trapping as described in Chapter 10.

## 11.1 GENERAL PRINCIPLES

The technique consists essentially of two components: the **lure**, which could consist of odours, visual cues or a combination of both, and an **affector**, which eliminates the attracted insect from the population. Table 11.1 describes the various lures and affectors that have been used to date. The odour sources that have been used could be a semiochemical such as a pheromone, a kairomone, an apneumone produced by a non-living substance such as a food or oviposition substrate, or an empirically derived attractant that has been found by screening candidate chemicals as attractants but that plays no known role in nature. **Affector** substances that have been used to date include conventional insecticides, insect growth regulators, sterilants or pathogenic organisms such as bacteria, viruses, fungi etc.

## 11.2 TEPHRITID FRUIT FLIES

The use of olfactory attractants in combination with insecticides has been a well established practice since the 1950s in the case of tephritid fruit flies such as the Mediterranean fruit fly, *Ceratitis capitata*, and the olive fly, *Bactrocera oleae*. The use of a protein hydrolysate–insecticide

**Table 11.1** Lures and affectors employed in attraction–annihilation (after Lanier, 1990)

**(a) Lures**

| *Type of lure* | *Examples* |
|---|---|
| Semiochemicals | |
| Pheromones: intraspecific attractants | |
| Kairomones: odours of a host or prey | |
| Apneumones: odours from non-living substances | Carrion attracts flies |
| Empirical attractants | Methyl eugenol attracts fruit flies (but its role in nature is uncertain) |
| Light source | UV attracts moths |
| Colours | Yellow attracts aphids and whiteflies |
| Objects | Large dark objects attract Tabanidae |

**(b) Affectors**

| *Mechanism* | *Examples* |
|---|---|
| Killing | Traps |
| | Sticky surface |
| | Container with restricted exit |
| | Container with insecticidal vapour |
| | Flight barrier (e.g. glass) over water or oil tray |
| | Electric grid |
| | Insecticide-treated surface |
| Indirect | Dislocation of insects from crop or material to be protected; pest succumbs to exhaustion and natural mortality factors |
| | Sterilization of insects |
| | Contamination with pathogens |
| | Harvesting or sanitation of trap crop |

mixture in a 'lure and kill' strategy allows the operator to spray a reduced surface area of crop, relying on the attractant to bring the insect to the treated area. Protein hydrolysates are used in many fruit-growing areas of the world for fruit fly control and have undoubtedly led to reduced environmental contamination by insecticides. Being a proteinaceous material, the hydrolysates are highly attractive to most insects, both good and bad, and as a consequence their use can significantly reduce natural enemy populations in the crops thus treated. The availability of species-specific semiochemical attractants has overcome this problem in many cases and some notable examples of their successful application are given below.

Lure and kill strategies for tephritid fruit flies fall into two groups: those that employ some form of **target device**; and those that rely on attracting the insect to a natural surface (e.g. host tree foliage) that has been treated with an attractant/insecticide mixture. This second technique will be referred to as **sprayable formulations**.

### 11.2.1 Target devices for male annihilation

Of all the parapheromones described in Chapter 10 for tephritid fruit flies, methyl eugenol stands out as being the most potent. Fibrous blocks containing methyl eugenol and an insecticide such as Naled™ (dibrom) have been used as target devices for male oriental fruit flies, *B. dorsalis*, in many eradication programmes. The first successful attempt at eradication of *B. dorsalis* by male annihilation was made by Steiner *et al.* (1965) on the island of Rota in the Mariana Islands. They used 5% by weight of an insecticide in methyl eugenol to saturate 5 cm fibreboard squares so that each held 24 g of the mixture. These fibreboard squares were then thrown out of an aircraft over uninhabited areas at given rates and re-application intervals to achieve eradication in about six months.

This technique was then adopted by the Japanese government in an ambitious eradication campaign against oriental fruit fly in the Ogasawara Islands. Over a period of 10 years, they were able to eradicate the fly from all the islands in the archipelago from Amami in the north to Okinawa in the south (Koyama *et al.*, 1984).

The success of this technique was undoubtedly based on the fact that methyl eugenol is very attractive indeed to *B. dorsalis* and is capable of attracting a sufficiently high percentage of males to the insecticide-treated target devices to leave a high percentage of the females unfertilized. None of the other parapheromones has been shown to produce the same effect when used alone as the lure; in most cases some other attractant has also been used which attracts females.

### 11.2.2 Target devices for male and female annihilation

The olive fruit fly, *Bactrocera (Dacus) oleae*, has been mentioned in the previous chapter as a species that can be controlled, at low population levels at least, using yellow sticky traps baited with sex pheromones and food attractants. However, such traps suffer from problems of trap catch saturation and detrimental effects on beneficials. Work in Greece over the last 10 years has been aimed at overcoming these problems through the development of target devices that carry an insecticide for killing the attracted flies instead of adhesives. The target devices used in most of the large-scale trials undertaken in Greece consisted of plywood rectangles (15 × 20 × 0.4 cm) dipped into appropriate concentrations of the pyrethroid insecticide Deltamethrin for a sufficiently long period of time to saturate the wood with the insecticide solution. These devices were baited with ammonium salt dispensers (food attractant) and one target in three or five was also baited with a sex pheromone dispenser.

Fewer sex pheromone dispensers than food attractants were thought necessary since their distance of attraction was shown in earlier experiments to be 60 – 80 m while that for the food attractants was only 15 – 20 m. Great logistical problems had to be overcome during the installation period of the devices in June and July, especially in years where over 2 million trees were treated. As the controlled release devices for the food and sex attractants became more advanced, it was possible to install the target devices during early summer and reasonably expect them to last until late autumn when olive fly populations reduce in importance because of decreasing temperatures and harvesting of the olives. Young agriculturalists monitored the effectiveness of the target devices throughout the periods of operation and they also monitored olive fly populations by the use of traps and by taking samples of olive fruit for periodic examination of damage levels. Samples of the target devices were taken back to the laboratory to verify by bioassay that the insecticide content of the plywood boards was still sufficient to kill the fly.

The results over five years from the area-wide application of these target devices can be summarized as follows. Fly populations as measured by McPhail traps were consistently lower in target-device treated areas compared with conventionally treated control areas. The average number of bait sprays that had to be used in target treated areas during the early years of the programme (1984/1985) was one as opposed to 2.5. No supplementary bait sprays were required in later years in the target-device treated areas. In most years, fruit infestation was lower than or equal to that in the controls where bait sprays were applied (Haniotakis *et al.*, 1991).

The target device method of controlling *B. oleae*, therefore, was very effective as a method of eliminating insecticide bait sprays, and significant increases in beneficial insect numbers were observed in target-device treated areas (Paraskakis, 1989). However, for the method to work to its greatest effect, it has to be applied on a large area. In small plots, large-scale adult movements over short distances can significantly override the effects of the devices. Similarly, when the system fails to contain pest populations, complementary measures are almost invariably required, significantly affecting the cost effectiveness of the technique. This method of managing *B. oleae* populations will nevertheless be pursued because legislative and environmental pressures will eventually restrict the broad-scale use of bait sprays.

### 11.2.3 Sprayable formulations

The wide-scale use of protein/insecticide bait sprays for controlling the olive fly has been mentioned several times in this chapter, together with several disadvantages to its use. Probably the most important problem with this technique is its lack of selectivity. Many important insect predators and parasites are known to be attracted by the protein hydrolysate component of the bait spray mix and this often leads to substantial reductions in their populations with continued use of this technique.

With the isolation and identification of the sex pheromone of *B. oleae*, a selective attractant became available which could substitute the protein hydrolysate. Trapping experiments showed the 1,7 dioxaspiro [5.5] undecane major component to be strongly attractive to the males (Jones *et al.*, 1983). Attempts were then made to disrupt the mating of *B. oleae* using techniques similar to those used for Lepidoptera (Chapter 12). However, instead of producing mating disruption, the wide-scale treatment of experimental olive groves with the pheromone produced only large immigrations of *B. oleae* adults – both male and female. It appeared that the pheromone, in addition to attracting the males in a clearly directed manner as seen with monitoring traps, was also attractive to females but not in such a strongly directed way. Attention was moved from trying to achieve mating disruption with the pheromone to using the pheromone as a substitute attractant in bait sprays, which were then much more selective.

A sprayable formulation of the pheromone was therefore required which would slowly release it over a period of time consistent with the effective life of the insecticide once applied in the field. The pheromone was encapsulated in polyurea-type micro-capsules and in polymer-entrapped micro-beads (both 5–10 μm in diameter) (Polycore SKL™) similar to those used for sprayable formulations of pheromones for lepidopteran mating disruption (Chapter 12). The formulated pheromone was then tank mixed with either malathion or dimethoate and applied aerially or from the ground.

In aerial sprays the plane delivers a 20 m wide swathe every 100 m of grove so that only 20% of the crop is treated. From the ground, the pheromone/insecticide mixture is applied either from a tractor-mounted sprayer, which applies the mixture to the south side of each row of

**Table 11.2** Results of lure-and-kill experiments carried out in southern Spain against *Bactrocera oleae* using a sprayable formulation of the sex pheromone and a conventional insecticide applied both from the ground and from the air

| Application | Year | Size of plot (ha) | | No. of applications | Infested fruits (mean %) | |
|---|---|---|---|---|---|---|
| | | Treated | Untreated control | | Treated plot | Control plot |
| Ground sprays | 1984 | 5 | 5 | 4 | 3.5 | 18.2 |
| | 1985 | 5 | 5 | 5 | 11.5 | 45.6 |
| Aerial sprays | 1986 | 200 | 10 | 3 | 10.2 | 43.5 |
| | 1987 | 1000 | 1000 | 3 | 4.3 | 7.0[a] |

[a] The control plot in this case was an area of olives treated with four applications of dimethoate mixed with protein hydrolysate applied in the same way as the Polycore SKL pheromone/insecticide mix.

trees; or, if a knapsack sprayer is used, only 1 $m^2$ of foliage, again on the south side of each tree, needs to be treated. Table 11.2 shows results obtained from trials using these various application methods.

## 11.3 HOUSE FLY

The common house fly, *Musca domestica*, is a cosmopolitan pest in domestic, industrial and animal-rearing premises. In animal-rearing buildings, the fly breeds in putrefying matter, especially excrement, and is particularly abundant in large intensive poultry units. In such units, large fly populations can sometimes build up which are not only a nuisance to the operators but can also lead to reduced egg production. Domestic premises located close to such units also suffer public health problems from the diseases transmitted by flies emanating from these units.

The traditional method of controlling such fly problems in animal-rearing units over the last 40 years is through the frequent application of insecticides as residual treatments or space sprays. Such concentrated use of insecticides has led to high selective pressure for the house fly populations in such units to develop resistance against the insecticides used. High levels of resistance have developed against most major groups of insecticides in countries where animals are intensively reared (Keiding, 1977). A more rational use of insecticides under such conditions is therefore required and one approach has been to reduce the total area treated with insecticides and use lure and kill formulations where an attractant brings the insect into contact with discrete amounts of insecticides. Both lure and kill baits and sprayable formulations have been developed and commercialized.

### 11.3.1 Toxic baits

Toxic baits, often called 'scatter baits' in the United States, consist of food impregnated with an insecticide and presented as a granular formulation. The insecticide is usually ingested and can overcome some of the resistance problems in that the contact toxicity of many insecticides is often reduced in resistant flies while its toxicity through oral uptake still remains high. As a result, toxic baits have remained an effective method of fly control even in flies showing multiple resistance to several classes of insecticide.

In an attempt to make such baits more effective, chemical attractants have been used to increase the number of flies that make contact with them. The house fly sex pheromone, (Z)-9-tricosene (common names Muscalure or Muscamone) (Carlson *et al.*, 1971) has also been widely studied as a chemical attractant in such bait granules. In laboratory tests, (Z)-9-tricosene was found to attract a larger number of males to the granules than did control granules with no pheromone (Uebel *et al.*, 1976; Adams and Holt, 1987) but with no significant effect on the females. In field tests, however, several authors have noted increased attraction of both males and females when the pheromone is added (Carlson and Beroza, 1973; Morgan *et al.*, 1974; Mitchell *et al.*, 1975). The precise reason why females are also attracted to (Z)-9-tricosene baits has still to be adequately resolved. Several explanations have been put forward, including the possibilities:

- that (Z)-9-tricosene is in fact an aggregation pheromone in the field (Fletcher, 1977);
- that (Z)-9-tricosene may potentiate the 'herding instinct'– Richter *et al.* (1976) reported that hydrocarbon cuticular extracts of female house flies lowered the threshold for optical cues stimulating aggregation of both sexes;
- that females may be responding to odours released by males attracted to the bait – Schlein and Galun (1984) suggested that male house flies secrete a sex pheromone from their testes and ejaculatory ducts;
- that females may be responding to a blend of (Z)-9-tricosene and the male-produced sex pheromone.

Whatever the reason for the female response, the net result is that large amounts of (Z)-9-tricosene are now being synthesised and incorporated in bait formulations such as Golden Malrin™ (based on methomyl) or Alfacron™ (based on azamethiphos). It is probably now one of the most widely used pheromones in terms of volume and was the first insect pheromone to be registered under the US Environmental Protection Agency (EPA). In late 1990, there were 10 different commercial products containing (Z)-9-tricosene registered for use in the United States (Ridgway and Inscoe, 1992).

### 11.3.2 Sprayable formulations

During studies on the best location for the toxic baits described above, Barson (1987) found that baits fixed to the walls gave better results in intensive animal-rearing units than baits scattered on the floor. This is due probably to the effects of low light intensities in animal-rearing houses which in turn influence the distribution of the flies; they tend to congregate on walls and structural supports. Formulations have therefore been developed which contain the pheromone and a knock-down insecticide that can be sprayed or painted on such surfaces directly or on wooden or plastic boards which are then attached to such surfaces. It has been shown that, when used in such a manner, the sex pheromone increases mortality of the flies on such devices by as much as 10-fold (P. Ridley, unpublished data).

## 11.4 TSETSE FLY

Tsetse flies (*Glossina* spp.) are serious pests of humans and their domesticated animals throughout sub-Saharan Africa. The tsetse fly feeds on the blood of these vertebrate hosts and, in doing so, transmits parasitic protozoan trypanosomes, the causative organism of sleeping sickness in humans and nagana in domesticated animals. The exclusively haematophagous habit of both sexes of adult tsetse and the relatively low energy value of blood as a food source means that flies must feed frequently, ingesting more than their own body weight at a single meal every few days. This behaviour, coupled with the relatively long life of individuals as adults, enhances the spread of trypanosomiasis among the vertebrate hosts upon which they feed.

Unlike most other insects that produce large numbers of eggs, the tsetse fly has developed an advanced form of adenotrophic viviparity. After mating, a mature egg is ovulated alternately from right and left ovaries from each of the paired ovarioles in each ovary in turn. Once fertilized with sperm from the single spermatotheca, the egg is retained in the oviduct and not laid. A first instar larva hatches from this and feeds on a nutritive secretion produced by highly modified accessory glands (milk glands) again within the female's body. The larva grows through feeding on this 'milk' and moults twice in the process. Eventually the female 'gives birth' to the fully grown larva and it is deposited on a suitably moist soil, where it buries itself to a depth of a few centimetres and pupates. About 30 days later a new adult emerges to complete the life cycle.

The unusual life cycle of tsetse presents one of the central problems for its control using conventional insecticides, since eggs and larvae are protected inside the parent female and the puparia are buried in the ground. However, the life cycle also means that females are capable of producing only a few offspring and any factor that puts an additional 2–4% mortality on the population can have a profound influence on population numbers. For most other insect species, the reproductive rate is 200 to 300 times higher and a much greater proportion of the population has to be killed in order to bring about effective population reduction.

The traditional method of controlling tsetse flies has involved the use of ground-spraying teams with pressurized knapsack sprayers delivering residual organochlorine insecticides to favoured tsetse resting sites on trees and bushes. The advent of aerial spraying techniques permitted the more widespread application of insecticides, but with little or no discrimination, and many non-target species were affected. Because of these problems, research was targeted at developing alternative methods of controlling this pest which were more cost effective and less harmful to the environment.

### 11.4.1 Target devices

Target devices were developed in Zimbabwe and West Africa which consisted of insecticide-coated cloth, usually about 1 $m^2$ in area, which provided a dark visual stimulus for the fly, mimicking its vertebrate host. The early version used in Zimbabwe was pivoted so that the face of the cloth was always at right angles to the wind direction, presenting the maximum visual target for flies approaching upwind. The target also had an invisible flanking net (Fig. 11.1), which was also treated with insecticide so that any flies colliding with it as they flew around the device also picked up a lethal dose of insecticide. From this design a simpler and more efficient version was developed with two flanking nets (Fig. 11.2) but without the roof (which was incorporated in the earlier design to protect the insecticide from the effects of the weather). As the efficiency of insecticide pick-up by the fly was improved over the years, the flanking nets were dispensed with and the target devices used today have only rectangular, darkly coloured cloths treated with suitable insecticides. The most attractive colour found was royal blue, followed by black; but yellows and greens were consistently unattractive (Green and Flint, 1986).

### 11.4.2 Olfactory attractants

As the design of target devices evolved, the possibility of enhancing the attraction of tsetse flies to them through the use of olfactory attractants

**Fig. 11.1** R-type target in Rifa Triangle, Zambezi.

was also investigated. The only sex pheromone described for *Glossina* spp. was a C-29 compound with contact chemostimulatory effects only and no distance attraction. Other olfactory attractants had to be found.

A series of experiments was undertaken in Zimbabwe which showed the importance of host odour in attracting tsetse flies. This involved enclosing host animals in underground pits and then extracting the air from those pits and releasing it near electrified nets which are invisible to insects but kill any that collide with them. The flies were observed to fly upwind to such an odour source (Bursell, 1987) and the number of flies attracted depended on the number of cattle enclosed in the underground pits. The possibility then existed of analysing this host odour for its chemical constituents in order to produce an artificial lure.

The method adopted was to carry out intensive chemical analysis of host odours and screen any compounds isolated in the laboratory using electrophysiological responses obtained from EAG preparations and behavioural responses from olfactometer tests as criteria for detecting

**Fig. 11.2** S-type target in Zambezi valley.

activity. Promising compounds and mixtures were then assayed in the field in Zimbabwe, using the standard capturing techniques.

Initially it was thought that the odours from host animal breath were of prime importance in the attraction of flies (Vale, 1974). Later buffalo urine was also found to be a powerful olfactory attractant for *G. pallidipes* in Kenya (Owaga, 1985). Various phenols were subsequently identified from the latter as comprising the most attractive components (Vale *et al.*, 1988a). A combination of 2 litres per minute of carbon dioxide, 5 mg of acetone per hour and 0.05 mg of 1-octen-3-ol per hour was found to be nearly as attractive as natural ox odour for *G. pallidipes* (Vale and Hall, 1985). Recent work has indicated that upwind flight in response to host odour occurs only in the presence of carbon dioxide and that acetone and octenol on their own appear to increase the responsiveness of flies to visual cues. (Torr, 1990).

Unfortunately, carbon dioxide is too expensive and inconvenient to use on a large scale in control operations. A number of the phenolic compounds are available commercially, however, and can be mixed in appropriate ratios and released with acetone and octenol in the field to provide a highly effective olfactory lure for use with target devices in control operations against tsetse fly.

### 11.4.3 Tsetse control with target devices

A number of experimental programmes have shown that tsetse control can be achieved by the deployment of target devices. Preliminary trials were carried out on isolated populations of *G. m. morsitans* and *G. pallidipes* on a 4.5 $km^2$ island in Lake Kariba, Zimbabwe (Vale *et al.* 1986). Early tests involved the use of traps to kill or sterilise the insect caught, but these were later replaced by 20 insecticide-treated targets baited with acetone and octenol. The targets killed about 2% per day of the *G. m. morsitans* and 5% per day of the *G. pallidipes*. The population declined rapidly and disappeared in 9 months and 11 weeks, respectively (Vale *et al.*, 1985). Following this, insecticide-treated targets were used to eliminate natural populations of these two species in the Rifa Triangle, an area of 600 $km^2$ in the Zambezi valley where re-invasion pressure from the neighbouring wildlife area was significant (Vale *et al.*, 1988b). Odour-baited traps and targets are now in widespread use for tsetse control in many parts of Africa.

A recent development in the use of target devices for controlling tsetse flies involves using insect sterilizing agents instead of insecticides on the target device itself. Langley and Weidhaas (1986) argued that sterilization of both sexes of tsetse in a population at a certain rate would be more effective at suppressing that population than simply killing them at the same rate. Their rationale for this goes as follows: killing females only would be as effective as killing both sexes, since it is only females that reproduce. It follows that to sterilize females only would be as effective as killing both sexes. Hence, the sterilization of males as well as females would be a bonus. They also calculated that the arc of influence of a sterilizing device should be greater than that of a target which simply kills flies. In theory also, a sterilizing device should reduce the risk of behavioural resistance developing against the device, since attracted sterilized flies would mate with unattracted individuals. The juvenile hormones and their mimics have long been known to disrupt metamorphosis in insects if applied to the larval stages. They are also able to disrupt embryogenesis and therefore prevent egg hatch. The juvenile hormone mimic, pyriproxifen, proved to be very useful as a sterilizing agent for tsetse fly females (Langley *et al.*, 1990). When a female makes contact with a target device treated with a suitable formulation of this compound it is absorbed through the female's cuticle and is transported to the larva '*in utero*'. The third instar larva appears to be normal but metamorphosis is disrupted after pupation. Only very small amounts of the insecticide pyriproxifen are required and a single treatment will ensure that the female is effectively sterilized for life. Laboratory studies have also shown that males making contact with a treated surface can transfer sterilizing doses to females when they mate (Langley *et al.*, 1990) Field trials in Zimbabwe have shown that pyriproxifen can be used as an alternative to conventional insecticides in target devices for tsetse control (Hargrove and Langley, 1990). Future research will show whether the same concept can be applied to other insect pests. If this proves to be the case, then a very exciting future lies in store for semiochemicals used in combination with juvenile hormone mimics.

## REFERENCES


Adams, T.S. and Holt, G.G. (1987) Effect of pheromone components when applied to different models on male sexual behaviour in the housefly, *Musca domestica. J. Insect Physiol.*, **33**, 9–18.

Barson, G.B. (1987) Laboratory assessment of different methods of applying a commercial granular bait formulation of methomyl to control adult houseflies (*Musca domestica* L.) in intensive animal units. *Pestic. Sci.*, **19**, 167–177.

Bursell, E. (1987) The effect of wind-borne odours on the direction of flight in tsetse flies, *Glossina* spp. *Physiological Entomology*, **12**, 149–156.

Carlson, D.A. and Beroza, M. (1973) Field evaluation of (Z)-9-tricosene, a sex pheromone of the house fly. *Environ. Entomol.*, **2**, 555–559.

Carlson, D.A., Mayer, M.S., Silhacek, D.L. *et al.* (1971) Sex attractant pheromone of the house fly: isolation, identification and synthesis. *Science*, **174**, 76–77.

Fletcher, B.S. (1977) Behavioural responses of Diptera to pheromones, allomones and kairomones. In *Chemical Control of Insect Behaviour: Theory and Application* (eds H.H. Shorey and J.J. McKelvey Jr). John Wiley and Sons, Inc., New York, pp. 129–148.

Green, C.H. and Flint, S. (1986) An analysis of colour effects in the performance of the F2 trap against *Glossina pallidipes* Austen and *G. morsitans morsitans* Westwood (Diptera: Glossinidae). *Bull. Ent Res.*, **76**, 409–418.

Haniotakis, G., Kozyrakis, M., Fitsakis, T. and Antonidaki, A. (1991) An effective mass-trapping method for the control of *Dacus oleae* (Diptera: Tephritidae). *J. Econ. Ent.* **84**, 564–569.

Hargrove, J.W. and Langley, P.A. (1990) Sterilising tsetse (Diptera: Glossinidae) in the field: a successful trial. *Bull. Ent. Res.*, **80**, 397–403.

Jones, O.T., Lisk, J.C., Longhurst, C. *et al.* (1983) Development of a monitoring trap for the olive fly, *Dacus oleae* (Gmelin) (Diptera, Tephritidae) using a component of its sex pheromone as lure. *Bull. Ent. Res.*, **73**, 97–106.

Keiding, J. (1977) Resistance in the housefly in Denmark and elsewhere. In *Pesticide Management and Insecticide Resistance* (eds D.L. Watson and A.W.A. Brown), Academic Press, New York, pp. 261–302.

Koyama, J., Teruya, T. and Tanaka, K. (1984) Eradication of the oriental fruit fly (Diptera: Tephritidae) from the Okinawa Islands by male annihilation *J. Econ. Ent.* **77**, 468–472.

Langley, P.A. and Weidhaas, D. (1986) Trapping as a means of controlling tsetse, *Glossina* spp. (Diptera: Glossinidae): the relative merits of killing and sterilization. *Bull. Ent. Res.*, **76**, 89–95.

Langley, P.A., Felton, T., Stafford, K. and Oouchi, H. (1990) Formulation of pyriproxyfen, a juvenile hormone mimic, for tsetse control. *Med. and Vet. Ent.*, **4**, 127–133.

Lanier, G.N. (1990) Principles of attraction–annihilation: mass trapping and other means. In *Behaviour-Modifying Chemicals for Insect Management* (eds R.L. Ridgway, R.M. Silverstein and M.N. Inscoe), Marcel Dekker Inc. New York and Basel, pp. 25–45.

Mitchell, E.R., Tingle, F.C. and Carlson, D.A. (1975) Effect of Muscalure on house fly traps of different color and location in poultry house. *J. Ga. Entomol. Soc.*, **10**, 169–174.

Morgan, P.B., Gilbert, I.H. and Fye, R.L. (1974) Evaluation of (Z)-9-tricosene for attractancy for *Musca domestica* in the field. *Fla. Entomol.*, **57**, 136–140.

Owaga, M.L. (1985) Observations on the efficacy of buffalo urine as a potent olfactory attractant for *Glossina pallidipes* Austen. *Insect Science and its Application*, **6**, 561–566.

Paraskakis, H.I. (1989) Results from the biological control of *Saissetia oleae* Oliv. on olive trees in Crete, Greece. In *Proceedings of the 2nd Panhellenic Congress of Entomology, 1987*, Entomological Soc. of Greece, Athens, pp. 274–284.

Richter, I., Krain, H. and Mangold, H.K. (1976) Long chain (Z)-9-alkenes are 'psychydelics' to houseflies with regard to visually stimulated sex attraction and aggregation. *Experientia*, **32**, 186–188.

Ridgway, R.L. and Inscoe, M.N. (1992) Insect behaviour-modifying chemicals: practical applications in the United States. In *Insect Pheromones and Other Behaviour-modifying Chemicals* (eds R.L.Ridgway, M.N. Inscoe and H. Arn), British Crop Protection Council Monograph No. 51, pp. 19–28.

Schlein, Y. and Galun, R. (1984) Male housefly (*Musca domestica* L.) genital system as a source of mating pheromone. *J. Insect Physiol.*, **30**, 175–177.

Steiner, L.F., Mitchell, W.C., Harris, E.J., *et al.* (1965) Oriental fruit fly eradication by male annihilation. *J. Econ. Ent.* **58**, 961–964.

Torr, S.J. (1990) Dose responses of tsetse flies (*Glossina*) to carbon dioxide, acetone and octenol in the field. *Physiol. Entomol.*, **15**, 93–103.

Uebel, E.C., Sonnet, P.E. and Miller, R.W. (1976) House fly sex pheromone: enhancement of mating strike activity by combination of (Z)-9-tricosene with branched saturated hydrocarbons. *Environ. Entomol.*, **5**, 905–908.

Vale, G.A. (1974) New field methods for studying the responses of tsetse flies (Diptera: Glossinidae) to hosts. *Bull. Ent. Res.*, **64**, 199–208.

Vale, G.A. and Hall, D.R. (1985). The role of 1-octen-3-ol, acetone and carbon dioxide in the attraction of tsetse flies, *Glossina* spp. (Diptera: Glossinidae), to ox odour. *Bull. Ent. Res.*, **75**, 209–217.

Vale, G.A., Bursell, E. and Hargrove, J.W. (1985) Catching-out the tsetse fly. *Parasitology Today*, **1**, 106–110.

Vale, G.A., Hargrove, J.W., Cockbill, G.F. and Phelps, R.J. (1986) Field trials of baits to control populations of *Glossina morsitans morsitans* Westwood and *G. pallipides* Austen (Diptera: Glossinidae) in Zimbabwe. *Bull. Ent. Res.*, **76**, 179–193.

Vale, G. A., Hall, D.R. and Gough, A.E.J. (1988a) The olfactory responses of tsetse flies, *Glossina* spp. (Diptera: Glossinidae) to phenols in the field. *Bull. Ent. Res.*, **78**, 293–300.

Vale, G.A., Lovemore, D.F., Flint, S. and Cockbill, G.F. (1988b). Odour baited targets to control tsetse flies, *Glossina* spp. (Diptera: Glossinidae) in Zimbabwe. *Bull. Ent. Res.*, **78**, 31–49.

# 12

# Mating disruption

Conventional crop protection using insecticides relies on hitting the insect pest with a lethal dose of insecticide through contact, ingestion or systemic action, at its most vulnerable stage of development. For most insect pest groups, and especially lepidopteran pests such as bollworms, budworms, leafworms, shoot borers and stem borers, this means the first instar larva. Its low body mass, coupled usually with an exposed position on the foliage following eclosion (hatching) from the egg, makes the first instar larva especially vulnerable to a relatively low dose of insecticide and particularly so with biological pesticides such as *Bacillus thuringiensis* (Mabbett *et al.*, 1980).

The success of this control strategy relies on insecticide sprays carefully synchronized and targeted at the onset of larval hatching as decided by 'scouting' returns (counts of eggs or small larva) (Mabbett and Nachapong, 1983) or as timed by pheromone traps (Chapter 9). Failure to pin-point accurately the insects pests' presence in the crop and their time of hatching leaves survivors to feed on and damage the crop. Failure is all the more likely with pests such as the pink bollworm, *Pectinophora gossypiella*, and the rice stem borers, *Chilo* and *Scirpophaga* spp., which bore into cotton bolls and rice stems, respectively immediately after hatching, thus presenting a very narrow window of control opportunity.

## 12.1 DISRUPTION OF MATING

The identification, development and commercialization of pheromones has opened the doors for a pre-emptive strategy referred to as 'mating disruption', in which egg-laying by the female is minimized or prevented by interfering with the behavioural processes that determine successful mating between male and female moths. Application timing is still important but not so critical, provided the controlled release formulations used are deployed prior to the onset of moth activity and before the crop becomes susceptible to pest damage. Once deployed, the controlled release formulation releases minute (though adequate) amounts of synthetic pheromone for weeks or even months to protect the crop from succeeding pest generations and throughout its period of susceptibility to pest damage.

Pheromones are now being used to disrupt mating in a wide range of lepidopteran pests in agriculture, horticulture and forestry and offer farmers, growers and foresters season-long, user-friendly and environment-friendly protection for their crops.

## 12.2 HOW IT WORKS

Male moths find their prospective mates by following a scent trail or pheromone plume given out by a 'calling' virgin female moth of the same species, as described in Chapter 3. The active ingredients are usually long chain and unsaturated esters, alcohols or aldehydes with more than one chemical component typically comprising the pheromone 'blend' of a particular species. Individual components may differ from each other both structurally (e.g. chain length) and geometrically as isomers or showing different positions of the double-bond(s) (Tumlinson, 1990).

The mechanisms and modes of action by which natural pheromones ensure mating, and the way pheromone control products disrupt it, are still under investigation. Dosage and distribution of point sources are generally thought to determine the variety of ways in which mating is disrupted by the use of synthetic pheromones (Weatherston, 1990).

Confusion, trail-masking and false-trail following are the three modes of action currently proposed as role players in the mating disruption process (Campion *et al.*, 1989). Confusion is caused by the constant exposure of the males to a high concentration of pheromone fog, causing adaptation of the antennal receptors and/or habituation of the central nervous system. These direct neuro-physiological effects prevent the male moth from responding to normal levels of the natural pheromone stimulus, which are defined as the amount of pheromone being emitted by calling virgin female moths (Weatherston, 1990).

Trail-masking, on the other hand, is though to happen if the natural pheromone plume is obliterated by synthetic pheromone, making trail-following impossible. False-trail following is thought to occur following the deployment or broadcasting of discrete point sources of pheromone at a sufficiently high concentration to create and present the insect with many false trails. This mode of mating disruption is heavily dependent for its success not only on the strength of pheromone emitted at and from the point source but also the number, intensity and deployment pattern of the pheromone dispensers.

Indeed, as Sanders (1981) points out, it is the chemical competition of the 'hot point sources' of synthetic pheromone with the pheromone emitted by calling females that ultimately determines the reduction in ability to locate mates and thus the success of mating disruption as an effective tool in crop damage limitation. Any increase in the ratio of virgin females to synthetic sources, whether by an influx of unmated females exploiting a deficiency in the initial deployment or by the premature exhaustion of pheromone in the dispenser, will reduce the level of mating disruption achieved.

The relative importance of these three modes of action and their role in the mating disruption process has been discussed at length by several different authors. Bartell (1982) suggests that the actual mechanism prevailing in a particular situation is determined mostly by the technique of pheromone release used, while Campion (1986) believes that the overall mode of action through which mating disruption is achieved at the practical field level may comprise one or any combination of the trio of mechanisms outlined above. Others, including Weatherston (1990). have concluded that there is a single mode of action or mechanism that operates at any one given time and that is the prominent or 'preferred' mode because of the prevailing set of conditions, including the pheromone release technique employed.

Clearly, the object of a successful mating disruption strategy is to release the pheromone from a sufficiently close number of point sources to 'cover' the whole zone in which the moths are sexually active and release it at a rate that provides an atmospheric concentration to which the insect responds. At the same time, the release profile of the pheromone active ingredients must ensure a duration of activity sufficient to protect the crop throughout that period when it is susceptible to attack and damage. In areas where crops of varying maturities are grown side by side, or where it is important to break the generation succession of a pest, mating disruption treatments may have to last beyond a crop's own specific period of damage susceptibility.

## 12.3 DELIVERY SYSTEM

The development of delivery systems which meet all the chemical, biological and agronomic requirements for economic control has taken several years to achieve. The need to synthesize all the components of multi-component pheromone blends not only specific for a particular species but also specific for species/geographical location was one of the first hurdles to overcome.

For instance, early work with the red bollworm showed its sex pheromone to consist of up to five components (Moorhouse *et al.*, 1969; Nesbitt *et al.*, 1973), while up to seven chemical components have been identified in the pheromone blends isolated from the cotton bollworms belonging to the genus *Helicoverpa* (= *Heliothis*) (Klun *et al.*, 1979;

Tumlinson *et al.*, 1982). Furthermore, the Egyptian cotton leafworm, *Spodoptera littoralis*, had two components in the pheromone blend from Crete but four and five, respectively, in those insects captured in Israel and Egypt (Campion *et al.*, 1980). Much of the early work on mating disruption, therefore, concentrated on finding out which of the many components of a pheromone blend were essential for inclusion in formulations for mating disruption.

Chemical instability presents another major problem because isomerization of double-bonds is readily caused by both photochemical (UV light) and thermal decomposition of the molecule, the resulting isomers inhibiting attraction in the case of *Pectinophora gossypiella* (Jacobson and Beroza, 1963; Bierl *et al.*, 1974) and *Earias vitella*, the spotted bollworm (Cork *et al.*, 1985).

Change in chemical structure and loss of activity may occur through oxidative decomposition, pH-induced hydrolysis and polymerization caused by metal ions, early in the manufacturing stage as well as during packaging, storage and most commonly in the field (Weatherston, 1990). These highly volatile and chemically unstable pheromone active ingredients need to be formulated in appropriate polymeric matrices which maintain the attractancy and chemical integrity of the pheromone blend for the necessary field lifetime of the formulation (Leonhardt *et al.*, 1990).

Instability of the pheromone molecule may be inherent and often related to the level of unsaturation and type of functional group in the molecule (Campion *et al.*, 1989). For instance, short chain aldehydes are usually more reactive and therefore harder to stabilize than longer chained acetates. Similarly, conjugated dienes such as (*E*,*Z*)-7,9-12:Ac, the major component of *Lobesia botrana* pheromone, or (*Z*,*E*)-9,11-14:Ac, the major component of *Spodoptera littoralis* pheromone, are also inherently unstable because of the proximity of the two double-bonds. This and other problems related to oxidative and photo-induced instability have been overcome by incorporating ultraviolet screeners and antioxidants into pheromone formulations (Hall and Maars, 1989).

In Chapter 9, the need to release a constant amount of pheromone from any particular trap design was stressed. Similarly, in devices used for the controlled release of pheromones in mating disruption, devices that release the pheromones at a constant rate (zero order kinetics) over a long period of time (Fig. 9.1) are preferable to ones that release large amounts of pheromone soon after field application and then tail off (first order kinetics). The latter is rather wasteful in terms of active ingredient but is often the only release kinetics system possible in most small point source, sprayable systems.

It is important to know also the release rates of pheromones from various dispensers under field conditions. Traditionally this has been measured by taking dispensers from the field at various time intervals following application and the quantity of pheromone remaining is then estimated gravimetrically or by gas chromatographic analysis. This, however, does not tell the user what quantities are actually released into the air from the dispenser or what concentrations of pheromones are actually achieved in the crop.

Unstable pheromones have been known to degrade on the external surfaces of dispensers and concentrations achieved within crops will depend greatly on the amount of crop foliage and the wind movement through it. Measurements of concentrations of pheromone in the crop have been made (Plimmer *et al.*, 1978; Caro *et al.*, 1980; Wiesner *et al.*, 1980; Meighen *et al.*, 1982) using chemical techniques. Recently, a new technique has been used in grapes and cotton to measure the concentration of pheromones of *Lobesia botrana* and *Pectinophora gossypiella*, respectively. The technique is based on a quantitative evaluation of electroantennogram signals recorded from a mobile, hand-held, electroantennograph which draws air over a mounted antenna of the insect concerned. Having first calibrated the system for known pheromone concentrations, measurements are then made at various points in the crop canopy and the values obtained are then read off the calibration curve (Sauer *et al.*, 1992). Such information about pheromone distribution in the crop canopy is very useful in determining the optimal distribution of dispensers.

With the physico-chemical properties related to blend ratio, stability, longevity and release rate thus achieved, it is the turn of crop and pest biology to be considered together with the agronomic requirements and limitations. For instance, where application needs to be made before the crop starts to foliate (e.g. in vines) sprayable formulations are clearly inappropriate and hand-applied dispensers that can be fixed to the branches easily and conveniently are required. Alternatively, sprayable formulations for mating disruption of rice stem borers, although they may be preferred by the farmers, are unsuitable in practice because of the absence of foliage in the early stages of growth of the crop and the difficulty of re-application in flooded rice. In such cases, hand-applied devices are the only option. Sprayable formulations are susceptible to weathering from rainfall or overhead irrigation. Sprayable formulations for the pink bollworm are therefore not suitable in areas which rely on overhead irrigation, but hand-applied devices – such as Selibate PBW rings placed over the central growing point of the plant in early season, where they stay secure as the crop grows – are often a better option.

Sprayable formulations are clearly versatile. They represent the most sensible and economic option when mating disruption of a specific pest is part of an integrated control programme aimed at a pest complex, some of which can only be controlled by spraying insecticides. Tank mixing of pheromone with insecticide can then be utilized. This practice has found favour in managing pink bollworm populations in the desertic southwest cotton growing areas of the United States. In addition, sprayable formulations offer the option of aerial application to very large areas of crop in which labour would be insufficient or too expensive for other forms of pheromone deployment. However, hand-applied dispensers such as 'twist-tie' ropes or the 'double ampoule' dispenser, which can be loaded with several different pest-specific pheromones, are now commercially available, thereby bringing some of the versatility and convenience of sprayables to the more environmentally secure hand-applied dispenser. In addition, polymeric matrix systems such as Selibate™ can be extruded into a variety of user-friendly shapes appropriate to the requirements of the crop and farmer, including rings for cotton, tubes and 'shoelace' strings tied to stakes and deployed above rice crops.

Figure 12.1 shows the variety of formulation types that have been developed in response to these varying needs. For mating disruption these include the widely spaced dispensers, usually deployed by hand and each releasing a relatively large amount of pheromone (e.g. twist-tie ropes, trilaminate ribbons and tapes and polymer rings and ties), and those that involve the application of large numbers of very small dispensers (e.g. hollow fibres, trilaminate confetti flakes, microcapsules and liquid flowables), some of which can be applied by conventional spraying equipment (Jones, 1991).

Controlled release technology as applied to pheromones has constituted the subject matter of many publications and the books by Jutsum and Gordon (1989) and Ridgway *et al.* (1990) provide several chapters on the subject.

## 12.4 EVALUATION OF MATING DISRUPTION TREATMENTS

Once the mating disruption formulation has been applied to the crop, it is necessary to monitor how well it is working. One simple way is to place monitoring traps within the treated field, baited with the normal monitoring dose of the pheromone, and compare catches of moths in those traps with others placed outside the treatment field and sufficiently far away from it so as not to be affected. If the dispensers are releasing sufficient pheromone to depress the catch in the traps within the disruption plot by 98 to 100%, then it is good early indication that the system is working. Certainly, if the percentage trap catch depression falls to below 95% it is indicative that the pheromone concentration within the disruption plot is reaching levels that might allow mating.

A more reliable means of proving that mating disruption is being achieved is to look for mating behaviour within the treatment plot, with

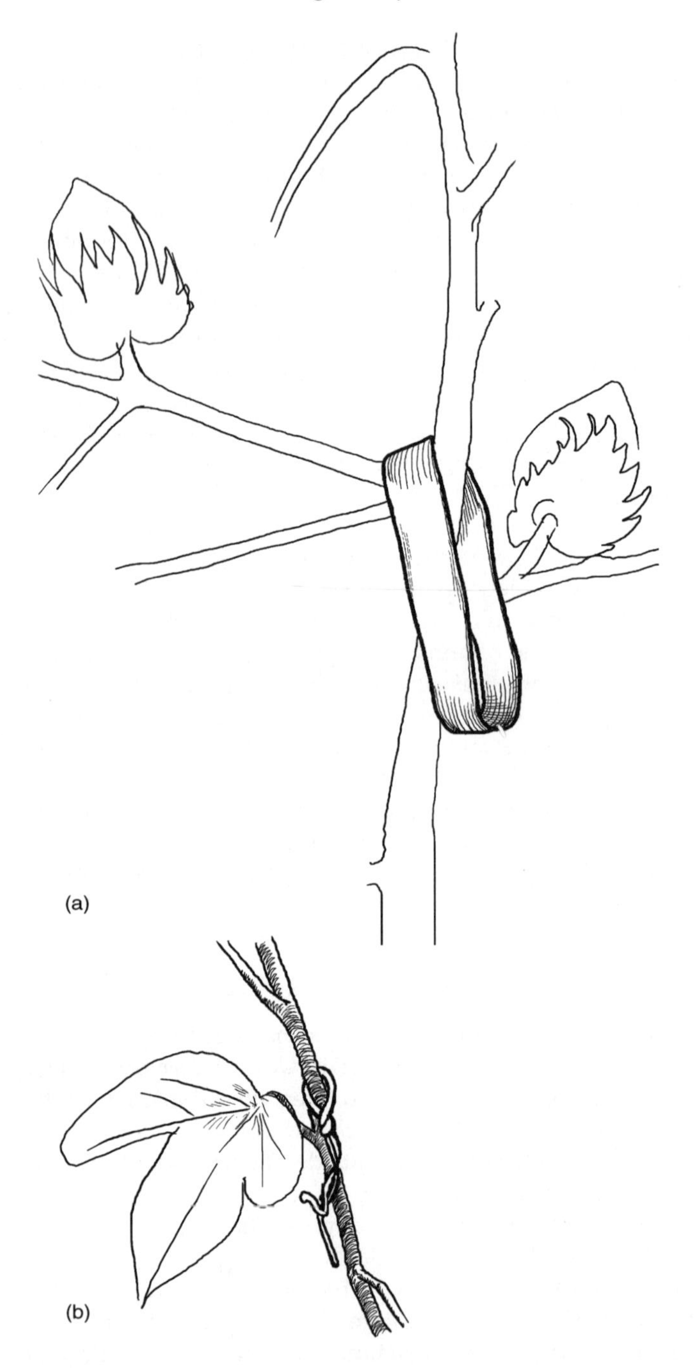


**Fig. 12.1** Containers for pheromone of *Pectinophora gossypiella*: (a) Selibate ring; (b) twist-tie.

night vision equipment if the species concerned is nocturnal. Researchers in this field have also tethered virgin female moths within the crop and collected them after one or more nights' exposure in the crop to see if they have been mated. Mated females will carry a spermatophore which can be seen from abdominal dissection of the insect. The ultimate proof of the technique's efficacy, however, comes from the damage assessments and yield data on the crop concerned, as will be seen from the many examples quoted blow.

## 12.5 INSECT PESTS OF COTTON

Cotton, more than any other crop, is bombarded with insect pests from emergence through to picking. Failure to control cotton insect pests economically is the single most important factor persuading farmers, especially small ones, to move out of this potentially lucrative cash crop.

Insecticides on their own have often failed to control pest populations, especially of the strong-flying, polyphagous and genetically versatile bollworm pests such as *Helicoverpa (Heliothis)* spp. A combination of natural enemy destruction and insecticide resistance has to some extent sealed the fate of almost all the chemical groupings introduced into cotton as insecticides. Integrated pest management with an important role for pheromones is now regarded as one of the few rational ways to manage such pests.

### 12.5.1 Pink bollworm (*Pectinophora gossypiella*)

That pink bollworm should employ the most stable and persistent of the sex pheromones belonging to the Lepidopteran pests of cotton was indeed fortunate. Pink bollworm causes economic damage to both New and Old World cotton, whether it be grown in temperate or tropical zones, and it is the only truly cosmopolitan species of bollworm. The activities of the similarly distributed *Helicoverpa* bollworms are divided amongst several plant species, according to the geographical zone.

It is difficult, if not impossible, to control this particular bollworm pest with conventional insecticide applications: the first instar larvae, on hatching from the egg, bore almost immediately and rapidly into bolls, where they stay to feed and develop, fully protected against contact insecticide. No other bollworms enter bolls and stay there so early in the period of larval development.

Today almost all pink bollworm management and control is achieved by using the synthetic pheromone Gossyplure. Ironically the research that started the ball rolling took place with the para-pheromone (Z)-7 hexadecenyl acetate in the United States, where effective reduction in mating frequency and pink bollworm control was achieved (Shorey *et al.*, 1974). Just as important, it dispelled early fears that the use of semiochemicals would just attract large numbers of moths from surrounding fields. It did not and this paved the way for use of the more active synthetic pheromone, a 1:1 mixture of (Z,Z) and (Z,*E*)-7,11-hexadecadienyl acetate identified and elucidated by Hummel *et al.* (1973) and Bierl *et al.* (1974) and subsequently shown to be superior to the parapheromone in field trials in the southwest cotton region of the United States (Shorey *et al.*, 1976; Gaston *et al.*, 1977). This new technology caught on fast, with the area sprayed in southern California and Arizona doubling every year to reach 40 000 ha by 1981 (Baker *et al.*, 1990).

### *(a) Formulation and use*

Possessing just two components, Gossyplure is considerably more stable and persistent than most (if not all) of the other multi-component pheromones. This has enabled rapid progress in formulation development to produce effective slow-release formulations. Otherwise formulation development and use have been dictated by the nature of the pest complex attacking cotton in a particular country or region and the socio-agronomic conditions under which cotton is produced in that location.

The slow-release formulations developed and used for mating disruption control of pink bollworm can be broadly classified as follows.

- Hollow fibres and laminate flakes usually applied by air with glue to secure adhesion to the foliage.
- Micro-capsules and polymer beads, dispersible and miscible in water, and suitable for application with conventional spraying machinery and nozzles.
- Hand-applied solid polymer dispensers such as twist-tie polyethylene tubes reinforced with soft wire and PVC resin, which can be extruded into user-friendly shapes – e.g. rings for growing points of cotton plants (Campion *et al.*, 1989; Critchley *et al.*, 1989).

With the development of Gossyplure and commercial slow-release formulations, cotton-growing countries afflicted with pink bollworm seized eagerly on this new strategy for controlling one of the crop's 'oldest' pests.

Out of the many trials and programmes thus established, a distinct pattern of strategies emerged, each governed by the prevailing pest complex, the agronomic and socio-economic conditions under which cotton was produced and even political pressures from various lobbies, including governments, farmers and the public at large.

These control strategies (which have been highlighted by McVeigh *et al.*, 1990, and Baker *et al.*, 1990) offer a more useful and meaningful assessment of the current status and future prospects for pink bollworm control by mating disruption than by looking at geographical location *per se*.

*(b) Season-long control*

This strategy has proved possible and highly pertinent in Egypt, where pink bollworm is the only important insect in the prevailing pest complex that cannot be controlled effectively with established cultural and biological methods. Other potentially important pests, such as *Spodoptera littoralis* (cotton leafworm), are contained in the critical early season period by teams of children who pick the egg masses off the foliage. Early season insecticide sprays to control pink bollworm, besides having a limited effect on the target pest, obliterate growing populations of beneficial arthropods, thus aggravating other insects into secondary pest status – including *Helicoverpa* spp. (cotton bollworms) and *Bemisia tabaci* (cotton whitefly). Faced with the prospect of uncontrollable infestations of whitefly such as occurred in neighbouring Sudan, the Egyptian government gave its full backing and cooperation to the introduction of Gossyplure for pink bollworm control.

Now, some 10 years later, established commercial use of Gossyplure controls pink bollworm; it has marginalized the use of insecticides, removed the threat of secondary pests and directly boosted boll size and honey production by removing the pressure of insecticides on pollinating bees. The Egyptian Ministry of Agriculture, more than any other, has made a significant commitment to the introduction of pheromones for pink bollworm control, as demonstrated by the fact that 300 000 acres of cotton (about 36% of the country's cotton acreage) was treated with pheromone in 1994. Table 12.1 shows this overwhelming benefit of mating disruption techniques on beneficial insect populations in Egypt, possibly explaining the absence of any severe outbreaks of secondary pests in the areas treated with pheromones.

*(c) Alongside insecticides*

Further east, Pakistan cotton has more pests to contend with and less inherent protection to fight them. Compared with hairy *Gossypium barbadense* varieties traditionally grown in Egypt, the smooth-leafed varieties of *Gossypium hirsutum* are especially vulnerable to attack from sucking pests such as cotton jassids. In addition, both *Earias insulana* (spiny bollworm) and *E. vitella* (spotted bollworm) present themselves as major pests. Thus cotton farmers in Pakistan do not have the luxury of dispensing with insecticides almost entirely, as do their counterparts

**Table 12.1** Mean numbers of predatory insects ha$^{-1}$ in cotton sampled by D-Vac suction apparatus following the first application of insecticide in pheromone- and insecticide-treated areas, Central Egypt, 7th July, 1983 (after Campion *et al.*, 1989)

| *Genus* | | *Insecticide treated* | | *Pheromone treated* | *Ratio (insecticides : pheromone)* | |
|---|---|---|---|---|---|---|
| | | *Area 1* | *Area 2* | *Area 3* | *1 : 3* | *2 : 3* |
| *Coccinellida* | adults | 33 | 250 | 3717 | 1 : 112.6 | 1 : 14.9 |
| *Paederus* | adults | 417 | 17 | 1717 | 1 : 4.1 | 1 : 101 |
| *Scymnus* | adults | 33 | 0 | 1184 | 1 : 35.9 | – |
| *Chrysoperla* | adults | 17 | 0 | 200 | 1 : 11.8 | – |
| *Orius* | adults | 0 | 0 | 583 | – | – |
| *Orius* | nymphs | 0 | 0 | 183 | – | – |
| Total | | 500 | 267 | 7584 | 1 : 15.2 | 1 : 28.4 |

in Egypt, although the case for eliminating early season sprays targeted at pink bollworm is even stronger. With aerial spraying banned, focus has been extended to the deployment of multi-component formulations such as PVC resins impregnated with the pheromone actives for *Pectinophora gossypiella* and both species of *Earias* to give season-long control of all three bollworm pests while relieving pressure on beneficial arthropods (McVeigh *et al.*, 1990).

### *(d) IPM programme integration*

Peru has proved to be the pace-setter for this strategy. In response to the arrival of pink bollworm in 1983 and its rapid spread through the country's cotton-growing regions, pheromone control was introduced. Overuse of insecticides in the 1960s and early 1970s and the resulting transformation of normally minor pests into economic pests had brought about the establishment of an integrated pest management programme based on minimizing the use of conventional insecticides, into which pink bollworm control by mating disruption fitted neatly. By using pheromone for pink bollworm control, the use of pyrethroids – effective against the pest but unfriendly to the beneficials – is delayed until at least 100 days after sowing. This avoids the upsurge of sucking pests, such as jassids and whiteflies, experienced elsewhere.

### *(e) Mixed motives*

The southwest United States (California and Arizona) is where pink bollworm control by mating disruption started and while, at one time, it represented the area where Gossyplure was most widely used, the progression and efficacy of pheromone use has not been particularly smooth. This may be due to the background of decreasing cotton profitability and acreages in these states and the willingness of farmers to 'cut corners' in pheromone use in an effort to cut costs over and above those achieved by switching away from insecticides. Thus the motives and strategy for use had deviated from those initially recognized when mating disruption control took off in the late 1970s.

The real and realized fears of farmers in the 1970s, that the widespread aerial application of the then relatively new pyrethroid insecticides induced big outbreaks of whitefly and *Helicoverpa (Heliothis)*, set against general public unease over insecticides, led to a positive vote by farmers for mating disruption control of pink bollworm. A Cotton Pest Abatement Board was established in California which fined any farmer who did not use Gossyplure. In spite of its success, this formalized programme was abandoned on the basis that farmers were convinced and responsible enough to ensure that they used the pheromone, although this assumption proved to be wrong.

A further problem was that the introduction of pheromone for pink bollworm control occurred at a time of accelerating decreases in profitability and a dramatic reduction in cotton acreage. Those farmers remaining in cotton tried to cut costs by cutting corners in the programme, including delaying application, reducing dose rate and even mixing insecticide with the sticker added to sprayed pheromone formulations. Such strategies, rarely a success even with conventional insecticide programmes, proved to be a disaster with the more delicately balanced pheromone programme, which undeservedly got some bad publicity.

Overall, the percentage of cotton in southwest United States treated with pheromone has gone up steadily even though the total acreage grown has gone down. More satisfactory formulations such as the sprayable polymer beads and hand-applied solid polymer formulations, including both twist-ties and PVC resins, have been deployed on some smaller plots in Arizona, although these formulations are generally thought to be of limited commercial value in countries such as the United States where labour costs are so high (Baker *et al.*, 1990; Ridgway and Inscoe, 1992).

## 12.6 INSECT PESTS OF RICE

Rice is a clear target for pheromone use in pest control. Second only to cotton in the breadth and intensity of insect pest attack, this age-old cereal crop, for years in balance with its ecosystem, received a severe jolt with the widespread use of synthetic insecticides (Mabbett, 1993a).

Indiscriminate use of insecticides in rice caused all sorts of pest aggravations, including pest resurgence and secondary pest infestation brought about by upsetting the 'natural' balance between pests and their arthropod predators and parasites. Furthermore, the way lowland rice is grown in varying depths of water makes the application of pesticides both difficult and inefficient, with run-off and leaching into the rice 'water', which is often home to carp and other fish species reared by rice farmers to supplement their protein intake. Faced with these problems and pressures many governments, including key rice producers such as Indonesia, have banned the use of all but the most target-specific and environmentally benign insecticides.

### 12.6.1 Rice stem borers

In terms of behaviour and how it affects control options, the rice stem borers, including both *Chilo* and *Scirpophaga* spp., present the same problems for rice as pink bollworm does for cotton. On hatching from the egg, the first instar larvae bore swiftly into the stem and stay there through larval development and pupation, thus presenting only the tiniest window of control opportunity with foliar-applied contact insecticides.

Research work and commercial use of pheromones to control *Chilo suppressalis*, the most serious of the rice stem borer moths, has gathered pace since the 1970s. Most research has focused on this pest species in East and South East Asia, the traditional home of paddy rice, though some of the most extensive commercial application of pheromones for mating disruption control has occurred on rice in southern Europe, especially in Spain (Jones, 1991).

The female sex pheromone of the rice stem borer *Chilo suppressalis* was first identified as a mixture of (Z)-11-hexadecenal and (Z)-13-octadecenal (Nesbitt *et al.*, 1975; Ohta *et al.*, 1976) but failure of this blend to attract males to the same extent as live virgin females suggested the presence of other synergistic components, and subsequently (Z)-9-hexadecenal, a positional isomer of (Z)-11-hexadecenal, was identified (Tatsuki *et al.*, 1983). This enormously increased the attractant properties of the existing two-blend formulation to male moths in the field and since then the three-blend mixture has formed the basis of experimental and commercial development of controlled release formulations combining aldehyde-stablizing agents (Tatsuki, 1990).

Large-scale mating disruption trials in Japan have employed two distinct methods of pheromone dispenser deployment: the 'spray method' using micro-capsules deposited close together over the crop to obtain an almost uniform permeation of the air with disruptant synthetic pheromone vapour; and the 'point source' method, using hand-applied solid polymer dispensers with a high pheromone release rate and a correspondingly wider spacing within the target area. The presence of rain-supplied or irrigation 'rice water' and the availability of labour in many traditional rice-growing regions have led to the general adoption of the point source method using polyethylene capillary tubes, laminated films and, most recently, PVC resin shoe-lace ties as pheromone disruptant dispensers.

### *(a) Timing dosage and distance*

Practical and economic considerations dictate that the evaporation points for point source pheromone deployment strategies are placed as far apart as possible. Orientation disruption levels greater than 90% have been achieved at a pheromone dispenser spacing of 16 m and a release rate of 50 mg $ha^{-1}$ $d^{-1}$. For these levels to be achieved at 32 m spacing, release rates of 234 mg $ha^{-1}$ $d^{-1}$ were required and at these rates significant disruption was achieved at double this spacing, i.e. 64 m. Overall, several years of research in Japan with both polyethylene capillary dispensers and plastic laminate dispensers suggests that inter-evaporation point distances can be maximized if the release rate of pheromone is sufficiently high: the comparative costs of labour and products then determine the grid pattern used in practice (Kanno *et al.*, 1982).

Timing is a key factor in the success of mating disruption control of *Chilo suppressalis* in rice paddies in Japan because of markedly different behaviour of first flight and second flight moths. Difficulties in correlating results of monitoring studies (using tethered females) with mating disruption control, as indicated by mating success of feral females and rice stem damage, were encountered with first flight moths (those of the second or overwintering generation) but not with second flight moths (those of the first generation). Release rates of 0.8 g $ha^{-1}$ $d^{-1}$ offered satisfactory mating disruption effect in tethered females but not in feral first flight females, almost all of which had mated.

This can be explained by the movement of overwintering moths: fully grown larvae bore into rice straws, where they hibernate for the winter. This rice straw is subsequently moved around as an agricultural and horticultural commodity. As female moths usually mate near their place of emergence, the mated females in the paddies were almost certainly immigrants, indicating the limited use of mating disruption for controlling the reproduction of first flight moths. On the other hand, mating disruption is altogether more useful for controlling the reproduction of second flight moths because all the adults tend to emerge and mate throughout the paddies. Monitoring tests with tethered females correlate well with actual control and with a release rate of 1.2–1.6 g $ha^{-1}$ $d^{-1}$ it is possible to achieve good control from an orientation disruption in excess of 95% (Tanaka *et al.*, 1987; Tatsuki, 1990).

### (b) Rice field control

Parallel developments in South and South East Asia and southern Europe have focused on other stem borers too, including several species of *Scirpophaga*. Work in Spain has been brought to the greatest fruition, with successful trials leading to commercial application over 1500 ha of rice in both the Ebro Delta and Valencia rice-growing areas. In Spain, *Chilo suppressalis*, with up to three generations per year and damage levels exceeding 50%, has been successfully controlled for up to 100 days by a single deployment of PVC resin impregnated with the *Chilo* three-component pheromone blend and UV blockers and antioxidants to protect the active ingredients. The 5% a.i. (a.i. = active ingredient) polymer is cut into lengths and secured to the tops of canes at 75 cm above the ground and well clear of the crop flood irrigation water.

Deployed at the rate of 2500 polymer dispensers $ha^{-1}$ (8 mg active pheromone each) translating into 20 g $ha^{-1}$, they performed as well as conventional insecticide (fenitrothion) by controlling stem borers down to a damage level of 0.2% (Table 12.2). Confusion efficacy rates were approaching 100%. Similarly high rates and control levels were achieved by reducing the point sources to 500 or 100 $ha^{-1}$, depending on infestation history, and using, respectively, 1.6 g or 8 g pieces of polymer to give a total dosage of 40 g $ha^{-1}$, a system which is now recommended commercially in Spain.

**Table 12.2** Mating disruption experiments with *Chilo suppressalis* in Valencia, Spain

| *Year* | *Area treated (ha)* | *Dispensers* $ha^{-1}$ | *Pheromone rate* $ha^{-1}$ *(g* $ha^{-1}$*)* | *Mean moth trap catch* | *Trap catch suppression index* | *Percentage damage* |
|---|---|---|---|---|---|---|
| 1988 | 60 | 2500 | 20 | 2 | 99.7 | 0.195 |
| | 1500 | Insecticide* | 0 | 692 | | 0.04 |
| 1989 | 60 | 625 | 30 | 15 | 93.1 | 5.04 |
| | 1500 | Insecticide* | 0 | 218 | | 0.3 |
| 1990 | 1200 | 156 | 40 | 9.3 | 97 | 0.26 |
| | 30 | 100 | 40 | 1.4 | 99.5 | 0.46 |
| | 1500 | Insecticide* | 0 | 295 | | 0.07 |

Insecticide*: Control plots were treated at least two times with 60% fenitrothion at 1.8 litres per ha or 70% Tetrachlorvinphos at 1 litre per ha.

Developments in South and South East Asia are well on the way to commercial application for control of *Scirpophaga incertulas* and *S. innotata*. The yellow stem borer, *Scirpophaga incertulas*, is the most important pest of rice in India. It is invariably present and there is little genetic basis for control by crop resistance. The females are attracted to the rice fields and both mating and oviposition take place at night, the females sitting on top of the rice plants during the day. Eggs are laid on the leaves in two or three masses of 50–80 and hatch within seven days. The first instar larvae disperse from the egg mass by means of silken threads and invade the stems, with hardly any surface feeding, 2–4 days after hatching. They go in at the top, one larva per tiller, and bore downwards, proceeding through the larval instar stages within 20–25 days, before pupating–still inside the stem.

In India this stem borer has two generations per season: the first at the tillering stage and the second when the panicles are forming. 'Dead heart' is the symptom caused by the first generation – young larvae bore into the tillers of nursery and seed bed plants and kill them. Stem borer attacks at this time significantly reduce the number of tillers per hill (normally 12–15) and even destroy the entire crop. Replanting is invariably impractical and if farmers are working to a tight rain or irrigation schedule there is insufficient time for the replanted crop to mature. Second generation attacks are economically even more damaging because the grains fail to fill, causing 'whiteheads' and an irrecoverable loss in yield.

Contact insecticides are inappropriate for stem borer control because of the narrow window of contact opportunity (48 hours from egg hatch to entry) and their drastic effect on natural enemies. Systemic insecticides such as carbofuran and phorate in granular formulations applied to the soil are the best insecticide option, but these are ineffective in the wet season because of rapid leaching from the soil.

Pheromones are target specific, unlike insecticides which can wipe out beneficial insects just as well as the target. Field trials on 20 ha of rice in west Bengal showed pheromone control to be equal to that achieved with insecticide regimes. The PVC resin formulation is deployed 1 m above the water level, attached to sticks, at a specific stage in crop development (10–14 days after transplanting) to coincide with moth activity and crop susceptibility. Lengths of PVC resin (20 cm) are tied to stakes at a rate of 625 pieces $ha^{-1}$, representing a dosage rate of 40 g active ha. This gives a mating disruption effect that lasts for 60–80 days.

Work has moved to Indonesia to evaluate pheromone blends for mating disruption control of *S. innotata*. This pest poses a threat to Indonesia's status as a rice exporter. It is endemic and capable of causing near total loss, with farmers not bothering to harvest what remains. The Indonesian government has a progressive policy of banning the use of all but the most target-specific and environmentally benign insecticides. The pheromone for this species has been identified and trials have begun on the island of Java. Future plans include the use of pheromones for mating disruption of key rice stem borer pests in China, Korea and Thailand (Hall *et al.*, 1991; Mabbett, 1993a).

## 12.7 FOREST INSECT PESTS

Commerical forest crops have a long generation time, a relatively low value $ha^{-1}$ and a delicate environmental balance. For these reasons the use of expensive and ecologically disruptive chemical insecticides has increasingly been adopted as a last resort only. Integrated pest management programmes with pheromones as a vital and integral part have been established throughout the world. Indeed the typical forest pest, with its narrow host range, is usually very amenable to management with pheromones With the development of sprayable pheromone formulations foresters now have the option to extend pheromone deployment using a mating disruption strategy over large areas of pest-susceptible forest by aerial application methods.

### 12.7.1 Gypsy moth (*Lymantria dispar*)

At first glance the gypsy moth, *Lymantria dispar*, which devastates huge areas of hardwood forest and shade and ornamental trees across North America and Europe, would appear to be an unpromising candidate for control by mating disruption. With a host range encompassing no less than 500 species and a damage record that includes 13 million acres defoliated with US$350 million losses in one year in the United States alone, the temptation to blanket spray huge areas of trees with chemical insecticides from the air is overwhelming. But an accelerating awareness of the damage caused by insecticides to public health and environmental integrity has made the development of an alternative strategy for gypsy moth control top priority for the last 25 years (Kolodny-Hirsch and Schwalbe, 1990).

Specific aspects of gypsy moth biology make it a more suitable candidate for control by mating disruption than its general physiological and ecological profile would indicate. A combination of polygamous males and highly fecund females (300–1200 eggs per mass) means that mating must be reduced to a greater extent, compared with most other forest lepidopteran pests, before the population can be significantly brought down and an economic impact on damage levels achieved.

Since female moths are wingless, the spread of the insect is achieved by windborne movements of the young larvae on their silk threads

(Campion *et al.*, 1989). The flightless status of the females means that they are unable to move far from their site of emergence, thus allowing specific areas to be treated without fear or threat of mass adult invasion. Similarly, males have little need to disperse widely from their emergence site to locate a mate, a fact which mark–recapture studies confirm (Schwalbe, 1981; Elkinton and Carde, 1980).

These aspects of insect behaviour, together with the appearance of males in a distinct and predictable flight over a relatively short six-week period, mean that pheromone applications for mating disruption can be safely synchronized with male moth activity. Contrary to previously accepted theory, it now appears that communication between sexes during mating is completely mediated by olfactory cues (Kolodny-Hirsch and Schwalbe, 1990), but random encounters play an important part in mate finding at higher population densities. It would appear, therefore, that control of this pest by mating disruption is best suited to low population densities at which random contacts between the sexes are minimal (Beroza and Knipling, 1972; Knipling, 1979).

*Disparlure and its development*

With the discovery of disparlure, (Z)-7,8-epoxy-2-methyl octadecanol, the female sex pheromone of *Lymantria dispar* (Bierl *et al.*, 1970), gypsy moth management was handed an important extra tool for both monitoring and control. Early tests indicated that the optically active (+) enantiomer of disparlure, synthesized in Japan (Iwaki *et al.*, 1974), was a superior attractant compared with the racemic mixture (Plimmer *et al.*, 1977). However, the much lower cost of the latter, coupled with field trials that showed it to be just as good as the optically active form (Plimmer *et al.*, 1982) for mating disruption control, led to its widespread adoption for almost all subsequent work.

Since the earliest trials in 1971 (Stevens and Beroza, 1972) the inherent and intrinsic activity of disparlure has never been in doubt. Diverse trials using micro-encapsulated formulations applied by air at 5–15 g a.i. $ha^{-1}$ (Cameron *et al.*, 1974; Schwalbe *et al.*, 1974) or knapsack mistblower at 18 g a.i. $ha^{-1}$ (Granett and Doane, 1975) and plastic laminate flakes and hollow fibres at 5 to 50 g a.i. $ha^{-1}$ (Schwalbe *et al.*, 1983) all gave significant control by mating disruption.

It is the inverse relationship between incidence of mating disruption to population density, long suspected and recently confirmed by Webb *et al.* (1988), that has constrained use of mating disruption in an industry that is geared to the control of insects at high population density. There is strong evidence to suggest that acceptable levels of mating disruption could be achieved at high population levels with a corresponding increase in dosage and formulations, with more rapid release rates. Current commercial use is restricted to hand application of laminated tape dispensers with a recently developed polymer bead formulation set to establish a sprayable option.

## 12.8 FRUIT AND VEGETABLE PESTS

Fruit and vegetable crops are obvious beneficiaries of insect pest control with pheromones. They are typically attacked by a wide range of pests, including many Lepidoptera. These include crop-specific species such as *Keiferia lycopersicella* (tomato pin worm), polyphagous species like *Helicoverpa (=Heliothis) armigera* and others such as *Cydia pomonella* (codling moth) which attack a range of taxonomically related fruit tree crops. These factors, together with a myriad of sucking pests (aphids, whiteflies, thrips and jassids), offer ideal conditions for pest explosions caused by beneficial insect destruction and acquired insecticide resistance due to misplaced and mistimed insecticide applications.

Fruit and vegetable growers are rapidly sucked into a spiral of decreasing insecticide efficacy and increasing pesticide use (higher dosages and reduced spray intervals) so that spraying every 2–3 days becomes necessary. With fruits and vegetables invariably consumed fresh, often in the 'raw' condition and picked almost daily, as in the case of tomatoes, growers soon run into severe residue problems, especially if their produce is earmarked for export to tightly controlled markets such as those in the United States and European Union. The introduction of pheromones into horticulture has helped to break this vicious spiral of insecticide use and abuse and bring stable, long-lasting and environmentally friendly control to fruit and vegetable production. The Chilean export fruit industry provides a good example where this is increasingly being achieved (Mabbett, 1994).

The idea of mating disruption was conceived over 30 years ago, although it took another 15 years before sufficient quantities of synthetic pheromone became available to deploy on crops for this purpose. Research has proceeded most rapidly in the sophisticated horticultural production and marketing areas of the globe including North America, Japan, Australasia and western Europe. Scientists and agriculturists have concentrated their efforts on pests of temperate top fruit and stone fruit, vine pests and pests of widely grown and high value vegetables such as tomato.

### 12.8.1 Codling moth (*Cydia pomonella*)

This fairly ubiquitous pest of apples and pears reduces the value of crops in most of the major top-fruit producing areas of the world: North America, Europe, South America and Australia.

The main component of the female sex pheromone, (*E,E*)-8,10-dodecadien-1-ol (*E,E*-8,10-12:OH), commonly called codlemone, was identified more than quarter of a century ago (Roelofs *et al.*, 1971a). When applied for mating disruption it is thought to camouflage the female's natural emission, thus pushing the males into false-trail following (Bartell, 1982). For over 11 years (1976–1987) Switzerland was the focus of intensive research into the use of codlemone for mating disruption, initially using rubber tubing dispensers holding 1.3 mg codlemone/$mm^{-1}$ and later Hercon dispensers (40 mg per unit). These were placed both on the borders of the orchard and inside at 40–200 point sources per hectare, depending on the size and shape of the orchard (Charmillot, 1990).

A decade of research has brought the mating disruption technique for codling moth to the point of widespread commercial application. Provisional registration was in place in Switzerland in 1988/89, followed by Germany and Austria (Minks and Deventer, 1990). However, the specific mode of action of codlemone on mating disruption and the inherent dispersal capacity of codling moths has placed a number of ecological and topographical restrictions on its deployment commercially (Charmillot, 1990).

Male moths move very long distances inside and outside the orchard (Mani and Windbolz, 1977). Female migration, though less significant, can encompass movement of 50 m to more than 100 m inside the orchard, with a tendency for aggregation on the borders (Windbolz and Baggiolini, 1959; White *et al.*, 1973). Thus isolated untreated apple trees outside the orchard, or orchards with a high population density and separated from each other by less than 100 m, offer real infestation sources (Myburgh *et al.*, 1975). For this reason, current recommendations are tempered by a list of restrictions including use only for isolated orchards of no less than 3 ha, isolated from external infestation sources by at least 100 m. Orchard borders must be protected by small dispensers no more than 4–5 m apart, although inside the orchard larger and more widely spaced dispensers can be used. When orchard borders are especially irregular or are too elongated, the border length (where matings occur) becomes too important relative to orchard and the use of hand-applied dispensers for mating disruption control is not advised.

These practical constraints, together with the labour required for hand application of dispensers, may have discouraged the commercial adoption of this technique in many areas, but, as our understanding of how the various systems actually achieve mating disruption increases, the control of *Cydia pomonella* through mating disruption technology is gradually being adopted in area-wide programmes in the United States and South Africa.

### 12.8.2 Oriental fruit moth (*Grapholitha molesta*)

*Grapholitha molesta* (oriental fruit moth–OFM), a ubiquitous lepidopteran pest of stone fruit, is one of the few examples where identification of the sex pheromone followed by relatively short trials programmes has led to the widespread commercial application of a mating disruption product. Given the commercial value of peach and nectarine crops and traditionally intensive use of organophosphate insecticides to control OFM, it was fortunate indeed that the breakthrough with mating disruption control in the stone fruit industry should have happened with this particular pest. Mating disruption programmes are now being used in almost all the major peach and nectarine production areas of the world, albeit in different strategies and ways depending on the nature of the entire pest complex (Vickers, 1990).

#### *(a) Australia*

Following early identification of the major component of the female sex pheromone, (Z)-8-dodecenyl acetate (Z8-12:Ac), in 1969 (Roelofs *et al.*, 1969) Australian scientists were quick to exploit its potential. With a stone fruit industry worth Aus $13.4 million annually and severe secondary pest problems with the mite *Tetranychus urticae* from the intense use of OFM-targeted organophosphate insecticide sprays, the economic and environmental inducements were in place for this work.

Trials began just one year later with this newly discovered compound (Rothschild, 1975), using material contaminated with 2–3% of the *E* isomer. This 'contaminant' was later shown to be an essential part of the natural pheromone blend (Cardé *et al.*, 1979) and is believed to have facilitated, albeit by accident, the rapid commercialization of the pheromone for mating disruption. Using various dispensers and release rates, researchers quickly pin-pointed polyethylene micro-centrifuge tubes (400 μl) loaded with 50 μl pheromone for release at 6 mg $ha^{-1}$ $h^{-1}$ to provide the most effective system available. Efficacy was subsequently enhanced by addition of both (Z)-8-dodecenol (Z8-12:OH) and dodecanol (12:OH) which have since been identified as components of the pheromone blend (Cardé *et al.*, 1979).

By 1974–75 the trials were already showing mating disruption to be as effective as the standard insecticide programme, but comments by researchers about the potentially uneconomic status of mating disruption control, coupled with a decline in the Australian peach industry, prevented any more immediate progress at that time (Rothschild, 1975).

Resumption of the work in 1980–81, with its expansion into entire peach growing districts (Vickers *et al.*, 1985), finally laid the firm foundations required for commercial deployment of the system. The use of a combination of pheromone trap catches, index of infestation (shoot and fruit damage combined) and assessment of the mating status of females caught in feeding lures showed conclusively that the system was a success. In particular the mating status of wild females, which is considered to be the most meaningful indicator of mating disruption, gave confidence to the results (Rothschild, 1981)

The 1984–85 seasons saw the first commercial deployment of mating disruption facilitated by a Japanese design and automated production of a dispenser and pheromone unit complete with a wire for hanging on the trees. Just one year later the system was offering an equivalent and reduced cost control (Aus \$267 versus Aus \$389 $ha^{-1}$) compared with standard insecticide programmes using azinphos methyl and cyhexatin, the former for OFM and the latter for mite explosions caused by azinphos-methyl.

*(b) North America*

Encouraged by the Australian success and the need to protect an even more valuable peach industry (Can \$17.8 million in 1985) the Australian mating disruption product for OFM was marketed in Canada. OFM in Canada is just one member of a lepidopteran pest complex that includes the peach tree borers (*Synanthedon pictipes* and *S. exitosa*) and that is routinely controlled by two to four sprays of organophosphate insecticide per season.

Trials conducted in Ontario during 1987 indicated good, if not spectacular, control and highlighted specific problems in Canada with very high infestations of OFM and the antagonistic effect of relaxing insecticide sprays on other lepidopteran pests, such as peach tree borers. The use of mating disruption as part of an integrated control programme for OFM management would appear to be the best way forward with initial insecticide sprays to reduce very high OFM infestations to levels with which mating disruption can cope (Vickers, 1990).

The OFM problem in the United States is different in both the West Coast and East Coast peach production areas and complicated by the widespread attack by OFM of many other stone fruit and top-fruit trees including apples, pears, plums, apricots and cherries. As in Canada, OFM is part of a lepidopteran pest compex that attacks the hugely valuable peach crop (Rice and Kirsch, 1990).

Early trials in Georgia (Gentry *et al.*, 1982), which highlighted both the potential and the problem of mating disruption control in the United States, were followed by large-scale trials in 1985 in California covering all the major peach and nectarine growing areas of the San Joaquin Valley (Rice and Kirsch, 1990).

Using dispensers sourced from Australia, each containing a 75 mg blend of (Z)-8-dodecenyl acetate (93%), (*E*)-8-dodecenyl acetate (6%) and (Z)-8-dodecenol (1%), point sources were located in the upper third of the tree canopy (2.5–4.0 m) and deployed at 1000 ha to give a release rate of 30–35 μg/dispenser per hour over a three-month period.

They are applied twice, one in late February/early March (full bloom) synchronized with emergence of the overwintering population, and exactly 90 days later to cope with the next generation. Assessments of female mating success and fruit damage over a three-year period showed mating disruption with pheromones to be a feasible alternative to conventional insecticide sprays provided that measures were taken (e.g. winter sprays and additional in-season sprays) to maintain control of the peach twig borer (*Anarsia lineatella*) and the omnivorous leaf roller (*Platynota stultana*).

Similarly encouraging trials were conducted in Virginia (eastern United States) in 1986 and 1987 with the added advantage of lesser appleworm (*Grapholitha prunivora*) mating disruption as the pest responds to the same pheromone components as OFM (Pfeiffer and Killian, 1986).

#### *(c) South Africa*

OFM was first reported in South Africa in the late 1980s and it rapidly became a problem that threatened the continued success of the stone fruit industry. Rather than immediately implementing control measures based on organophosphate insecticides, the industry coordinated an area-wide application of hand-applied pheromone dispensers in the entire Tulbagh Valley in 1991/2 which had approximately 2000 ha of mixed stone fruit. Despite recording of 85–100% tip damage in the last generation of the previous season, complete control was achieved in a single season (Kirsch and Lingren, 1993) with no supplementary applications of insecticide. The Tulbagh Valley pheromone treatment rates were halved in 1992/3 and populations again did not recover to damaging levels, thus showing once again the wisdom of carrying out mating disruption programmes on an area-wide basis.

### 12.8.3 Tomato pinworm

Owing to its day-neutral flowering status, tomato is grown in just about every corner of the world, albeit under protection where low temperatures, arid atmosphere and soil or tropical monsoon rainfalls are limiting. By and large, tomato crops escape the worst ravages of insect pests but they are more susceptible to fungal and bacterial pathogens. Where they do occur insect pests, and especially Lepidoptera such as

*Keiferia lycopersicella* (tomato pinworm–TPW), wreak economic havoc. TPW is a primary pest of tomato (*Lycopersicon esculentum*) in North, Central and South America, with perhaps Mexico being the hardest hit (Mabbett, 1993b).

Everything about the tomato crop and its crop protection requirements points towards the use of biological control. Tomato is an ideal export cash crop for many so-called developing countries such as Mexico, which is now a major supplier of fresh tomatoes to the almost insatiable North American market. When marketed in this way, the fruits have to be picked almost daily, making them prone to the accumulation of unacceptably high insecticide residues which are quickly picked up by zealous import inspectors, who invariably reject the consignment.

In addition, the behaviour of *Keiferia lycopersicella*, like that of its gelechiid relative *Pectinophora gossypiella* (pink bollworm–PBW), means that the larvae, deep inside the host tissue, are effectively protected against broad spectrum residual insecticides. Furthermore, overuse of this type of insecticide on tomato to control TPW invariably makes the situation worse by encouraging the development of insecticide-resistant strains of TPW and aggravating secondary pests such as *Helicoverpa zea* and *Spodoptera exigua* into causing economic loss (Jenkins *et al.*, 1990)

### *Mating disruption of TPW*

Following the identification of the TPW sex pheromone (96 : 4 mixture of (*E*)- and (*Z*)-4-tridecenyl acetate) in 1979, mating disruption trials and associated definitive research were initiated. Dispensers consisting of hollow fibres gave the best catches and have subsequently been used as a standard lure to assess mating disruption levels.

Again using hollow fibre dispensers, but with a higher dose of pheromone, TPW mating disruption trials were initiated in Florida. These trials proved inconclusive, but follow-up work in Mexico using two applications, 10 g a.i. $ha^{-1}$ followed by 20 g a.i. $ha^{-1}$ 20 days later, proved more successful and recorded mean TPW infestation on fruits of just 2% in pheromone-treated plots, compared with 34% in the control.

During the following year (1981) mating tables were used in Florida which showed just 4.2% mated females in pheromone-treated plots against 52.6% in the control plots. Comparisons of different hollow fibre formulations (polyethylene fibres on sticky tape, Celconese™ plastic fibres on sticky tape and Celconese fibres applied with polybutene sticker) gave similar and not significantly different high levels of control by mating disruption.

Large demonstration trials in Mexico covering 600 ha of tomato using hand-applied hollow fibres in sticker at 10 g a.i. $ha^{-1}$ with 36 fibres per point and 1300 point sources $ha^{-1}$ gave positive results. Fewer conventional insecticide applications were needed (all pests) in pheromone-treated areas; one pheromone application replaced 2.5 insecticide sprays and fewer infested fruit were recovered from pheromone-treated fields.

In spite of this general success, hand application was calculated to be uneconomic (Van Steenwyk and Oatman, 1983) but with labour accounting for the lion's share of the total cost, the socio-economic conditions under which tomatoes are grown are clearly more important than agronomic factors in determining economic viability of hand-applied mating disruption formulations such as hollow fibres on tape or in sticker.

A TPW mating disruption system is now in commercial use in California, Florida and Mexico using clusters of fibres and adhesive applied to the leaves or stakes supporting the tomatoes. Given the huge differences in labour costs between the United States and Mexico, the need to adopt mating disruption control by pheromones at the expense of insecticides, due to residues, resistance and secondary pests, has caused a shift in tomato production from the United States to Mexico (Buckley *et al.*, 1986). Mexican growers are well motivated towards the production of clean, minimal residue tomatoes following previous experiences when the USDA-FDA prevented the entry of large consignments because of pesticide levels exceeding the laid-down tolerance level.

### 12.8.4. Vine pests

Because of the limitations of space, it has not been possible to include in this chapter a great deal of work that has been done on the mating disruption of lepidopteran pests of grape vines. The following references give much valuable information on the subject: Arn *et al.*, 1976, 1981; Buschmann *et al.*, 1987; Buser *et al.*, 1984; Charmillot *et al.*, 1985; Dennehy *et al.*, 1990; Descoins, 1990; Leonhardt *et al.*, 1990; Neumann, 1990; Roehrich *et al.*, 1979; Roehrich and Carles, 1982; Roelofs *et al.*, 1971b, 1973; Saglio *et al.*, 1977; Taschenberg and Roelofs, 1977.

## REFERENCES

Arn, H., Rauscher, S., Buser, H.R. and Roelofs, W.L. (1976) Sex pheromone of *Eupoecilia ambiguella: cis*-9-Dodecenyl acetate as a major component. *Z. Naturforsch*, **31C**, 499–503.

Arn, H., Rauscher, S., Schmid, A. *et al* (1981) Field experiments to develop control of the grape moth, *Eupoecilia ambiguella*, by communication disruption. In *Management of Insect Pests with Semiochemicals* (ed. E.F. Mitchell), Plenum Publishing, New York, pp. 327–838.

Baker, T.C., Staten, R.T. and Flint, H.M. (1990) Use of pink bollworm pheromone in the southwestern United States. In *Behaviour-modifying Chemicals for Insect Management: Applications of Pheromones and Other Attractants* (eds R.L. Ridgway, R.M. Silverstein and M.N. Inscoe), Marcel Dekker Inc., New York, pp. 417–436.

Bartell, R.J. (1982) Mechanisms of communication disruption by pheromone in the control of Lepidoptera: a review. *Physiol. Entomol.*, **7**, 353–364.

Beroza, M. and Knipling, E.F. (1972) Gypsy moth control with the sex attractant pheromone. *Science*, **177**, 19–27.

Bierl, B.A., Beroza, M. and Collier, C.W. (1970) A potent sex attractant of the gypsy moth: its isolation, identification and synthesis. *Science*, **170**, 87–89.

Bierl, B.A., Beroza, M., Staten, R.T. *et al.* (1974) The pink bollworm sex attractant. *J. Econ. Entomol.*, **67**, 211–216.

Buckley, K.C., VanSickle, J.J., Bredahl, M.E. *et al* (1986) Florida and Mexico competition for the winter fresh vegetable market. *USDA ESA Agric. Econ. Rept*, **556**, 101 pp.

Buschmann, E., Seufert, W. and Kreig, W. (1987) A new synthetic attractant for the European grape vine moth (*Lobesia botrana*) (Z)-9-dodecadecen-7-yn-1-yl acetate. Paper presented at the Euchem Conference Semiochemicals in Plant and Animal Kingdom, Angers, October 12–16, 1987.

Buser, H.R., Rauscher, S. and Arn, H. (1984) Sex pheromone of *Lobesia botrana. E7,Z9-*dodecadienyl acetate in the female grape vine moth. *Z. Natur. Forsch*, **29**, 731–748.

Cameron, E.A., Schwalbe, C.P., Beroza, M. and Knipling, E.F. (1974) Disruption of gypsy moth mating with microencapsulated disparlure. *Science*, **183**, 972–973.

Campion, D.G. (1986) Survey of pheromone uses in pest control. In *Techniques in Pheromone Research* (eds H.E. Hummel and T.A. Miller), Springer Verlag, New York, pp. 405–449.

Campion, D.G., Hunter-Jones, P., McVeigh, L.J. *et al.* (1980) Modifications of the attractiveness of the primary pheromone component of the Egyptian cotton leafworm. *Spodoptera littoralis* (Boisduval) (Lepidoptera: Noctuidae), by secondary pheromone components and related chemicals. *Bull. Entomol. Res.*, **70**, 417–434.

Campion, D.G., Critchley, B.R. and McVeigh, L.J. (1989) Mating disruption. In *Insect Pheromones in Plant Protection* (eds A.R. Jutsum and R.F.S Gordon), John Wiley & Sons, Chichester, pp. 89–122.

Cardé, A.M., Baker, T.C. and Cardé, R.T. (1979) Identification of four-component sex pheromone of the female oriental fruit moth, *Grapholitha molesta* (Lepidoptera: Tortricidae). *J. chem. Ecol.* **5**, 423–427.

Caro, J.H., Glotfelty, D.E. and Freeman, H.P. (1980) (Z)-9-Tetradecen-1-ol formate: distribution and dissipation in the air within a corn crop after emission from a controlled-release formulation. *J. Chem. Ecol.*, **6**, 229–239.

Charmillot, P.J. (1990) Mating disruption technique to control codling moth in Western Switzerland. In *Behaviour-modifying Chemicals for Insect Management: Applications of Pheromones and Other Attractants* (eds R.L. Ridgway, R.M. Silverstein and M.N. Inscoe), Marcel Dekker Inc., New York, pp. 65–182.

Charmillot, P.J., Bloesch, B., Schmid, A. and Newmann, U. (1985) Essais de lutte contre cochylis *Eupoecilia ambiguella* Hbn. par la technique de confusion sexuelle. In *Mediateurs chimiques et biosystematique chez les lepidopteres*. Colloques INRA, Valence, Decembre 13–14 1985.

Cork, A., Beevor, P.S., Hall, D.R., *et al.* (1985) A sex attractant for the spotted bollworm, *Earias vitella. Tropical Pest Management*, **31**, 158.

Critchley, B.R., Campion, D.G. and McVeigh, L.J. (1989) Pheromone control in the integrated pest management of cotton. In *Pest Management in Cotton* (eds M.B. Green and D.J. de B. Lyon), Ellis Horwood, Chichester, pp. 83–92.

Dennehy, T.J., Roelofs, W.L., Taschenburg, E.F. and Taft, T.N. (1990) Mating disruption for control of grape berry moth in New York vineyards. In *Behaviour-modifying Chemicals for Insect Management: Applications of Pheromones and Other Attractants* (eds R.L. Ridgway, R.M. Silverstein, and M.N. Inscoe), Marcel Dekker Inc., New York, pp. 223–240.

Descoins, C. (1990) Grape berry moth and grape vine moth in Europe. In *Behaviour-modifying Chemicals for Insect Management: Applications of Pheromones and Other Attractants* (eds R.L. Ridgway, R.M. Silverstein and M.N. Inscoe), Marcel Dekker Inc., New York, pp. 213–222.

Elkinton, J.S. and Cardé, R.T. (1980) Distribution, dispersal and apparent survival of the male gypsy moth as determined by capture in pheromone-baited traps. *Environ. Entomol.*, **9**, 729–737.

Gaston, L.K., Kaae, R.S., Shorey, H.H. and Sellers, D. (1977) Controlling the pink bollworm by disrupting sex pheromone communication between adult moths. *Science*, **196**, 904–905.

Gentry, C.R., Yonce, C.E. and Bierl-Leonhardt, B.A. (1982) Oriental fruit moth: mating disruption trials with pheromone. In *Insect Suppression with Controlled Release Pheromone Systems*, Vol. II. (eds A.F. Kydonieus and M. Beroza), CRC Press, Boca Raton, pp. 107–115.

Granett, J. and Doane, C.C. (1975) Reduction of gypsy moth mating potential in dense populations by mistblower sprays of microencapsulated disparlure. *J. Econ. Entomol.*, **68**, 435–437.

Hall, D.R. and Marrs, G.J. (1989) Microcapsules. In *Insect Pheromones in Plant Protection* (eds A.R. Jutsum and R.F. S. Gordon), John Wiley & Sons, Chichester, pp. 89–122.

Hall, D.R., Beevor, P.S., Cork, A. *et al.* (1991) The use of pheromones for monitoring and control of insect pests of rice. In *Symposium Proceedings of Rice Research–New Frontiers*, directorate of Rice Research, Hyderabad, India, 15–19 November, 1990, pp. 226–234.

Hummel, H.E., Gaston, L.K., Shorey, H.H. *et al.* (1973) Clarification of the chemical status of the pink bollworm sex pheromone. *Science*, **181**, 873–875.

Iwaki, S., Marumo, S., Saito, T. *et al.* (1974) Synthesis and activity of optically active disparlure. *J. Am Chem. Soc*, **96**, 7842–7844.

Jacobsen, M. and Beroza, M. (1963) Chemical insect attractants. *Science*, **140**, 1367–1373.

Jenkins, J.W., Doane, C.C., Schuster, D.J. *et al.* (1990) Development and commercial application of sex pheromone for control of the tomato pinworm. In *Behaviour-modifying Chemicals for Insect Management: Applications of Pheromones and Other Attractants* (eds R.L. Ridgway, R.M. Silverstein and M.N. Inscoe), Marcel Dekker Inc., New York, pp. 269–280.

Jones, O.T. (1991) Monitoring and control of insect pests with pheromones and other semiochemicals. *Agro-Industry Hi-Tech.*, **2** (4), 27–32.

Jutsum, A.R. and Gordon, R.F.S (eds) (1989) *Insect Pheromones in Plant Protection.* John Wiley & Sons, Chichester, 369 pp.

Kanno, H., Hattori, M., Sato, A., *et al* (1982) Release rate and distance effects of evaporators with Z-11-hexadecenal and Z-5-hexadecene on disruption of male orientation in the rice stem borer moth, *Chilo suppressalis* Walker (Lepidoptera: Pyralidae). *Appl. Entomol. Zool.*, **17**, 432–438.

Kirsch, P. and Lingren, B. (1993) Commercial advancement on pheromone related monitoring and control technology. In *Insect Pheromones, Proceedings of an IOBC/WPRS Working Group Meeting, 11–14 May 1993, Chatham, UK*, Bulletin OILB/SROP, **16**, pp. 121–127.

Klun, J.A., Plimmer, J.R., Bierl-Leonhardt, B.A. *et al* (1979) Trace chemicals: the essence of chemical communications in *Heliothis* species. *Science*, **204**, 1328–1329.

Knipling, E.F. (1979) *The basic principles of insect population suppression and management*, Agric. Handbk. 512, USDA.

Kolodny-Hirsch, D.M. and Schwalbe, C.P. (1990) Use of Disparlure in the management of the gypsy moth. In *Behaviour-modifying Chemicals for Insect Management: Applications of Pheromones and Other Attractants* (eds R.L. Ridgeway, R.M. Silverstein and M.N. Inscoe), Marcel Dekker Inc., New York, pp. 363–385.

Leonhardt, B.A., Cunningham, R.T., Dickerson, W.A. *et al* (1990) Dispenser design and performance criteria for insect attractants. In *Behaviour-modifying Chemicals for Insect Management: Applications of Pheromones and Other Attractants* (eds R.L. Ridgway, R.M. Silverstein and M.N. Inscoe, Marcel Dekker Inc., New York, pp. 113–130.

Mabbett, T.J. (1993a) Stem borer breakthrough for pheromones. *Far Eastern Agriculture*, July/August 1993, pp. 14–15.

Mabbett, T.J. (1993b) Feromonas en Latinoamerica. *Agriculturas de las Americas*, **42 (3)**, 18–24.

Mabbett, T.J. (1994) Fruta chilena: exito fenomenal. *Agriculturas de las Americas*, **43 (1)**, 5–12

Mabbett, T.H. and Nachapong, M. (1983) Some aspects of oviposition by *Heliothis armigera* pertinent to cotton pest management in Thailand. *Tropical Pest Management*, **29(2)**, 159–165.

Mabbett, T.H., Dareepat, M. and Nachapong, M. (1980) Behaviour studies on *Heliothis armigera* and their application to scouting techniques for cotton in Thailand. *Tropical Pest Management*, **26(3)**, 268–273.

Mani, E. and Windbolz, T. (1977) The dispersal of male codling moths (*Laspeyresia pomonella L*). in the Upper Rhine Valley. Z. *Angew. Entomol*, **83**, 161–168.

McVeigh, L.J., Campion, D.G. and Critchley, B.R. (1990) The use of pheromones for the control of cotton bollworms and *Spodoptera* spp. in Africa and Asia. In *Behaviour-modifying Chemicals for Insect Management: Applications of Pheromones and Other Attractants* (eds R.L. Ridgway, R.M. Silverstein, and M.N. Inscoe), Marcel Dekker Inc., New York, pp. 407–416.

Meighen, E.A., Slessor, K.N. and Grant, G.G. (1982) Development of bioluminescence assay for aldehyde pheromones of insects. I. Sensitivity and specificity. *J. Chem. Ecol.*, **8**, 911–921.

Minks, A.K. and van Deventer, P. (1990) Practical applications: the European scene. In *Insect Pheromones and Other Behaviour Modifying Chemicals* (eds R.L. Ridgway, M. Inscoe and H. Arn), British Crop Protection Council Monograph No. 51, pp. 9–18.

Moorhouse, J. E., Yeadon, R., Beevor, P.S. and Nesbitt, B.F. (1969) Method for use in studies of insect communication. *Nature (Lond.)*, **223**, 1174–1175.

Myburgh, A.C., Madsen, H.F., Rust, D.J. and Bosman, I.P. (1975) Codling moth (Lepidoptera: Olethreutidae): sex attractant traps as an adjunct to control programmes. *Proc. I. Congr. Entomol. Soc. South Afr.*, 1975, pp. 99–108.

Nesbitt, B.F., Beevor, P.S., Cole, R.A., *et al* (1973) Sex pheromones of two noctuid moths. *Nature New Biol.*, **224**, 208–209.

Nesbitt, B.F., Beevor, P.S., Hall, D.R. *et al.* (1975) Identification of the female sex pheromone of the moth *Chilo suppressalis. J. Insect Physiol.*, **21**, 1883–1886.

Neumann, U. (1990) Commercial development: mating disruption of the European grape berry moth. In *Behaviour-modifying Chemicals for Insect Management: Applications of Pheromones and Other Attractants* (eds R.L. Ridgway, R.M. Silverstein, and M.N. Inscoe), Marcel Dekker Inc., New York, pp. 539–546.

Ohta, K., Tatsuki, S., Uchiumi, K. *et al.* (1976) Structures of sex pheromones of rice stem borer. *Agric. Biol. Chem.*, **40**, 1987–1899.

Pfeiffer, D.G. and Killian, J.C. (1986) Pheromone disruption for the control of oriental fruit moth and lesser peach tree borer. *Proc. 62nd Cumberland-Shenandoah Fruit-Workers Conf.*, 1986, 3 pp.

Plimmer, J.R., Schwalbe, C.P., Paszek, E.C. *et al.* (1977) Contrasting effectiveness of (+) and (–) enantiomers of disparlure for trapping native populations of gypsy moth in Massachusetts. *Environ. Entomol.*, **6**, 518–522.

Plimmer, J.R., Caro, J. H. and Freeman, H.P. (1978) Distribution and dissipation of aerially applied disparlure under a woodland canopy. *J. Econ. Ent.*, **71**, 155–157.

Plimmer, J.R., Leonhardt, B.A. and Webb, R.E. (1982) Management of gypsy moth with its sex attractant pheromone. In *Insect Pheromone Technology: Chemistry and Application* (eds B.A. Leonhardt and M. Beroza), ACS Symp. Series. No. 190, American Chemical Society, Washington, D.C., pp. 231–242.

Rice, R.E. and Kirsch, P. (1990) Mating disruption of oriental fruit moth in the United States. In *Behaviour-modifying Chemicals for Insect Management: Applications of Pheromones and Other Attractants* (eds R.L. Ridgway, R.M. Silverstein and M.N. Inscoe), Marcel Dekker Inc., New York, pp. 193–212.

Ridgway, R.L., and M.N. Inscoe, (1992) Insect behaviour-modifying chemicals: practical applications in the United States. In *Insect Pheromones and Other Behaviour-modifying Chemicals* (eds R.L.Ridgway, M.N. Inscoe, and H. Arn), British Crop Protection Council Monograph No. 51, pp. 19–28.

Ridgway, R.L., Silverstein, R.M. and M.N. Inscoe, (eds) (1990) *Behaviour-modifying Chemicals for Insect Management: Applications of Pheromones and Other Attractants*, Marcel Dekker Inc., New York, 761 pp.

Roehrich, R. and Carles, J.P. (1982) Essais de 'confusion sexuelle' en vignoble centre l'Eudemis de la vigne, *Lobesia botrana* Schiff. In *les Mediateurs Chiminiques Agissant sur le Comportement des Insectes*, Colloques INRA No. 7, pp. 365–371.

Roehrich, R., Carles, J.P., Tresor, C. and De Vathaire, M.A. (1979) Essais de 'confusion sexuelle' contre les tordeuses de la grappe, l'Eudemis *Lobesia botrana* Den. et Schiff et la Cochylis *Eupoecilia ambiguella Tr. Ann. Zool. Ecol. Anim*, **11**, 654–675.

Roelofs, W.L., Comeau, A. and Selle, R. (1969) Sex pheromone of the oriental fruit moth. *Nature*, **224**, 723.

Roelofs, W.L., Comeau, A., Hill, A. and Milicevic, G. (1971a) Sex attractants of the codling moth: characterization with electroantennogram technique. *Science*, **174**, 297–299.

Roelofs, W.L., Tette, J.P., Taschenberg, E.F. and Comeau, A. (1971b) Sex pheromone of the grape berry moth: identification by classical and electro-antennogram methods and field tests. *J. Insect Physiol*, **17**, 2235–2243.

Roelofs, W.L., Kochansky, J., Cardé, R. *et al.* (1973) Sex attractant of the grape vine moth, *Lobesia botrana. Mitt Schwez. Entomol.Ges*, **46**, 71–73.

Rothschild G.H.L. (1975) Control of the oriental fruit moth *Cydia molesta* (Busck) (Lepidoptera: Tortricidae) with synthetic pheromone. *Bull. Entomol. Res*, **65**, 473–490.

Rotshchild, G.H.L. (1981) Mating disruption of lepidopterous pests: current status and future prospects. In *Management of Insect Pests with Semiochemicals* (ed. E.R. Mitchell), Plenum Press, New York, pp. 207–228.

Saglio, P., Descoins, C., Gallois, M. *et al.* (1977) Etude de la pheromone sexuelle de la cochylis de la vigne. *Eupocilia (Clysia) ambiguella* Hb. Lepidoptere Tortricoidea Cochylidae. *Ann. Zool. Ecol. Anim.*, **9**, 553–562.

Sanders, C.J. (1981) Disruption of spruce budworm mating – the state of the art. In *Management of Insect Pests with Semiochemicals* (ed. E.R. Mitchell), Plenum Press, New York, pp. 339–349.

Sauer, A., Karg, G., Koch, U.T. *et al.* (1992) A portable EAG system for measurement of pheromone concentrations in the field. *Chemical Senses*, **17**, 543–553.

Schwalbe, C.P. (1981) Disparlure-baited traps for survey and detection. In *The Gypsy Moth: Research Toward Integrated Pest Management*, (eds C.C. Doane and M.L. McManus), USDA Technical Bulletin 1584, pp. 542–548.

Schwalbe, C.P., Cameron, E.A., Hall, D.J. *et al.* (1974) Field tests of micro-encapsulated disparlure for suppression of mating among wild and laboratory reared gypsy moths. *Environ. Entomol*, **3**, 589–592.

Schwalbe, C.P., Paszek, E.C., Bierl-Leonhardt, B.A. and Plimmer, J.R. (1983) Disruption of the gypsy moth (Lepidoptera: Lymantriidae) mating with disparlure. *J. Econ. Entomol.*, **76**, 84–844.

Shorey, H.H., Kaae, R.S. and Gaston, L,K. (1974) Sex pheromones of Lepidoptera, development of a method for pheromonal control of *Pectinophora gossypiella* in cotton. *J. Econ. Entomol.*, **67**, 347–350.

Shorey, H.H., Gaston, L.K. and Kaae, R.S. (1976) Air-permeation with gossyplure for control of the pink bollworm. In *Pest Management with Insect Sex Attractants* (ed. M. Beroza), ACS Symposium Series 23, American Chemical Society, Washington, DC, pp. 67–74.

Stevens, L.J. and Beroza, M. (1972) Mating inhibition field tests using disparlure, the synthetic gypsy moth sex pheromone. *J. Econ. Entomol.*, **65**, 1090–1095.

Tanaka, F., Yabuki, S., Tatsuki, S. *et al.* (1987) Control effect of communication disruption with synthetic pheromones in paddy fields in the rice stem borer, *Chilo suppressalis* (Walker) (Lepidoptera: Pyralidae). *Jpn. J. Appl. Entomol. Zool.* **31**, 125–133.

Taschenberg, E.F. and Roelofs, W.L. (1977) Mating disruption of the grape berry moth, *Paradobesia viteana*, with pheromone released from hollow fibres. *Environ. Entomol.*, **6**, 761–763.

Tatsuki, S. (1990) Application of the sex pheromone of the rice stem borer moth, *Chilo suppressalis*. In *Behaviour-modifying Chemicals for Insect Management: Applications of Pheromones and Other Attractants* (eds R.L. Ridgway, R.M. Silverstein and M.N. Inscoe), Marcel Dekker Inc., New York, pp. 387–406.

Tatsuki, S., Kurihara, M., Usui, K. *et al.* (1983) Sex pheromone of the rice stem borer, *Chilo suppressalis* (Walker) (Lepidoptera: Pyralidae): the third component, Z-9-hexadecenal. *Appl. Entomol. Zool.*, **18**, 443–446.

Tumlinson, J.H. (1990) Chemical analysis and identification of pheromones. In *Behaviour-modifying Chemicals for Insect Management: Applications of Pheromones and Other Attractants* (ed. R.L. Ridgway, R.M. Silverstein and M.N. Inscoe), Marcel Dekker Inc., New York, pp. 73–92.

Tumlinson, J.H., Heath, R.R. and Teal, P.E.A. (1982) Analysis of chemical communications systems of lepidoptera. In: *Insect Pheromone Technology: Chemistry and Applications* (eds B.A. Leonhardt and M. Beroza), ACS Symposium Series 190, American Chemical Society, Washington D.C., pp. 1–25.

Van Steenwyk, R.A. and Oatman, E.R. (1983) Mating disruption of tomato pinworm (Lepidoptera: Gelechiidae) as measured by pheromone trap, foliage, and fruit infestation. *J. Econ. Entomol*, **76**, 80–84.

Vickers, R.A. (1990) Oriental fruit moth in Australia and Canada. In *Behaviour-modifying Chemicals for Insect Management: Applications of Pheromones and Other Attractants* (eds R.L. Ridgway, R.M. Silverstein and M.N. Inscoe), Marcel Dekker Inc., New York, pp. 183–192.

Vickers, R.A. Rothschild, G.H.L. and Jones, E.L. (1985) Control of the oriental fruit moth, *Cydia molesta* (Busck) (Lepidoptera: Tortricidae), at a district level by mating disruption with synthetic female pheromone. *Bull. Entomol. Res.*, **75**, 625–634.

Weatherston, I. (1990) Principles of design of controlled-release formulations. In: *Behaviour-modifying Chemicals for Insect Management: Applications of Pheromones and Other Attractants* (eds R.L. Ridgway, R.M. Silverstein and M.N. Inscoe), Marcel Dekker Inc., New York, pp. 93–112.

Webb, R.E., Tatman, K.M., Leonhardt, B.A. *et al.* (1988) Effect of aerial applications of racemic disparlure on male trap catch and female mating success of gypsy moth (Lepidoptera: Lymantriidae). *J. Econ. Entomol.*, **81**, 268–273.

White, L.D., Hutt, R.B. and Butt, B.A. (1973) Field dispersal of laboratory reared fertile female codling moths and population suppression by release of sterile males. *Environ. Entomol.*, **2**, 66–69.

Wiesner, C.J., Silk, P.J., Tan, S.H. and Fullarton, S. (1980) Monitoring of atmospheric concentrations of the sex pheromone of the spruce budworm, *Choristoneura fumiferana* (Lepidoptera: Tortricidae). *Can. Ent.*, **112**, 333–334.

Windbolz, T. and Baggiolini, M. (1959) Uber das Mass der Ausbreitung des Apfelwicklers wahrend der Eiablageperiode. *Mitt Schweiz. Entomol. Ges.*, **32**, 241–257.

# 13

# Other uses of semiochemicals

As the chemical ecology of insect pests slowly becomes unravelled, new opportunities have arisen to exploit semiochemicals in different ways in order to achieve the goal of managing insect pest problems. The semiochemicals used often differ substantially both in chemical structure and in biological function from lepidopteran sex pheromones. Only a few examples will be offered in this chapter but as time goes on new analytical breakthroughs will produce even more varieties of semiochemicals which will find new uses in IPM.

## 13.1 ANTI-AGGREGATION PHEROMONES

The semiochemicals involved in the attack of bark beetles on living trees (Chapter 10) were obviously crucial in attracting the required number of beetles for the mass attack to succeed. However, when the population reaches a sufficiently high level, the beetles already attracted begin to release anti-aggregation pheromones that interrupt the response of other attacking beetles to the attractive semiochemicals. There is obviously a survival advantage to this mechanism in that it prevents the build-up of beetles on any particular tree from reaching such a high level that successful reproduction could be endangered. Synthetic versions of these chemicals have potential as a means of preventing colonization of high value trees, protecting infested stands from attack, or alternatively, of reducing infestation levels in stands that have already suffered some attack.

One of the earliest anti-aggregation pheromones to be discovered was methylcyclohexenone (MCH), a potent anti-aggregant for Douglas-fir beetle, *Dendroctonus pseudotsugae*, and spruce beetle, *D. rufipennis* (McGregor *et al.*, 1984; Lindgren *et al.*, 1989a). Another compound reported to reduce mass attack of mountain pine beetle, *D. ponderosae* (Ryker and Yandell, 1983; Lindgren *et al.*, 1989b), western pine beetle, *D. brevicomis* (Bedard *et al.*, 1980a), and southern pine beetle, *D. frontalis* (Rudinsky, 1973), is verbenone. In the case of *D. brevicomis*, racemic ipsdienol has an additive inhibitory effect when used with verbenone in a 14% to 86% mixture (Paine and Hanlon, 1991).

Anti-aggregants are released from many small point sources, roughly 100 000 $ha^{-1}$, or from reservoir release devices called 'bubble caps' at 100–150 $ha^{-1}$. Typical dosages range from 80 to 150 g ha. Both MCH and verbenone have been field tested with promising results (Furniss *et al.*, 1977; McGregor *et al.*, 1984; Amman *et al.*, 1989; Lindgren *et al.*, 1989a,b) and have proved particularly useful in preventing the build-up of beetles on isolated windblown trees. The anti-aggregants are applied from the air on to the fallen trees using helicopters, and the build-up of beetles on them is avoided until ground crews get to the site some weeks later for their eventual removal. Anti-aggregant pheromones continue to be investigated for their ability to disrupt mass attacks, thereby reducing beetle survival and reproduction.

## 13.2 OVIPOSITION-DETERRING PHEROMONES

Female cabbage white butterflies (*Pieris brassicae* and *P. rapae*) have been shown to add an oviposition-deterring pheromone to their eggs during egg-laying. This is produced by glands at the tip of the abdomen (Behan and Schoonhoven, 1978) and it has been shown to inhibit further egg-laying by conspecific females. Blaakmeer *et al.* (1994) isolated and identified three novel alkaloids (miriamides), the combined action of which could explain the oviposition-deterring effects of crude egg washes.

Similar oviposition-deterring pheromones have been demonstrated in the cherry fruit fly, *Rhagoletis cerasi*, and the apple maggot fly, *R. pomonella*. Both species have been described marking the fruit around the oviposition puncture by dragging the ovipositor across the surface. Prokopy *et al.* (1982) showed that the pheromone is produced in the midgut of *R. pomonella* females and is released with other gut contents during the marking process. Egg-laying *R. pomonella* females showed a marked avoidance of fruits marked with a water extract of oviposition-deterring pheromone (Averill and Prokopy, 1987). In the case of *R. cerasi*, Hurter *et al.* (1987) identified the oviposition-deterring pheromone as *N*(15-(beta glucopyranosyl)oxy-8-hydroxy)taurine. Field trials with faecal extracts had already demonstrated that the deterrent action can be achieved in the field (Katsoyannos and Boller, 1976, 1980) but only after considerable efforts in synthetic chemistry was enough material produced to verify those earlier results. The results proved very satisfactory but, because of the complex nature of the chemical, it is doubtful whether an economically viable synthetic route for the compound can be developed.

Oviposition-deterring pheromones have been shown in other insect orders, e.g. the spotted cow pea bruchid, *Callosobruchus maculatus* (Messina *et al.*, 1987), but to date none has been developed to the point of commercial sales. This is an area where we might see significant advances in the future, provided the chemistry of the compounds concerned does not prove to be too complex and costly.

## 13.3 SEX AND ALARM PHEROMONES IN APHIDS AND MITES

In temperate countries, aphids are probably the most important insect pests, causing damage to their host plants not only by their feeding activities but also through their role as vectors of plant viruses. Aphids have been shown to produce sex pheromones but usually only when they are on their primary or winter hosts. The sex pheromones of a number of pest species, including the black bean aphid (*Aphis fabae*), the pea aphid (*Acyrthosiphon pisum*), the green bug (*Schizaphis graminum*), the peach potato aphid (*Myzus persicae*) and the damson aphid (*Phorodon humuli*), have now been shown to comprise nepetalactol I and/or nepetalactone II or nepetalactol III (Fig. 13.1) (Dawson *et al.*, 1987, 1988, 1989; Campbell *et al.*, 1990). Attempts are currently being made to incorporate the use of these chemicals in an IPM package for these species, but because it is the asexual forms attacking the secondary or summer host that cause the economic damage it is these stages that need to be targeted to achieve control. For obvious reasons, the sex pheromones have no practical value during the secondary host period but other pheromones have shown some promise.

When aphids are attacked by predators, they produce an alarm pheromone, the major component of which is (*E*)-β-farnesene (Bowers *et al.*, 1972). When the alarm pheromone is released, it causes other aphids nearby to stop feeding, move away and sometimes drop off the plant.

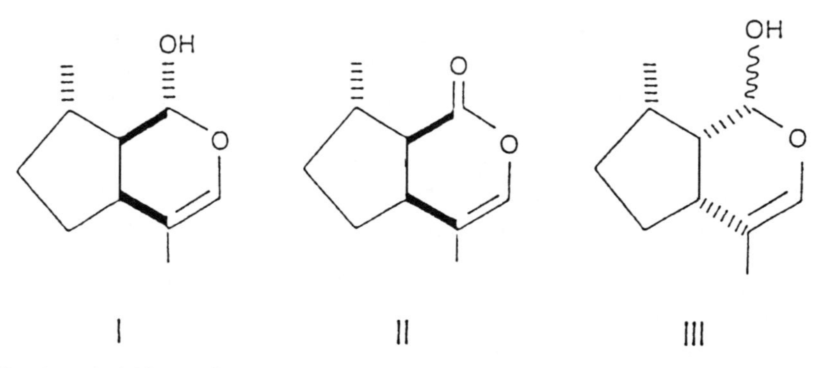


**Fig. 13.1** Aphid sex pheromone component structures.

Attempts to use the alarm pheromone alone for aphid control met with failure. However, Griffiths and Pickett (1987) showed that when synthetic alarm pheromone was formulated in a hydrocarbon propellant and applied electrostatically, it was highly effective in increasing aphid mobility in the field. When the above formulation was used with an insecticide such as permethrin or bendiocarb, increased pick-up of the insecticide was observed which led to increased aphid mortality (Pickett *et al.*, 1986). Similarly it has been shown that the alarm pheromone can be used to improve the efficacy of biological agents. Under greenhouse conditions, aphids such as *Aphis gossypii* have become almost completely resistant to several families of conventional insecticides but they are still susceptible to the fungal pathogen *Verticillium lecanii*, provided enough fungal spores can be picked up by the aphid over a relatively short period of time. Use of the alarm pheromone improved pick-up of spores by *A. gossypii* even though it responds rather weakly to the synthetic pheromone (Griffiths and Pickett, 1987).

Work on mite alarm pheromones has shown that citral and isopiperitenone (Kuwahara *et al.*, 1987) are the principal constituents of the alarm pheromones of at least five species of mites. The use of these compounds or chemical analogues in synergizing the effects of acaricides has also given positive results, which has led to commercial products.

The commercialization of (*E*)-β-farnesene as a spray adjuvant to improve knock-down of aphids by conventional insecticides and the use of farnesol in the same way with acaricides against mites are currently being undertaken by at least two industrial companies and the use of these pheromones is projected to increase significantly during the coming years.

## 13.4 STIMULO-DETERRENT DIVERSIONARY STRATEGY OR PUSH–PULL STRATEGIES

The discovery by Gibson and Pickett (1983) that glandular hairs on leaves of the wild potato, *Solanum berthaultii*, release (*E*)-β-farnesene to repel the aphid *Myzus persicae* makes it clear that it is not sufficient to study semiochemical interactions between insects in isolation. It is important to consider not only the possible synergistic effects of host plant volatile on insect responses to pheromones from their conspecifics but also the fact that, as Gibson and Pickett (1983) showed, plants can actually synthesize insect pheromones, or their analogues, and use them for their own defence (as allomones), as in the case of the wild potato above, or as a kairomone in the case of wild orchids which release exact copies of hymenopteran sex pheromone for pollination (Kaiser, 1993).

There are now good examples from many insect orders in which a pheromone is synergized by plant-derived volatiles – synergistic effects of tree volatiles on the aggregation pheromones of bark beetles being a good example. In aphids, too, there is evidence that volatiles from the primary host synergize the attraction of males to the sex pheromone of the damson-hop aphid, *Phorodon humuli* (Pickett *et al.*, 1992). Earlier in this chapter we noted the deterrent effects of other semiochemicals such as the antiaggregation pheromones of bark beetles and the oviposition-deterrent pheromones of Tephritidae, among others. There are attempts currently under way to combine these attraction and repulsion effects of semiochemicals to form robust 'push–pull' systems of pest management (Pyke *et al.*, 1987) or 'stimulo-deterrent diversionary strategies', as they have been termed by Miller and Cowles (1990). In essence, semiochemicals are used to divert insect pests from the crop being protected and are then aggregated on other parts of the crop, or even on non-host plants, where they can be destroyed using conventional insecticides or, better still, biological ones. Early indications are that the sum of the two effects is not just additive but even synergistic. The demonstration by Nottingham *et al.* (1991) that non-host volatiles can significantly deter aphids from settling on host plants suggests that the phenomenon is probably applicable even to aphids that apparently have a much greater ability to register olfactory cues from both host and non-host plants than was previously thought (Pickett *et al.*, 1992).

Involatile semiochemicals such as antifeedants could also be used in these strategies. Polygodial, obtained from the weed *Polygonum hydropiper*, shows antifeedant activity against a broad range of coleopteran and lepidopteran pests and against aphids that are virus vectors. In a field trial to reduce barley yellow dwarf virus transmission by *Rhopalosiphum padi*, yields were increased by over 1 tonne $ha^{-1}$ in the polygodial-treated crop compared with untreated controls, a result equivalent to the control obtained using a pyrethroid insecticide (Pickett *et al.*, 1987). In other experiments to control the pea and bean weevil, *Sitona lineatus*, Smart *et al.* (1994) have used an antifeedant formulation of neem oil from the neem tree, *Azadirachta indica* to 'push' weevils from parts of a field bean crop, and the insect's aggregation pheromone, 4-methyl-3,5-heptanedione, to 'pull' the weevils on to other parts of the crop treated with a pyrethroid insecticide.

An armoury of insect- and plant-derived olfactory repellents and deterrents is therefore being elucidated which could be combined with insect- or plant-derived attractants to arrive at a push–pull strategy which is robust enough to be accepted in practice.

## 13.5 SEMIOCHEMICALS INFLUENCING THE BEHAVIOUR OF PARASITOIDS AND PREDATORS

### 13.5.1 Parasitoids

In searching for an insect host, a parasitoid may first of all locate its host's habitat before it searches out the host itself. Plant volatiles emanating from the host's food or foodplant have been shown to be important cues in host habitat location for a number of hymenopteran parasitoids (Herrebout, 1969a,b). Camors and Payne (1972) showed that α-pinene alone attracted *Heydenia unica*, a parasitoid of the bark beetle *Dendroctonus frontalis*, but that host-derived compounds also influenced this response. Similarly, both male and female *Diaeretiella rapae*, an aphid parasitoid, are attracted to fresh collard leaves and to very low concentrations of mustard oil (allyl isothiocyanate), which occurs in collard leaves (Read *et al.*, 1970).

To locate the host, once in the host's location, a parasitoid may use a number of odour cues. Some insects, for instance, release mandibular gland secretions during feeding which may aid host location by a parasitoid. The mandibular gland secretion of larvae of the flour moth, *Ephestia kühniella*, for example, contains an epideitic pheromone that elicits oviposition movements in the hymenopteran parasite *Venturia canescens* (Corbet, 1971).

In some species, the parasitoids employ the host insect's own sex pheromone as a kairomone for host location. This has been shown to be the case with *Trichopoda pennipes*, a parasite of stink bugs (Mitchell and Man, 1970), parasitoids of *Ips confusus* and other bark beetle species (Rice, 1969; Vite and Williamson, 1970; Dixon and Payne, 1979; Greany and Hagen, 1980) and scale insect parasitoids (Sternlicht, 1973).

The kairomonal influences on host location by the parasitoids of the genus *Trichogramma* have been studied extensively. Chemicals released by the eggs of *Heliothis zea* have been shown to increase the rate at which *T. evanescens* found and parasitized them (Lewis *et al.*, 1971). Lewis *et al.* (1972) showed that the host seeking response of *T. evanescens* was stimulated by compounds emanating from the scales of *H. zea* adults, and Jones *et al.* (1973) identified tricosane as the most active of several chemicals. Detailed studies of the response of the parasitoid to tricosane showed conclusively that the kairomone functions primarily by releasing and maintaining host-seeking responses rather than by attracting and serving as a steering mechanism (Lewis *et al.*, 1975a,b). Perhaps the most exciting aspect relating to parasitoid kairomones that has been discovered recently is that parasitoids can be conditioned or 'fixed' to a certain olfactory substance while they are being bred and this can later be used to manipulate their behaviour in the field by applying the same olfactory cue to an area where host-seeking behaviour by the parasitoid is desirable. This phenomenon, if it can be developed to the point of wide-scale application, has the potential to revolutionize the outdoor use of parasitoids which, to date, have been used predominantly in greenhouses, for reasons of containment.

### 13.5.2 Predators

Host selection behaviour of parasitoids and prey selection behaviour of predators are generally very similar, with habitat location and host location being exhibited by both entomophagous groups. The coccinellid beetle, *Anatis ocellata*, for instance, a predator of aphids found on pine needles, is attracted to infested trees by chemicals found in pine needles (Kesten, 1969). Predaceous *Chrysopa carnea* larvae have also been shown to respond to kairomones from the scales of *H. zea* adults by initiating prey-seeking behaviour (Lewis *et al.*, 1977) but acceptance of *H. zea* eggs is stimulated by an additional kairomone, probably originating in the accessory gland secretion which is associated with egg deposition (Nordlund *et al.*, 1977). Some predators respond to chemicals from both the prey and its habitat. *Enoclerus lecontei* and *Temnochila chloridia*, two beetle predators of the bark beetle pest, *Dendroctonus brevicomis*, have been shown to arrive during and very shortly after the mass arrival of their prey. *T. chlorodia* responds very specifically to *exo*-brevicomin (prey-derived), while its response decreases when (*E*)-verbenol and verbenone (host tree compounds modified by beetles) are also present (Bedard *et al.*, 1969, 1980b). *E. lecontei*, on the other hand, which is a significant predator of *D. brevicomis*, responds to the host tree volatiles, but not significantly to the pheromone of *D. brevicomis*, although it does respond to the pheromone of *Ips typographus*, another of its prey species (Lanier *et al.*, 1972; Wood, 1972).

There are very few documented cases of the applied uses of predator manipulations in the field by semiochemicals but the manipulation of *Chrysopa carnea* through the use of tryptophan is well documented (Hagen *et al.*, 1976; McEwen *et al.*, 1994) and the attraction of *Rhizophagus grandis* using a kairomone produced by *Dendroctonus micans* in a trap to monitor the predator's distribution in the field (Wainhouse *et al.*, 1991, 1992) serves also as an indicator of what is possible in terms of natural enemy manipulation through semiochemicals.

## REFERENCES

Amman, G.D., Their, R.W., McGregor, M.D. and Aschmitz, R.F. (1989) Efficacy of verbenone in reducing lodgepole pine infestation by mountain pine beetles in Idaho. *Canadian Journal of Forest Research*, **19**, 60–64.

Averill, A.L. and Prokopy, R.J. (1987) Residual activity of oviposition-deterring pheromone in *Rhagoletis pomonella* (Diptera: Tephritidae) and female response to infested fruit. *J. Chem. Ecol.*, **13**, 167–177.

Bedard, W.D., Tilden, P.E., Wood, D.L. *et al* (1969) Western pine beetle: field response to its sex pheromone and a synergistic host terpene myrcene. *Science*, **164**, 1284–1285.

Bedard, W.D., Tilden, P.E., Lindahl, K.Q.Jr *et al.* (1980a) Effects of verbenone and *trans*-verbenol on the response of *Dendroctonus brevicomis* to natural and synthetic attractant in the field. *J. Chem. Ecol.*, **6**, 997–1013.

Bedard, W.D., Wood, D.L., Tilden, P.E. *et al.* (1980b) Field responses of the western pine beetle and one of its predators to host- and beetle-produced compounds. *J. Chem. Ecol.*, **6**, 625–641.

Blaakmeer, A., Stork, A., Van Veldhuizen, A. *et al.* (1994) Isolation, identification and synthesis of miriamides, new host markers from eggs of *Pieris brassicae* (Lepidoptera: Pieridae). *J. Nat. Prod.*, **57**, 90–99.

Bowers, W.S., Nault, L.R., Webb, R.E. and Dutky, S.R. (1972) Aphid alarm pheromone: isolation, identification, synthesis. *Science*, **177**, 1121–1122.

Campbell, C.A.M., Dawson, G.W., Griffiths, D.C. *et al.* (1990) The sex attractant pheromone of the damson-hop aphid *Phorodon humuli* (Homoptera, Aphididae). *J. Chem. Ecol.*, **16**, 3455–3456.

Camors, F.B. and Payne, T.L. (1972) Resistance of *Heydenia unica* (Hymenoptera: Pteromalidae) to *Dendroctonus frontalis* (Coleoptera: Scolytidae) pheromones and a host tree terpene. *Ann. Entomol. Soc. Am.*, **65**, 31–33.

Corbet, S.A. (1971) Mandibular gland secretion of larvae of the flour moth, *Anagasta kuehniella*, contains an epideictic pheromone and elicits oviposition movement in a hymenopteran parasite. *Nature*, **232**, 481–484.

Dawson, G.W., Griffiths, D.C., Janes, N.F. *et al.* (1987) Identification of an aphid sex pheromone. *Nature*, **325**, 614–616.

Dawson, G.W., Griffiths, D.C., Merritt, L.A. *et al.* (1988) The sex pheromone of the greenbug, *Schizaphis graminum. Entomologia Experimentalis et Applicata*, **48**, 91–93.

Dawson, G.W., Janes, N.F., Mudd, A. *et al.* (1989) The aphid sex pheromone. *Pure and Applied Chemistry*, **61**, 555–558.

Dixon, W.N. and Payne, T.L. (1979) Aggregation of *Thanasimus dubius* on trees under mass-attack by the southern pine beetle. *Environ. Entomol*, **8**, 178–181.

Furniss, M.M., Young, J.W., McGregor, M.D. *et al.* (1977) Effectiveness of controlled-release formulations of MCH for preventing Douglas-fir beetle (Coleoptera: Scolytidae) infestation in felled trees. *Canadian Entomology*, **106**, 1063–1069.

Gibson, R.W. and Pickett, J.A. (1983) Wild potato plant repels aphids by release of aphid alarm pheromone. *Nature (London)*, **302**, 608–609.

Greany, P.D. and Hagen, K.S. (1980) Prey selection. In *Semiochemicals: Their Role in Pest Control* (eds D.A. Nordlund, R.L. Jones and W.J. Lewis), John Wiley, New York.

Griffiths, D.C. and Pickett, J.A. (1987) Novel chemicals and their formulation for aphid control. *Proceedings of the 14th International Symposium on Controlled Release of Bioactive Materials*, **14**, 243–244.

Hagen, K.S., Greany, P., Sawall, E.F. Jr and Tassan, R.L. (1976) Tryptophan in artificial honeydews as a source of an attractant for adult *Chrysopa carnea. Environ. Entomol.*, **5**, 458–468.

Herrebout, W.M. (1969a) Some aspects of host selection in *Eucarcelia rutilla* Vill. *Neth. J. Zool.*, **19**, 1–104.

Herrebout, W.M. (1969b) Habitat selection in *Eucarcelia rutilla* Vill. II. Experiments with female of known age. *Z. Angew. Entomol.*, **63**, 336–349.

Hurter, J., Boller, E.F., Staedtler, E. *et al.* (1987) Oviposition-deterring pheromone in *Rhagoletis cerasi L.*: purification and determination of the chemical constitution. *Experientia*, **43**, 157–164.

Jones, R.L., Lewis, W.J., Beroza, M. *et al.* (1973) Host-seeking stimulants (kairomones) for the egg-parasite *Trichogramma evanescens. Environ. Entomol*, **2**, 593–596.

Kaiser, R. (1993) *The Scent of Orchids. Olfactory and Chemical Investigations*, Elsevier Science Publishers, Amsterdam, 264 pp.

Katsoyannos, B.I. and Boller, E.F. (1976) First field application of oviposition-deterring marking pheromone of European cherry fruit fly. *Environ. Entomol*, **5**, 151–152.

Katsoyannos, B.I. and Boller, E.F. (1980) Second field application of oviposition-deterring pheromone of the European cherry fruit fly, *Rhagoletis cerasi L.* (Diptera: Tephritidae). *Z. Angew. Ent*, **89**, 278–281.

Kesten, U. (1969) Zur Morfologie und Biologie von *Anatis ocellata* (L.) (Coleoptera, Coccinellidae). *Z. Angew. Entomol.*, **63**, 412–445.

Kuwahara, Y., Akimoto, K., Leal, W.S. *et al.* (1987) Isopiperitenone: a new alarm pheromone of the acarid mite, *Tyrophagus similis* (Acarina: Acaridae). *Agric. Biol. Chem*, **51**, 3441–3442.

Lanier, G.N., Birch, M.C., Schmitz, R.F. and Furniss, M.M. (1972) Pheromones of *Ips pini* (Coleoptera: Scolytidae): variation in response among three populations. *Can. Ent*, **104**, 1917–1923.

Lewis, W.J. and Jones, R.L. (1971) Substance that stimulates host-seeking by *Microplitis croceipes*, a parasite of *Heliothis* species. *Ann. Ent. Soc. Am.*, **64**, 471–473.

Lewis, W.J., Sparks, A.N., Jones, R.L. and Barras, D.J. (1972) Efficiency of *Cardiochiles nigriceps* as a parasite of *Heliothis virescens* on cotton. *Environ. Entomol.*, **1**, 468–471.

Lewis, W.J., Jones, R.L., Nordlund, D.A. and Gross, H.R. Jr (1975a) Kairomones and their use for management of entomophagous insects. II. Mechanisms causing increase in the rate of parasitization by *Trichogramma* spp. *J. Chem. Ecol.*, **1**, 349–360.

Lewis, W.J., Jones, R.L., Nordlund, D.A. and Sparks, A.N. (1975b) Kairomones and their use for management of entomophagous insects. I. Evaluation for increasing rates of parasitization by *Trichogramma* spp. in the field. *J. Chem. Ecol.*, **1**, 343–348.

Lewis, W.J., Nordlund, D.A., Gross, H.R. Jr *et al.* (1977) Kairomones and their use for management of entomophagous insects. V. Moth scales as a stimulus for predation of *Heliothis zea* (Boddie) eggs by *Chrysopa carnea* Stephens larvae. *J. Chem. Ecol*, **3**, 483–487.

Lindgren, B.S., McGregor, M.D., Oakes, R.D. and Meyer, H.E. (1989a) Suppression of spruce beetle attacks by MCH released from bubble caps. *Western Journal of Applied Forestry*, **4** (2), 49–52.

Lindgren, B.S., Borden, J.H., Cushon, G.H. *et al.* (1989b) Reduction of mountain pine beetle (Coleoptera: Scolytidae) attacks by verbenone in lodgepole pine stands in British Columbia. *Canadian Journal of Forest Research*, **19**, 65–68.

McEwen, P.K., Jervis, M.A. and Kidd, N.A.C. (1994) Use of a sprayed L-tryptophan solution to concentrate numbers of the green lacewing *Chrysoperla carnea* in olive tree canopy. *Entomol. Exp. Appl.*, **70**, 97–99.

McGregor, M.D., Furniss, M.M., Oakes, R.D. *et al.* (1984) MCH pheromone for preventing Douglas-fir beetle infestation in windthrown trees. *Journal of Forest History*, **82**, 613–616.

Messina, F.J., Barmore, J.L. and Renwick, J.A.A. (1987) Oviposition deterrent from eggs of *Callusobruchus maculatus*: spacing mechanism or artefact? *J. Chem. Ecol*, **13**, 219–226.

Miller, J.R. and Cowles, R.S. (1990) Stimulo-deterrent diversion: a concept and its possible application to onion maggot control. *J. Chem. Ecol.*, **16**, 3197–3212.

Mitchell, W.C. and Man, R.F.L. (1970) Response of the female southern green stink bug and its parasite *Trichopoda pennipes* to male stink bug pheromones. *J. Econ. Entomol.*, **64**, 856–859.

Nordlund, D.A., Lewis, W.J., Jones, R.L. *et al.* (1977) Kairomones and their use for management of entomophagous insects. VI. An examination of the kairomones for the predator *Chrysopa carnea* Stephens at the oviposition sites of *Heliothis zea* (Boddie). *J. Chem. Ecol*, **3**, 507–511.

Nottingham, S.F., Hardie, J., Dawson, G.W. *et al.* (1991) Behavioural and electrophysiological responses of aphids to host and non host volatiles *J. Chem. Ecol.*, **17**, 1231–1242.

Paine, T.D. and Hanlon, C.C. (1991) Response of *Dendroctonus brevicomis* and *Ips paraconfusus* (Coleoptera: Scolytidae) to combinations of synthetic pheromone attractants and inhibitors verbenone and ipsdienol. *J. Chem. Ecol.*, **17**, 2163–2176.

Pickett, J.A., Cayley, G.R., Dawson, G.W. *et al.* (1986) Use of the alarm pheromone and derivatives against aphid mediated damage. *Abstracts 6th International Congress Pesticide Chemistry*, Ottawa 1986, 2C–08.

Pickett, J.A., Dawson, G.W., Griffiths, D.C. *et al.* (1987) Development of plant-derived antifeedants for crop protection. In *Pesticide Science and Biotechnology* (eds R. Greenhalgh and T.R. Roberts), Blackwell Scientific Publications, Oxford, pp. 125–128.

Pickett, J.A., Pye, B.J., Wadhams, L.J. *et al.* (1992) Potential applications of semiochemicals in aphid control. In *Insect Pheromones and Other Behaviour-Modifying Chemicals* (eds R.L. Ridgeway, M. Inscoe and H. Arn), BCPC Monograph No 51, pp. 29–33.

Prokopy, R.J., Averill, A.L., Bardinelli, C.M. *et al.* (1982) Site of production of an oviposition-deterring pheromone component in *Rhagoletis pomonella* flies. *J. Insect Physiol*, **28**, 1–10.

Pyke, B., Rice, M., Sabine, B. and Zaluchi, M. (1987) The push–pull strategy–behavioural control of *Heliothis. Aust. Cotton Grower*, May–July 1987, 7–9.

Read, D.P., Feeny, P.P. and Root, R.B. (1970) Habitat selection by the aphid parasite *Diaeretiella rapae. Can. Entomol*, **102**, 1567–1578.

Rice, R.E. (1969) Response of some predators and parasites of *Ips confusus* (leC.) to olfactory attractants. *Contrib. Boyce Thompson Inst.*, **24**, 189–194.

Rudinsky, J.A. (1973) Multiple functions of the southern pine beetle pheromone verbenone. *Environmental Entomology*, **2**, 511–514.

Ryker, L.C. and Yandell, K.L. (1983) Effect of verbenone on aggregation of *Dendroctonus ponderosae* Hopkins (Coleoptera: Scolytidae) to synthetic attractant. *Zeitschrift fur Angewandte Entomologie*, **96**, 452–459.

Smart, L.E., Blight, M.M., Pickett, J.A. and Pye, B.J. (1994) Development of field strategies incorporating semiochemicals for the control of the pea and bean weevil, *Sitona lineatus* L. *Crop Protection*, **13(2)**, 127–135.

Sternlicht, M. (1973) Parasitic wasps attracted by the sex pheromone of the coccid host. *Entomophagae*, **18**, 339–342.

Vite, J.P. and Williamson, D.L. (1970) *Thanasimus dubius*: prey perception. *J. Insect Physiol.*, **16**, 233–239.

Wainhouse, D., Wyatt, T., Phillips, A. *et al.* (1991) Response of the predator *Rhizophagus grandis* to host plant derived chemicals in *Dendroctonus micans* larval frass in wind tunnel experiments (Coleoptera: Rhizophagidae, Scolytidae). *Chemoecology*, **2**, 55–63.

Wainhouse, D., Beech-Garwood, P.A., Howell, R.S. *et al.* (1992) Field response of predator *Rhizophagus grandis* to prey frass and synthetic attractants. *J. Chem. Ecol.*, **18**, 1693–1705.

Wood, D.L. (1972) Selection and colonisation of ponderosa pine by bark beetles. In *Insect/Plant Relationships* (ed. H.F. Van Emden), *Symp. R. Ent. Soc. Lond*, **6**, 101–117, Blackwell, Oxford.

# Index